T0251643

DRUG DELIVERY
APPROACHES AND
NANOSYSTEMS

Volume 2: Drug Targeting Aspects of
Nanotechnology

DRUG DELIVERY APPROACHES AND NANOSYSTEMS

Volume 2: Drug Targeting Aspects of Nanotechnology

Edited by

Raj K. Keservani, MPharm
Anil K. Sharma, MPharm
Rajesh K. Kesharwani, PhD

APPLE
ACADEMIC
PRESS

Apple Academic Press Inc.
3333 Mistwell Crescent
Oakville, ON L6L 0A2 Canada

Apple Academic Press Inc.
9 Spinnaker Way
Waretown, NJ 08758 USA

© 2018 by Apple Academic Press, Inc.

First issued in paperback 2021

Exclusive worldwide distribution by CRC Press, a member of Taylor & Francis Group
No claim to original U.S. Government works

ISBN 13: 978-1-77-463113-3 (pbk)
ISBN 13: 978-1-77-188584-3 (hbk)

Library and Archives Canada Cataloguing in Publication

Drug delivery approaches and nanosystems / edited by Raj K. Keservani, MPharm, Anil K. Sharma, MPharm, Rajesh K. Kesharwani, PhD.
Includes bibliographical references and indexes.
Contents: Volume 1. Novel drug carriers -- Volume 2. Drug targeting aspects of nanotechnology.
Issued in print and electronic formats.
ISBN 978-1-77188-583-6 (v. 1 : hardcover).--ISBN 978-1-77188-585-0
(set : hardcover).--ISBN 978-1-77188-584-3 (v. 2 : hardcover).--
ISBN 978-1-31522537-1 (v. 1 : PDF).--ISBN 978-1-31522-536-4 (v. 2 : PDF)
1. Drug delivery systems. 2. Nanotechnology. 3. Nanostructures. I. Kesharwani, Rajesh Kumar, 1978-, editor II. Sharma, Anil K., 1980-, editor III. Keservani, Raj K., 1981-, editor
RS420.D78 2017   615.1'9   C2017-902014-5   C2017-902015-3

Library of Congress Cataloging-in-Publication Data

Names: Keservani, Raj K., 1981- editor. | Sharma, Anil K., 1980- editor. | Kesharwani, Rajesh Kumar, 1978- editor.
Title: Drug delivery approaches and nanosystems / editors, Raj K. Keservani, Anil K. Sharma, Rajesh K. Kesharwani.
Other titles: Novel drug carriers. | Drug targeting aspects of nanotechnology. Description: Toronto ; New Jersey : Apple Academic Press, 2017. | Includes bibliographical references and index.
Identifiers: LCCN 2017012335 (print) | LCCN 2017013149 (ebook) | ISBN 9781315225371 (ebook) | ISBN 9781771885836 (hardcover ; v. 1 : alk. paper) | ISBN 9781771885843 (hardcover ; v. 2 : alk. paper) | ISBN 9781771885850 (hardcover ; set : alk. paper) | ISBN 9781315225371 (eBook)
Subjects: | MESH: Drug Delivery Systems | Nanotechnology | Nanostructures
Classification: LCC RS420 (ebook) | LCC RS420 (print) | NLM QV 785 | DDC 615.1/9--dc23
LC record available at https://lccn.loc.gov/2017012335

Apple Academic Press also publishes its books in a variety of electronic formats. Some content that appears in print may not be available in electronic format. For information about Apple Academic Press products, visit our website at **www.appleacademicpress.com** and the CRC Press website at **www.crcpress.com**

The Present Book Is Dedicated To
Our Beloved
Aashna,
Atharva
Vihan
&
Vini

CONTENTS

LIST OF CONTRIBUTORS

Yukio Ando
Department of Neurology, Graduate School of Medical Sciences, Kumamoto University, 1-1-1 Honjo, Chuo-ku, Kumamoto 860-8556, Japan

Hidetoshi Arima
Program for Leading Graduate Schools "HIGO (Health Life Science: Interdisciplinary and Glocal Oriented) Program," Kumamoto University, 5-1 Oe-honmachi, Chuo-ku, Kumamoto 862-0973, Japan

Kantha Deivi Arunachalam
Center for Environmental Nuclear Research, SRM University, Tamilnadu, India

Roberta Cassano
Department of Pharmacy, Health and Nutritional Sciences, University of Calabria, Arcavacata di Rende, Cosenza, Italy

Lakshmi Kiran Chelluri
Transplant Biology, Immunology, Stem Cell Unit, Global Hospitals, Hyderabad, India

Giuseppe Cirillo
Department of Pharmacy, Health and Nutritional Sciences, University of Calabria, Rende (CS), Italy

Tuba Demirci
Institute for Biochemical Engineering and Analytics, University of Applied Sciences, Giessen, Germany

Merve Erginer
IBSB-Industrial Biotechnology and Systems Biology Research Group, Bioengineering Department, Marmara University, Istanbul, Turkey

Sweta K. Gupta
Department of Chemical Engineering, University of Rhode Island, Kingston, USA

Silke Hampel
Leibniz Institute for Solid State and Materials Research Dresden, Dresden, Germany

Yuya Hayashi
Research Fellow of Japan Society for the Promotion of Science, Department of Physical Pharmaceutics, Graduate School of Pharmaceutical Sciences, Kumamoto University, 5-1 Oe-honmachi, Chuo-ku, Kumamoto 862-0973, Japan

Juergen Hemberger
Institute for Biochemical Engineering and Analytics, University of Applied Sciences, Giessen, Germany

Taishi Higashi
Department of Physical Pharmaceutics, Graduate School of Pharmaceutical Sciences, Kumamoto University, 5-1 Oe-honmachi, Chuo-ku, Kumamoto 862-0973, Japan

Hirofumi Jono
Department of Clinical Pharmaceutical Sciences, Graduate School of Pharmaceutical Sciences, Kumamoto University, 1-1-1 Honjo, Chuo-ku, Kumamoto 860-8556, Japan

Jun-ichi Kadokawa
Department of Chemistry, Biotechnology, and Chemical Engineering, Graduate School of Science and Engineering, Kagoshima University, 1-21-40 Korimoto, Kagoshima 890-0065, Japan

Yoshiharu Kaneo
Laboratory of Biopharmaceutics, Faculty of Pharmacy and Pharmaceutical Sciences, Fukuyama University, Fukuyama, Hiroshima, Japan

Sabyasachi Maiti
Department of Pharmaceutics, Gupta College of Technological Sciences, Asansol, West Bengal, India

Sami Makharza
Department of Chemistry, Faculty of Science & Technology, Hebron University, Hebron, Palestine
Leibniz Institute for Solid State and Materials Research Dresden, Dresden, Germany

Megha Marwah
Department of Pharmaceutics, Bombay College of Pharmacy, Mumbai–400098, India

Samantha A. Meenach
Department of Chemical Engineering, University of Rhode Island, Kingston, USA

Yamuna Mohanram
Transplant Biology, Immunology, Stem Cell Unit, Global Hospitals, Hyderabad, India

Keiichi Motoyama
Department of Physical Pharmaceutics, Graduate School of Pharmaceutical Sciences, Kumamoto University, 5-1 Oe-honmachi, Chuo-ku, Kumamoto 862-0973, Japan

Mangal Nagarsenker
Department of Pharmaceutics, Bombay College of Pharmacy, Mumbai–400098, India

Preethi Naik
Department of Pharmaceutics, Bombay College of Pharmacy, Mumbai–400098, India

Ebru Toksoy Oner
IBSB-Industrial Biotechnology and Systems Biology Research Group, Bioengineering Department, Marmara University, Istanbul, Turkey

Gopal Singh Rajawat
Department of Pharmaceutics, Bombay College of Pharmacy, Mumbai–400098, India

Raghavendra Ramalingam
Center for Environmental Nuclear Research, SRM University, Tamilnadu, India

Elisa A. Torrico-Guzmán
Department of Chemical Engineering, University of Rhode Island, Kingston, USA

Sonia Trombino
Department of Pharmacy, Health and Nutritional Sciences, University of Calabria, Arcavacata di Rende, Cosenza, Italy

Meenakshi Venkataraman
Department of Pharmaceutics, Bombay College of Pharmacy, Mumbai–400098, India

Orazio Vittorio
Children's Cancer Institute Australia, Lowy Cancer Research Centre, University of New South Wales, Sydney, Australia

Zimeng Wang
Department of Chemical Engineering, University of Rhode Island, Kingston, USA

Songul Yasar Yıldız
IBSB-Industrial Biotechnology and Systems Biology Research Group, Bioengineering Department, Marmara University, Istanbul, Turkey

LIST OF ABBREVIATIONS

ABC	ATP-binding cassette
AD-PEG	adamantine-PEG
AD-PEG-Tf	AD-PEG-transferrin
ADEPT	antibody-directed enzyme prodrug therapy
ADR	adriamycin
AF	asialofetuin
AM	adriamycin
AmB	amphotericin B
AMS	antimicrobial solutions
APC	antigen presenting cell
APIs	active pharmaceutical ingredients
apoB	apolipoprotein B
AR	all-trans retinol
ASGP-R	asialoglycoprotein receptor
ASGPR1	asialoglycoprotein receptor 1
AV	aloe vera
BD	biodistribution
BE	branching enzyme
BLG	beta lactoglobulin
BMP2	bone morphogenetic protein 2
BSA	bovine serum albumin
CA	cellulose acetate
CDex	cluster dextrin
CDI	N, N'-carbonyldiimidazole
CDP	CyD polymer
CF	cystic fibrosis
CFTR	cystic fibrosis transmembrane conductance regulator
CLSM	confocal laser scanning microscopy
CLT	clotrimazole
CMC	critical micellation concentration

CME	clathrin-dependent endocytosis
CMP	carboxymethylpullulan
CNT	carbon nanotubes
COPD	chronic obstructive pulmonary disease
CP	cisplatin
CPT	camptothecin
CS	chitosan
CS-EDTA	chitosan–ethylenediaminetetraacetic acid
CU	curcumin
CVD	chemical vapor deposition
CyD	cyclodextrin
DADA	diisopropylamine dichloroacetate
DCA	dichloroacetate
DDS	drug delivery systems
DEX-MPS	dextran-methylprednisolone succinate
DLS	dynamic light scattering
DMSO	dimethyl sulfoxide
DNa	diclofenac sodium
DNM	daunomycin
DOPE	dioleoylphosphatidylethanolamine
DOX	doxorubicin
DP	degree of polymerization
DPI	dry powder inhalers
DSC	differential scanning calorimetry
ECN	econazole nitrate
EGF	epidermal growth factor
EPC	egg yolk phosphatidylcholine
EPR	enhanced permeability and retention
F-SLNs	fluorescent solid lipid nanoparticles
FA	ferulic acid
FA	folic acid
FAP	familial amyloid polyneuropathy
FD	flurbiprofen-dextran
FDA	Food and Drug Administration
FEV1	forced expiratory volume over 1-second
FITC	fluorescein isothiocyanate

FU-P	5-fluorouracil-pectin
GDEPT	gene-directed enzyme prodrug therapy
GI	gastrointestinal
GO	graphene oxide
GP	galactosylated pullulan
GRAS	generally recognized as safe
GTP	green tea polyphenols
HA	hyaluronic acid
Hap	hydroxyapatite
HDFs	human dermal fibroblasts
hEGF	human epidermal growth factor
HENC	corneal endothelial cells
HES	hydroxyethyl starch
HPMA	N-(2-hydroxypropyl)methacrylamide
HPSEC	high-performance size exclusion chromatography
HRP	horse radish peroxidase
HSP	heat shock protein
KDR	kinase domain receptor
KOL	kollicoat IR
LANH$_2$	lactobionic amine
LMW	low molecular weight
LV	lamivudine
MA	mycolic acid
MATE	multi drug toxic compound extrusion
MB	melt blowing
MDI	metered-dose inhalers
MDR	multi drug resistance
MEC	minimal effective concentration
MFD	microfiber disc
MFS	major facilitator superfamily
MFS	microfiber sheet
MIC	minimal inhibitory concentration
MLX	meloxicam
MMAD	median aerodynamic diameter
MN	miconazole nitrate
MP	maleilated pullulan

MPS	mononuclear phagocytic system
MRI	magnetic resonance imaging
MTX	methotrexate
MW	molecular weight
MWCNT	multiwalled carbon nanotube
nCmP	nanocomposite microparticles
NCS	neocarzinostatin
NE	neem
NP	nanoparticle
ODN	oligonucleotide
PA	pectin-adriamycin ester
PA	poly alcohols
PAD	polyaldehyde dextran
PAM	pectin–adriamycin
PAMAM	polyamidoamine
PANCA	poly-acrylonitrile-co-acrylic acid
PBS	phosphate buffered saline
PC	prednicarbate
PCL	poly ε-caprolactone
pDNA	plasmid DNA
PEG	polyethylene glycol
PEG 400 MS	polyethylene glycol 400 monostearate
PEI	polyethyleneimine
PEO	polyethylene oxide
PG	propylene glycol
PK	pharmacokinetics
PLA	poly lactic acid
PLGA	poly(lactide-coglycolide) acid
PLLA	poly L-lactide
PMMA	poly-methyl methacrylate
PNA	N-phenyl-1-naphthylamine
PNIPAAm	poly-N-isopropylacrylamide
PSS	poly-sodium 4-styrenesulfonate
PTT	poly tri-methylene terephthalate
PTX	paclitaxel
PU	pullulan

PU-ALD	pullulan aldehyde
PVA	poly vinyl alcohol
PVB	poly vinyl butyral
PVC	poly vinyl chloride
PVP	poly vinylpyrrolidone
QCh	quaternized chitosan
QRTPCR	quantitative real time PCR
RES	reticulo-endothelial system
rGO	reduced graphene oxide
RJS	rotary jet spinning
RME	receptor mediated endocytosis
RNAi	RNA interference
RND	resistance-nodulation-cell division
ROS	reactive oxygen species
RRM2	ribonucleotide reductase
RT	room temperature
SA	sodium alginate
SA	stearyl alcohol
SC	stratum corneum
SC	succinylcurcumin
SCF	supercritical fluid
SD	suprofen-dextran
SELEX	systematic evolution of ligands by exponential enrichment
SF	silk fibroin
SF	stearyl ferulate
SFD	spray-freeze drying
SFD	submicrofiber disc
SFS	submicrofiber sheet
SHCC	secondary hepatic carcinoma
siRNA	small-interfering RNA
SLN	solid lipid nanoparticles
SMA	styrene maleic acid
SMR	small multi drug resistance
sPEG	O-methyl-O'-succinyl polyethylene glycol
SPION	super paramagnetic iron oxide nanoparticles
SWCNT	single walled carbon nanotube

SX	Sialyl Lewis^x
TAT	trans activator transcriptor
TDC	titanocene dichloride
TIP	tobramycin-inhaled powder
TOB	tobramycin
TP	triptolide
TPP	tripolyphosphate
TRN	tretinoin
TTR	transthyretin
Tween 80	Polysorbate 80
URCP	urocanylpullulan
UVR	ultraviolet radiation
WP	whey proteins
XXDR	extremely drug resistant tuberculosis
ZnPP	zinc protoporphyrin IX
α-CDE	α-cyclodextrin
βCDPs	β-CyD-containing polycations

PREFACE

This edited book, *Drug Delivery Approaches and Nanosystems*, is comprised of two volumes—*Volume I: Novel Drug Carriers* and *Volume II: Drug Targeting Aspects of Nanotechnology*. The volumes present a full picture of the state-of-the-art research and development of actionable knowledge discovery in the real-world discovery of drug delivery systems using nanotechnology and its applications.

The book is triggered by the ubiquitous applications of nanotechnology or nano-sized materials in the medical field, and the real-world challenges and complexities to the current drug delivery methodologies and techniques.

As we have seen, and as is often addressed though many methods have been used, very few of them have been validated in medical use.

A major reason for the above situation, we believe, is the gap between academia and research and the gap between academic research and real-time clinical applications and needs.

This book, *Drug Delivery Approaches and Nanosystems: Drug Targeting Aspects of Nanotechnology*, includes 11 chapters that contain, information in particular about the targeting facet of drug delivery systems. Targeting is a focused maneuver to achieve desired goals. Similarly the drug could be escorted to specific locations where activity is desired. By this the normal cells/tissues/organs are spared, and only the ailing ones receive the drug treatment. The targeting has now been successfully achieved in several diseases/disorders; however, its role is noteworthy in cancer treatment where chemotherapy is the main course of approach. Nanotechnology-based products have great potential by virtue of their inherent features.

This edited book provides a detailed application of nanotechnology in drug delivery systems in health care systems and medical applications. The book discusses general principles of drug targeting, materials of construction and technology concerns of nanoparticles, and different drug delivery systems and their preparation thereof.

The role of nanoscience to provide site-specific delivery of active pharmaceutical ingredients is given in Chapter 1, *Drug Targeting: General Aspects and Relevance to Nanotechnology*, written by Preethi Naik and associates. The authors have attempted to describe the diverse approaches to achieving drug targeting with special attention to treatment of infectious diseases and cancer, focusing on the trending delivery systems being formulated.

The targeting to tumor by a variety of nanocarriers has been explained in Chapter 2, *Nano-Oncotargets and Innovative Therapies*, written by Yamuna Mohanram and Lakshmi Kiran Chelluri. The multidisciplinary approach has been described for innovative approaches in treating the cancer. The way the nano-particles have driven into current applications are elucidated. Various materials to make them bio-compatible for human applications have been described to help advance our understanding of this highly exploratory science.

Chapter 3, *Dendrimer Conjugates as Novel DNA and siRNA Carriers*, written by Yuya Hayashi and colleagues, focuses on the potential of dendrimers to be exploited for delivery of genetic substances. The potential use of various polyamidoamine (PAMAM) starburst™ dendrimer conjugates with α-cyclodextrin (α-CDE) as a novel DNA and siRNA carriers (vectors) has been demonstrated. This review reports on the recent findings on hepatocyte-targeted α-CDE for DNA and siRNA delivery.

The synthesis of polysaccharides via agency of enzymes has been discussed in Chapter 4, *Precise Preparation of Functional Polysaccharide Nanoparticles by an Enzymatic Approach*, written by Jun-ichi Kadokawa. Polysaccharides are one of the major classes of biological macromolecules, which often have very complicated structures composed of a wide variety of monosaccharide units and glycosidic linkages. This chapter mainly deals with phosphorylase- and sucrase-catalyzed synthesis of functional polysaccharides with controlled nanostructures such as nanoparticles.

The promise exhibited by polysaccharides in drug targeting has been provided in Chapter 5, *Polysaccharide-Drug Nanoconjugates as Targeted Delivery Vehicles*, written by Sabyasachi Maiti. The polymeric prodrug conjugates are being widely investigated to solve different bio-performance related issues for a variety of drugs. This overview focuses on some water-soluble natural polymers such as alginate, pullulan, dextran and pectin; and their drug conjugates.

The profile and applications of glycans that are linear or branched polymer molecules composed of repeating sugar units adjoined by glycosidic bonds is given in Chapter 6, *Glycan-Based Nanocarries in Drug Delivery*, written by Songul Yasar Yıldız and colleagues. The various properties of glycans, such as biocompatibility, solubility, potential for modification, and innate bioactivity, offer great potential for their use in drug delivery systems (DDS). There is an observed increase in investigations involving use of glycans to formulate nanosystems. In this chapter, after a brief introduction to glycans and glycan-based DDS, various glycans, including chitosan, hyaluronic acid, alginate, dextran, hydroxyethyl starch, levan and mannan that are used in drug delivery, are summarized.

The clinical aspects of nanotechnology, in particular of carbon nanotubes and graphene, have been described in Chapter 7, *Physiological and Clinical Considerations of Drug Delivery Systems Containing Carbon Nanotubes and Graphene*, written by Sami Makharza and colleagues. This chapter focuses on understanding the physiological and clinical issues concerning the shape, surface chemistry, and dimensionality of Nanosized particles. Moreover, the chapter discusses the cellular paradigm (*in vitro* and *in vivo*) in order to have a complete overview of the applicability of these engineered nanoparticles.

Chapter 8, *Nanofibers: General Aspects and Applications*, written by Raghavendra Ramalingam and Kantha Deivi Arunachalam, explains the characteristics of nanofibers relevant to drug delivery applications. The chapter begins with a classification of these nanostructures and stretches to applications in treatment of variety of diseases/disorders.

The nCmP formulations have been described in Chapter 9, *Nanocomposite Microparticles (nCmP) for Pulmonary Drug Delivery Applications*, written by Zimeng Wang and associates. Over the past few years, research in the field of pulmonary drug delivery has gained momentum for the treatment of pulmonary and non-pulmonary diseases. The nCmP preparations are 'intelligent' systems widely used to deliver small molecules, genes, protein/peptides or drug nanoparticles to various targeted locations in the body.

Chapter 10, *Solid Lipid Nanoparticles for Topical Drug Delivery*, written by Sonia Trombino and Roberta Cassano highlights the role of solid lipid nanoparticles (SLNs) as topical drug carriers of antimicotic, antiinflammatory, antibiotic and antioxidant molecules. In recent years various strategies have been emerging to optimize the delivery across the skin. In particular, formulations that enhance the penetration of drugs through the stratum corneum are essential to improve their efficacy as topical agents. In this context the SLNs are attracting great attention.

The applications of hydrophilic drug carriers are provided in Chapter 11, *Hydrophobized Polymers for Encapsulation of Amphotericin B in Nanoparticles*, written by Yoshiharu Kaneo, summarizes the usefulness of self-assembled nanoparticles using a variety of synthetic polymers and polysaccharides. The driving force for the development of such systems is poor water solubility of a number of drugs for instance amphotericin B.

The book also provides detailed information on the application of nanotechnology in drug delivery systems in health care systems and medicine with an emphasis on inherent targeting potential of nanocarriers.

This volume provides a wealth of information that will be valuable to scientists and researchers, faculty, and students.

ABOUT THE EDITORS

Raj K. Keservani, MPharm
Faculty of Pharmaceutics, Sagar Institute of Research and Technology-Pharmacy, Bhopal, India

Raj K. Keservani, MPharm, has more than seven years of academic (teaching) experience from various institutes of India in pharmaceutical education. He has published 25 peer-reviewed papers in the field of pharmaceutical sciences in national and international journals, 15 book chapters, two co-authored books, and two edited books. He is also active as a reviewer for several international scientific journals. Mr. Keservani graduated with a pharmacy degree from the Department of Pharmacy from Kumaun University, Nainital (UA), India. He received his Master of Pharmacy (MPharm) (specialization in pharmaceutics) from the School of Pharmaceutical Sciences, Rajiv Gandhi Proudyogiki Vishwavidyalaya, Bhopal, India. His research interests include nutraceutical and functional foods, novel drug delivery systems (NDDS), transdermal drug delivery/ drug delivery, health science, cancer biology, and neurobiology.

Anil K. Sharma, MPharm
Delhi Institute of Pharmaceutical Sciences and Research, University of Delhi, India

Anil K. Sharma, MPharm, is working as a lecturer at the Delhi Institute of Pharmaceutical Sciences and Research, University of Delhi, India. He has more than seven years of academic experience in pharmaceutical sciences. He has published 25 peer-reviewed papers in the field of pharmaceutical sciences in national and international journals as well as 12 book chapters. He received a bachelor's degree in pharmacy from the University of Rajasthan, Jaipur, India, and a Master of Pharmacy degree from the School of Pharmaceutical Sciences, Rajiv Gandhi Proudyogiki Vishwavidyalaya,

Bhopal, India, with a specialization in pharmaceutics. His research interests encompass nutraceutical and functional foods, novel drug delivery systems (NDDS), drug delivery, nanotechnology, health science/life science, and biology/cancer biology/neurobiology.

Rajesh K. Kesharwani, PhD
Faculty, Department of Biotechnology, NIET, NIMS University, Jaipur, India

Rajesh K. Kesharwani, PhD, has more than seven years of research and two years of teaching experience in various institutes of India, imparting bioinformatics and biotechnology education. He has received several awards, including the NASI-Swarna Jayanti Puruskar-2013 by The National Academy of Sciences of India. He has authored over 32 peer-reviewed articles and 10 book chapters. He has been a member of many scientific communities as well as a reviewer for many international journals. Dr. Kesharwani received a BSc in biology from Ewing Christian College, Allahabad, India, an autonomous college of the University of Allahabad; his MSc (Biochemistry) from Awadesh Pratap Singh University, Rewa, Madhya Pradesh, India; and MTech-IT (specialization in Bioinformatics) from the Indian Institute of Information Technology, Allahabad, India. He earned his PhD from the Indian Institute of Information Technology, Allahabad, and received a Ministry of Human Resource Development (India) Fellowship and Senior Research Fellowship from the Indian Council of Medical Research, India. His research fields of interest are medical informatics, protein structure and function prediction, computer-aided drug designing, structural biology, drug delivery, cancer biology, and next-generation sequence analysis.

CHAPTER 1

DRUG TARGETING: GENERAL ASPECTS AND RELEVANCE TO NANOTECHNOLOGY

PREETHI NAIK, MEGHA MARWAH, MEENAKSHI VENKATARAMAN, GOPAL SINGH RAJAWAT, and MANGAL NAGARSENKER*

Department of Pharmaceutics, Bombay College of Pharmacy, Mumbai–400098, India
E-mail: mangal.nagarsenker@gmail.com

CONTENTS

ABSTRACT

Over time immemorial the fight against diseases has warranted development of drugs and drug delivery systems for achieving better disease management and improved survival. One such example is the method of preferentially accumulating active drug molecules at the relevant pharmacological sites, referred to as drug targeting. A journey that started with the concept of magic bullet by Dr. Paul Ehrlich, traversed through development of different delivery system, today has culminated into an era of modern drug therapy encompassing nano-sized drug carrier systems and/or medical devices. In this chapter the authors have attempted to describe the diverse approaches to achieving drug targeting with special attention to treatment of infectious diseases and cancer, focusing on the trending delivery systems being formulated. Whilst elaborating the advantages and application of this drug delivery boon, the authors have also touched upon the pitfalls and shortcomings of present day drug targeting systems.

1.1 INTRODUCTION

Through decades of intensive research and revolutionary discoveries in basic science coupled with the advancement in technology, the scientific community has achieved tremendous progress and transformed the landscape of drug discovery and drug development. Success of drug therapy and disease treatment lies in availability of drug at specific sites, at therapeutically effective concentration for sufficient time period. This can be achieved by designing targeted drug delivery systems, also referred to as smart delivery systems (Muller and Keck, 2004). These systems selectively and effectively localize drugs at predetermined target(s) in therapeutic concentration, while restricting its access to nontarget(s) normal cellular linings (Torchilin, 2010). In a general sense, drug targeting refers to a mode of specifically and quantitatively accumulating drug at relevant pharmacological sites relative to others, independent of route of administration. The specificity can be achieved by designing a formulation or device that facilitates drug administration, improves pharmacokinetics and biodistribution of drug while enhancing the efficacy and safety of the drug therapy.

1.2   A BRIEF INSIGHT INTO THE CHRONICLES THAT LED TO THE CONCEPT OF TARGETING

The battle of cancer has encumbered human race since times immemorial and is still the leading cause of death across the globe. The first incidence of cancer, reported in ancient Egypt, dates back to 1500 BC (Sudhakar, 2009). Investigating the underlying cause of cancer in addition to comprehending the complexity of this disease engendered theories and hypotheses in the early twentieth century. The humoral theory instituted by the Greek physician Hippocrates, who also coined the term 'cancer', (from the Greek word carcinos) prevailed through the middle ages for a span of 1300 years. The journey traversed from then to the present times, has witnessed a considerable growth in the conception of molecular and cellular pathways ensuing the onset of cancer.

Retrocession in time brings to light the inception of targeting with the concept of magic bullet, introduced by German Nobel laureate Paul Ehrlich, which served as the Holy Grail on the drug developmental frontier (Strebhardt and Ullrich, 2008). He envisaged that it was possible to specifically kill bacteria without harming body cells. In this attempt, he successfully developed Salvarsan, an arsenic derivative against bacteria causing Syphilis (Ehrlich, 1913; Strebhardt and Ullrich, 2008; Winau et al., 2004). This was the first targeted delivery system. Chemotherapy, made its presence felt at the time of World War II, when Louis Goodman and Alfred Gilman demonstrated the use of nitrogen mustard in tumor regression for the first time (Goodman et al., 1946). Subsequently, Hertz and Li established the application of methotrexate, a small molecule inhibitor, in the treatment of choriocarcinoma, which corroborated the hypothesis of Ehrlich (Li et al., 1958). The magic bullet theory thus revolutionized the preexisting chemotherapeutic era and continues to augment research efforts directed towards modern day therapy, particularly contributing to the ambit of disease management.

The journey from the naissance of the concept of magic bullet to the present in the development of targeted drug delivery has been delineated

TABLE 1.1 The Journey From the Naissance of the Concept of Magic Bullet to the Present in the Development of Targeted Drug Delivery

Sr. no.	Timeline	Milestone
1	1906–1910	Discovery of magic bullet and invention of chemotherapy with Salvarsan.
2	1928	Discovery of Penicillin by Dr. Fleming. Invention of antibacterial antibiotics.
3	1946	First anticancer chemotherapy using Nitrogen mustard by Goodman, Gilman and Linkog.
4	1958	Use of methotrexate in cancer treatment.
5	1964	Development of first controlled release device (Ocusert®) by Alza Corporation.
6	Late 1960s	Development of human vaccines against tetanus, diphtheria using nanoparticles by Dr. Speiser.
7	1970s	Advent of sustained release system.
8	1950–1960	Emergence of Penicillin resistance. Development of other class of antibiotics like Anthracyclins and Macrolids.
9	1960–1980	Two decades of tremendous progress in the field of nanotechnology-based drug delivery. Beginning of concept of PEGylation for improved circulation. Development of polymer-drug conjugates as targeting systems
10	1980s	Beginning of the era of Cephalosporins and Quinolones.
11	1984	Initiation studies on Enhanced Permeation and Retention effect in cancer therapy.
12	1990	First Amphotericin B liposomal preparation under brand name AmBisome® – Inception of liposomal drug delivery.
13	1992	FDA approval of Taxol®, parenteral formulation containing Paclitaxel.
14	1995	FDA approval of Doxil®, first PEGylated doxorubicin loaded liposome for cancer treatment.
15	1996	Imatinib (Gleevec®) – Dawn of targeted treatment.
16	1998	Commercial success of antibody guided magic bullet with Trastuzumab (Herceptin) for breast cancer treatment.
17	Early 2000	Nanoparticle-based drug delivery across Blood Brain Barrier.
18	Late 2000s	Antibiotic Resistance era and emergence of multi drug resistant Tuberculosis.
19	2000 – 2010	Remarkable advancement in drug delivery strategies. Advent of gene-based therapy.

TABLE 1.1 (Continued)

Sr. no.	Timeline	Milestone
20	2012	FDA approves Abraxane®, injectable suspension of protein bound Paclitaxel for metastatic cancer therapy.
21	Ongoing	Battle for improved cancer chemotherapy and treatment of infectious disease continues.

in Table 1.1 (Hoffman, 2008; Kreuter, 2007; Shi et al., 2011; Strebhardt and Ullrich, 2008).

As the knowledge bank of biology, physiology and disease etiology increased, the exciting journey of drug delivery has transformed from era of macro to micro to nano system and has presently culminated to an era of bio-controlled delivery systems.

It is a herculean task to discuss the continuously growing wide variety of targeted drug delivery approaches since each differs with respect to targeting method, target site and expected outcomes. In the following chapter, the authors have attempted to provide some general understanding of the field giving representative examples of current scientific interest and their application in fighting infectious diseases and cancer.

1.3 TARGETED DRUG DELIVERY: BASIC CONCEPTS

For many years, chemical molecules or drugs have bolstered the body in its fight against diseased state or disorder. The therapeutic response of a drug depends upon the interaction of drug molecules with cell in a concentration dependent manner. During transit from site of administration to site of action, the drug encounters biological barriers, different environments and enzymes. Traditional chemotherapy, wherein fate of drug was governed by molecular physicochemical properties, was plagued with therapeutic failure owing to inactivation or loss of drug in vivo and increased incidences of drug toxicity due to administration of larger doses. This marked the advent of carrier-based drug delivery which protected the drug against a hostile environment, helped reduce drug dose, improved pharmacokinetic and pharmacodynamic profile and consequently enhanced efficacy and safety of therapy (Gaumet et al., 2008).

The obvious advantages of targeting include:

1. Improved therapeutic efficacy due to sharp increase in drug concentration at target site(s).
2. Lower frequency of administration.
3. Decrease in adverse side effects and off target toxicity of drug.
4. Overcoming drug related problems like poor pharmacokinetics, enzymatic degradation, and low specificity.
5. Means of administering drug of low therapeutic index.
6. Decrease in the overall cost of therapy.

1.4 APPROACHES TO DRUG TARGETING

In a broad sense, targeting can be achieved by one or more of the following approaches detailed below:

1. Chemical modification of drug at molecular level.
2. Entrapment of drug into a carrier system which imparts targeting.
3. Controlled release of drug from the carrier system to the bioenvironment so as to ensure programmed and desired biodistribution.

1.4.1 CHEMICAL MODIFICATION OF DRUG AT MOLECULAR LEVEL

Direct coupling of targeting moiety to drug is the simplest way of drug targeting. This approach involves chemically modifying the structure of the drug either through derivatization into prodrugs or covalent conjugation with other agents (Juillerat et al., 2007). Prodrugs are biorevesible derivatives of drug molecules which undergo chemical or enzymatic transformation to the pharmacologically active parent molecule in vivo (Dahan et al., 2014). It is a powerful tool to improve physicochemical (Lang et al., 2014; Patel et al., 2014; Vig et al., 2013) and/or pharmacokinetic (Liang et al., 2013; Sheng et al., 2015) properties of the parent molecule. The greatest potential of prodrug is organ specific targeted delivery on account of site-specific bioconversion. Classical example is CNS targeting using Levodopa, which readily crosses BBB and upon biotransformation to dopamine gets trapped there, thus enabling improved pharmacodynamic effect (Rautio et al., 2008). Recent advances in tumor targeting approach employing prodrugs involves antibody-directed enzyme prodrug therapy (ADEPT) and gene-directed enzyme prodrug therapy (GDEPT) which have been discussed in the later part of chapter. Co-drug is a form of prodrug wherein two active moieties are chemically tethered together. Since the constituent drugs are indicated for the same disease and may function via disparate mechanisms of action, codrug approach can result in synergistic activity and superior therapy (Chen et al., 2012; Huang et al., 2014). Drug-polymer conjugation is another impressive chemical modification approach for improving therapeutic efficacy of drugs. The wide flexibility and variety in the properties of polymer imparts manifold applicability for such conjugates (Khandare and Minko, 2006; Kopeček, 2013; Pang et al., 2013). These systems encompass conjugates of various actives like chemical agents, proteins, enzymes with different polymers like poly lactic glycolic acid (PLGA), Polyethylene glycol (PEG), polysaccharides, N-(2-hydroxypropyl) methacrylamide (HPMA) to name a few. Bioconjugation is a novel field of covalent linkage of two moieties, at least one of which is a biomolecule. Classical examples include folate conjugated methotrexate (Zhang et al., 2010); transferrin conjugated Paclitaxel (Gan and Feng, 2010). It was envisaged that conjugation of antibodies imparted unique specificity in drug targeting, thus burgeoning the field of immunoconjugates (Alley et al., 2010). This approach offers a lethal cocktail of high specificity of monoclonal antibodies to target antigen combined with high potency of anticancer agents. The emergence of immunotoxins, wherein the active (toxic) moiety of a natural toxin was attached to an antibody, in the mid-1980s, has now transformed into promising clinical responses of immunoconjugates. For example, Brentuximab vedotin-Auristatin conjugate (ADCETRIS®) has been approved for Hodgkin lymphoma (Sievers and Senter, 2013).

1.4.2 ENTRAPMENT OF DRUG INTO A CARRIER SYSTEM WHICH IMPARTS TARGETING

Development of carrier systems for targeting drugs has been an old science which has evolved over years of research and development offering diverse advantages and applications. The principal strategies in the scheme of targeting are:

a. physical targeting;
b. passive targeting; and
c. active targeting.

A summary of different targeting modalities employing various nano-carriers has been depicted in Figure 1.1.

1.4.2.1 Physical Targeting

Physical targeting is by virtue of changes in physical factors in the bioen-vironment. These triggering factors may be endogenous like temperature, pH, and redox condition or exogenous like applied magnetic field, sound, and heat. The nanocarriers employed become active participants in ther-apy and are also referred to as stimuli responsive delivery systems (Mura et al., 2013). It is known that under certain pathological conditions the micro-bio-environment of the target site is appreciably different from normal physiological condition. These differences like low pH due to acidosis, low pH in tumor, high temperature due to hyperthermia and redox

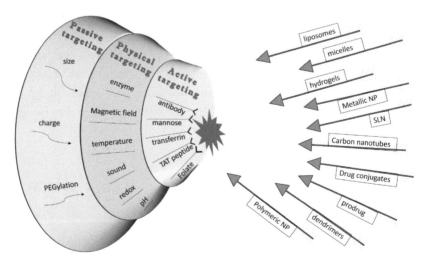

FIGURE 1.1   A summary of different targeting modalities employing various nanocarriers.

condition of tumor (Huo et al., 2014) are capitalized as triggers for drug accumulation and release. The unique benefit of such system is that even though the drug is distributed throughout the body, its therapeutic activity is elucidated only at the target site (Ganta et al., 2008). While polymeric particles are popular players in this approach, (Bawa et al., 2009), carriers like liposomes, micelles, hydrogels and dendrimers have also been reported (Mura et al., 2013). Similarly, triggered delivery systems responsive to exogenous stimuli include nanocarriers containing ferromagnetic properties that can be easily maneuvered under the influence of externally applied magnetic field thus offering targeting. The need for high strength magnetic field and blood flow velocity limits this targeting application. However, successful theronaustic systems have been reported (Yang et al., 2011). In the same way, vehicles responsive to sound have been employed for drug/gene delivery and theronaustic purposes. Ultrasound focusing has been found to disrupt intravascular endothelial cells creating pores for drug transport into the target tissue as observed with blood–brain barrier/ blood tumor barrier (Park et al., 2012). Major class of sound aided drug delivery vehicles include microbubbles, micelles, liposomes, and perfluorocarbon nanoparticles with noteworthy applicability in transdermal delivery, delivery to solid tumors, and delivery across the blood brain barrier, and enhanced thrombolysis (Ferrara, 2008).

1.4.2.2 Passive Targeting

Passive targeting is driven by virtue of particle properties of size, shape and surface and, abnormal physiological factors of cellular permeation under diseased condition (Devarajan and Jain, 2015). Enhanced permeation and retention (EPR) effect and uptake by Reticuloendothelial Systems (RES) are two important physiological considerations to design systems for passive targeting (Torchilin, 2010). EPR effect is characteristic of tumors, whereby, the tissue is marked by the presence of large gaps with fenestrations ranging from 100 to 1000 nm in the vascular endothelial cells as opposed to 10–20 nm of normal vascular endothelium, leading to improved permeation. Moreover, it lacks functional lymphatic drainage retaining the permeated material in the tumor (Duan and Li, 2013). In other words, EPR effect talks about high permeability due to leaky vasculature and high retention due to compromised lymphatic filtration associated with tumor tissue. Thus, passive targeting by taking advantage of EPR can be realized by tailoring size within appropriate range of 10–500 nm (Torchilin, 2010, 2011).

RES is a network of phagocytic cells like monocytes and macrophages, primarily accumulated in lymph nodes, liver, spleen, lungs and bone marrow, which function to eliminate foreign particles from the body as a part of the immune response. Uptake and clearance of administered particles by the RES is one of the major barriers to drug therapy. RES uptake depends on factors like size and surface characteristics and these properties determine how the body responds to, distributes and eliminates administered particles (Nie, 2010). Particles of small size, less than 20 nm, spread widely to different organs but are rapidly excreted by the kidneys while larger particles (e.g., 1 µm) are cleared by RES. Particles > 100 nm are promptly sequestered by sinusoids of spleen and fenestrations of liver. Thus, size range of 20–100 nm is preferred for avoidance of RES scavenging (Duan and Li, 2013). Particle recognition by RES is also governed by surface charge and hydrophilicity. It has been generally observed that surface coating with hydrophilic agents like PEG, evades RES uptake and prolongs circulation. Effect of physicochemical properties of size and surface on biodistribution of nanoparticles has been presented in Table 1.2 (Duan and Li, 2013).

1.4.2.3 Active Targeting

Active targeting is the most efficient carrier mediated targeting strategy. This approach is inspired from the natural process of ligand-receptor interaction, except that the ligand is appended to a drug loaded carrier. The conception of active targeting dates back to 1970s when Ringsdorf suggested conjugation of targeting unit to insoluble carrier loaded with drug. Since then, advancement in the fields of molecular biology and nanoparticle engineering technology has offered recognition of number of targets and ligands and development of targeted delivery systems. Some of the common ligands employed include antibodies, proteins, mono/oligo/polysaccharides, transferrin, folic acid and others (Byrne et al., 2008).

TABLE 1.2 Influence of Physicochemical Properties of Nanosystems and Their Significance in Tumor Targeting by EPR Effect

Sr. no.	Physicochemical property	Significance on tumor targeting by the EPR effect	References
1	Size	Smaller particles accumulate faster and deeper in tumor tissue, but also bear potential to diffuse back equally fast. Optimally sized particles can diffuse and retain in poorly permeable tumors.	(Cabral et al., 2011). (Dreher et al., 2006)
2	Charge	Overall charge on particles affects opsonization by the RES. Negative surface charge could increase, decrease or bear no effect on the circulation time of nanosystems, while positively charged systems have reduced blood circulation times. Positively charged nanosystems also favor interaction with endothelial cells, thereby bearing application in antiangiogenic therapy.	(Schmitt-Sody et al., 2003) (He et al., 2010) (Xiao et al., 2011) (Strieth et al., 2008)
3	Shape	Shapes other than spherical could affect extravasation into tumors owing to a higher aspect ratio irrespective of their size.	(Shukla et al., 2012)
4	Hydrophilicity	Increasing surface hydrophilicity of nanosystems with the aid of polymers like PEG confers the ability to circulate in blood for a longer span of time and reduce RES uptake for the same.	(Cho et al., 2008) (Decuzzi et al., 2009)

Active targeting can be classified into three levels of targeting as:

1. First order targeting or organ targeting: homing the particles into a particular organ.
2. Second order targeting or cellular targeting: cell surface recognition resulting in uptake by cells.
3. Third order targeting or subcellular targeting: internalization of particles within the cell and effect selective binding to specific cell organelles.

It is observed that two qualities viz. specificity and deliverability are essential for active targeted delivery system. The highly specific recognition and binding of ligand to the cell surface receptor on the target site not only ensures specificity but also significantly reduces off target toxicity. It is imperative that the ligand loaded carrier system is in the vicinity of target site for effective ligand-receptor interaction. Carrier-based factors like choice and nature of ligand, type of ligand carrier conjugation, carrier architecture, ligand density, and size together determine the efficacy of active targeting (Bertrand et al., 2014). The specific applications of active targeting to cancer and infectious diseases shall be dealt subsequently.

1.4.3 CONTROLLED RELEASE OF DRUG FROM THE CARRIER SYSTEM TO THE BIOENVIRONMENT TO ENSURE PROGRAMMED BIODISTRIBUTION

In this targeting modality, the design of carrier system is such that release of drug is controlled over longer duration of time. Thus, carrier system is localized, enabling effective drug concentrations at target site for prolonged periods (Weiser and Saltzman, 2014). Classical examples include mucoadhesive systems (Andrews et al., 2009; Mansuri et al., 2016), gastrorententive delivery systems (Pawar et al., 2011), hydrogel-based systems (Vashist et al., 2014), biodegradable biocompatible polymeric nanoparticles (Yuan et al., 2010) and in situ gelling systems (Bakliwal and Pawar, 2010).

1.5 ROLE OF DRUG TARGETING IN INFECTIOUS DISEASE

1.5.1 INFECTIOUS DISEASES IN MODERN WORLD

Infectious diseases have afflicted mankind since centuries with changing faces from plague, cholera in the eighteenth century to influenza, polio, and HIV/AIDS epidemics in the nineteenth century to hepatitis, malaria, dengue, tuberculosis and H1N1 flu in the twentieth century and Ebola and Zika viral diseases being the most recent addition to the list. Emergence and reemergence of infectious diseases has added to the epidemiological profile in which these diseases influence morbidity and mortality to a great extent. The twentieth century has seen remarkable advances in medical research and treatments, yet infectious diseases kill more people than any other single cause and remain among the leading causes of death worldwide. Epidemic can assume form of outbreak restricted to a smaller population or sometimes-pandemic affecting people worldwide due to globalization and movement of people across different geographic zones. The unpredictability of epidemics causes enormous suffering and socio-economic loss to population. Over the centuries such occurrences have impacted living and survival and defined the course of human civilization and development.

1.5.2 PHYSIOLOGICAL RESPONSE TO INFECTION

Infection is invasion and multiplication of pathogenic microorganisms, such as bacteria, viruses, parasites or fungi that are not normally present in the body and can be spread, directly or indirectly, from one person to another. Disease occurs when normal cells of the body get damaged due to infection and clinical symptoms of illness appear. In response to infection, the immune system gets activated and a cascade of activities involving white blood cells, antibodies, macrophages occur and culminate in the elimination of the foreign invader. A pathogen challenges the immune system in many ways and the physiological response to infection has been shown in Figure 1.2.

1.5.3 BACKSLIDE OF CONVENTIONAL THERAPY: NEED FOR TARGETING

Morbidity and mortality associated with infections has been curtailed to a significant extent since the advent of a plethora of antimicrobial agents. However, over these years, there has been indiscriminate use of antibiotics which has ultimately led to the emergence of resistance. It is spreading at an alarming rate particularly in hospitals which are hotspots for transfusion of resistance traits among microorganisms. Resistance to majority of the anti-tubercular drugs owing to the evolution from multi drug resistance (MDR) tuberculosis to more recently extremely drug

FIGURE 1.2 Physiological response to infection.

resistant tuberculosis (XXDR) depicts how gruesome the situation is at present Smart bugs can evade susceptibility to antimicrobial activity broadly by these mechanisms; poor uptake of drug molecules owing to reduced permeability, active efflux of drug molecules, enzymatic inactivation of molecules, alteration of target sites. The lipopolysaccharide barrier in Gram-negative bacteria is one of the prime reasons for impermeability of antibiotics (Kumar and Schweizer, 2005). This barrier exhibits reduced uptake of hydrophobic and cationic agents (Bengoechea et al., 1998; Moriyón and López-Goñi, 2010). *Efflux systems* identify and efficiently push out multiple antibiotics in an energy dependent manner thus attributing to multidrug resistance (Kumar and Schweizer, 2005). Multi drug efflux transporter proteins can be classified to 5 families: the Major Facilitator Superfamily (MFS) (Marger and Saier, 1993), the ATP-binding cassette (ABC) superfamily (Van Veen and Konings, 1998), the Small Multi drug Resistance (SMR) family (Paulsen et al., 1996), the Resistance-Nodulation-cell Division (RND) superfamily (Saier et al., 1994), and the Multi drug Toxic Compound Extrusion (MATE) family (Brown et al., 1999). MFS transporter proteins are widely prevalent in Gram-positive bacteria like *S. aureus* (Yoshida et al., 1990), *B. subtilis* (Neyfakh, 1992), *and Streptococcus pneumonia* (Tait-Kamradt et al., 1997). In many cases two mechanisms, reduced uptake and efflux pumps, work in consonance to result in intrinsic/acquired resistance especially in Gram-negative bacteria. *Enzymes* modify antibiotics by hydrolysis (Abraham and Chain, 1940; Barthelemy et al., 1984), group transfer (Leslie et al., 1988) or redox mechanisms (Yang et al., 2004) thus reducing the critical intracellular concentration mandatory for antimicrobial action. Mutations in target sites such as penicillin binding proteins (Kosowska et al., 2004), certain enzymes like DNA gyrase (Willmott and Maxwell, 1993), RNA polymerase (Mariam et al., 2004), are some of the strategies adapted by microbes to elude antibacterial. *Biofilms* are an assembly of surface associated microorganisms comprising exopolysaccharide in a polymeric matrix (van Tilburg Bernardes et al., 2015). Artificial joints, catheters, prosthetic heart valves are devices susceptible to biofilm formation (Smith, 2005). Aforementioned mechanisms, lack of penetration, efflux systems, and inactivating enzymes are possible reasons for biofilm resistance.

1.5.4 TARGETING MODALITIES AGAINST INFECTIOUS DISEASES

In the face of rising incidences of resistance and drug failure, a single targeting approach is seldom effective. The use of combinatorial approach not only extends the life of drug but also helps eradicate infections. In the following section, the authors have attempted to describe some of the strategies trending in the last decade with a view to overcome problems of conventional approach.

1.5.4.1 Physical Targeting

Triggered systems responsive to altered pathological microenvironment of pH, temperature and specific enzymes at the infection sites have been capitalized. Anaerobic fermentation and metabolic activity of bacteria coupled with host immune response results in localized acidity at infection site (Radovic-Moreno et al., 2012). pH responsiveness of nanocarriers imparted by surface charge switching, owing to selective protonation of functional groups like imidazole (Radovic-Moreno et al., 2012), oxime (Jin et al., 2011) and piperidine present in the carrier components, at low pH can potentially treat polymicrobial infection associated with acidity. Their application includes treatment of *H. Pylori* infection of stomach (Siddalingam and Chidambaram, 2014), ulcerative colitis (You et al., 2015) and vaginal candidiasis (Johal et al., 2016). Similarly, 'on demand' delivery vehicles responsive to enzymes like phospholipase and phosphatase secreted by bacteria have been developed for antibiotic delivery (Thamphiwatana et al., 2014).

1.5.4.2 Passive Targeting

Unlike general targeting strategies which would aim to escape RES, treatment of diseases like tuberculosis, leishmaniasis, malaria and hepatitis warrants RES uptake and targeting liver and spleen macrophages which act as hosts for invading parasites. Phagocytosis and internalization of drug carriers inside the host cell is imperative for killing such intracellular pathogens. Passive targeting solely may not be effective however it is complementary and imperative for active targeting. Passive targeting of silver nanoparticles by virtue of small size against number of parasites like HIV virus, Gram-negative bacteria, *S. Typhi* has been described in literature (Blecher et al., 2011). Infection by major pathogens like S. aureus, *P. Aeroginosa, Staphylococi and Klebsiella* have been found to induce EPR effect mediated through protease or Bradykinin activation and effectors like NO, VEGF, eicosanoids (Azzopardi et al., 2012). Researchers have demonstrated early and significant macromolecular accumulation at sites of infection attributed to post infection EPR (Laverman et al., 2001). However, the significance of EPR in infection has not yet been fully exploited in the passive targeted delivery of antibiotics but opens new opportunities for betterment of antibiotic therapy against infections.

1.5.4.3 Active Targeting

Designing nanosystems with their surface modified with ligands to specific receptors is an unfaltering approach for highly regulated site-specific delivery of antibiotics to organ/cellular/subcellular levels. These bioengineered nanoparticles insure increased receptor recognition, internalization and intracellular localization of antimicrobials thus altering biodistribution to annihilate pathogens. Ligand mediated targeted systems are proficient

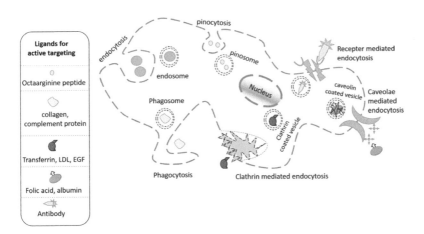

FIGURE 1.3 Uptake mechanism, intracellular trafficking pathways and ligands for active targeting of nanocarriers by macrophages.

TABLE 1.3 Examples of Active Targeting for Antiinfective Agents

Sr. No.	Nanocarrier	Ligand	Receptor	Disease	Therapeutic agent	Significant findings	Reference
1	Liposome	LFA-1 mAb	LFA-1 T cells Macrophages	HIV	siRNA	Reduction in plasma viral load in BLT mice.	(Kim et al., 2010)
2	Liposomes	Anti-HLA-DR (Fab' fragment)	HLA-DR	HIV	Indinavir	Increased concentration of Indinavir in lymphoid tissues.	(Gagné, Désormeaux et al., 2002)
3	Lipid nanoparticles	CD4 binding peptide	Lymphoid tissues	HIV	Indinavir	Higher drug concentration in lymph nodes. High antiviral activity in HIV infected macques.	(Kinman et al., 2006)
4	Gelatin nanoparticles	Mannan	Macrophage	HIV	Didanosine	Higher localization of drug in spleen, lymph node and brain.	(Kaur et al., 2008)
5	Liposomes	O-palmitoyl mannose	Lectin receptors on MPS	HIV	Stavudine	Significant level of drug in liver, spleen, lungs for upto 12h.	(Garg et al., 2006)
6	Polymeric nanoparticles	Trans activator transcriptor (TAT)	Brain	HIV	Ritonavir	Therapeutic levels of drug in brain and increased CNS bioavailability.	(Rao et al., 2008)
7	Albumin Nanoparticles	Transferrin	Brain	HIV	Azidothymidine	Enhanced localization in brain.	(Mishra et al., 2006)

TABLE 1.3 (Continued)

Sr. No.	Nanocarrier	Ligand	Receptor	Disease	Therapeutic agent	Significant findings	Reference
8	Liposomes	O-palmitoyl Mannan, O-palmitoyl Pullulan	Alveolar macrophages	Aspergilosis infection	Amphotericin B	Attain high concentration of drug in lungs for prolonged period of time (24 h).	(Vyas et al., 2005)
9	Liposomes	Tuftsin	Macrophages	Aspergilosis infection	Amphotericin B	Macrophage activation improves fungal chemotherapy.	(Owais et al., 1993)
10	Carbon nanotubes	Mannose	Macrophages	Leishmaniasis	Amphotericin B	Significant accumulation in macrophage rich organs.	(Pruthi et al., 2012)
11	Chitosan nanoparticles	Mannose	Macrophages	Leishmaniasis	Curcumin	Higher mean residence time of Curcumin with nanoparticle vis-a-vis Curcumin solution.	(Chaubey et al., 2014)
12	Nanoparticles	Folic acid	Folate receptors on macrophage	Tuberculosis	Rifampicin	Enhanced macrophage uptake.	(Patel et al., 2013)
13	Liposome	Antierythrocyte F(ab') 2	Erythrocyte	Malaria	Chloroquine	Enhanced liposomal binding to erythrocyte.	(Agrawal et al., 1987)

to exterminate deadly microbial reservoirs in macrophages as required in treatment of diseases like HIV (Duzgunes et al., 1999), tuberculosis (Nimje et al., 2009), influenza (Shiratsuchi et al., 2000), herpes simplex (Cheng et al., 2000). Cellular uptake of material occurs via multiple modes of receptor mediated endocytosis (RME) like clathrin mediated endocytosis, clathrin independent endocytosis, caveolae mediated endocytosis, macropinocytosis which exhibit different binding and internalization pathways as depicted in Figure 1.3. Following RME, the subsequent intracellular trafficking pathway governs the fate of engulfed material. Clathrin dependent and independent RME leads to accumulation in lysosomes and endosomes respectively (Bareford and Swaan, 2007). Active targeting molecules anchored in nanocarriers regulate internalization pathway. Cholesterol, albumin (Schnitzer, 2001) folic acid (Chang et al., 1992) facilitate uptake through caveolae mediated endocytosis (Bareford and Swaan, 2007). Ligands like collagen, glucose, mannan, mannose, fucose, for glycoreceptors promote clathrin-mediated endocytosis (Ezekovvitz et al., 1991; Greupink et al., 2005; Petros and DeSimone, 2010). Few examples of application of ligand coated nanocarriers in active targeting for antiinfective therapy has been enlisted in Table 1.3. Some of the noteworthy specialized targeting approaches which exemplify the eminence of ligand mediated active targeting in infectious diseases caused by three different parasites has been discussed below.

1.5.4.3.1 Malaria

Invasion and infection of RBCs and hepatocytes by *Plasmodium* species is the prime etiology of Malaria. Multiple drug resistance and lack of specific delivery to intracellular parasites with conventional malarial chemotherapy makes it an arduous task to localize drug in infected cells (Santos-Magalhães and Mosqueira, 2010). Ligand coated nanosystems have been designed targeting either infected erythrocytes in blood or hepatocytes in liver to combat parasites. Peptides, targeting receptors in hepatocytes, antibodies like MAb F10, antierythrocyte F(ab')2 targeting RBCs, carbohydrate molecules like pullulan that target asiologlycoprotein receptors overexpressed in hepatocytes, remarkably circumvent the issues with conventional systems

(Agrawal et al., 1987; Joshi and Devarajan, 2014; Longmuir et al., 2006; Owais et al., 1995; Singhal and Gupta, 1986). Liposomes' surface modified with amino acid sequences from *Plasmodium berghei* circumsporozoite protein with heparan sulfate proteoglycan binding sequence exhibited higher liver uptake in mice with extensive accumulation in hepatocytes and to some extent in nonparenchymal cells. This novel system with glycosaminoglycan recognition showed 100-fold higher liver uptake compared to heart, kidney, lung and more than 10-fold higher than spleen (Longmuir et al., 2006). Translocation of antimalarial drugs across blood brain barrier to treat cerebral malaria caused by invasion of parasites in brain is another formidable challenge. Transferrin (Tf), leptin, insulin, insulin like growth factors have high affinity to receptors in brain capillaries thereby forming primary candidates for brain targeted delivery. Gupta et al. have reportedly prepared Tf conjugated solid lipid nanoparticles (SLN) for localization of Quinine dihydrochloride, a polar drug prescribed in cerebral malaria with poor translocation across BBB owing to efflux pump, in brain tissues for treatment of cerebral malaria (Gupta et al., 2007). In vivo studies in albino rats revealed approximately 7 and 4 times higher brain uptake of the drug from Tf-SLN than free drug and unconjugated SLN, respectively.

1.5.4.3.2   HIV

HIV infection is highly progressive and pestilent wherein the virus targets immune cells *viz*, CD4+ T cells, monocytes, macrophages, and dendritic cells. Explicitly targeting the unique markers on these cells is a propitious strategy for prophylactic or therapeutic effect. A number of approaches have emerged for confining drug molecules into HIV infected cells. Glycoporteins gp120, gp41 present on viral envelope are exposed on infected host cells which makes them an attractive target for cell specific delivery (Ramana et al., 2014). Immunoliposomes' surface modified with ligand derived from Fab' fragment of HIV-gp120-directed monoclonal antibody F105 reportedly facilitates specific uptake by nonphagocytic HIV infected cells (Clayton et al., 2009). Aptamers can also be used in lieu of antibodies to avoid loss in specificity owing to conformational changes involved in formulation steps of immunoliposomes. They are small molecules of oligonucleotides or peptides developed using SELEX (systematic evolution of ligands by exponential enrichment) and exhibit selectivity like antibodies with excellent structural stability (Brody and Gold, 2000). Chemokine receptor, CCR5 (co receptor for HIV) is expressed on T cells, macrophages, microglial cells, and dendritic cells (Lee et al., 1999). Natural ligand CCR5 (He et al., 1997), synthetic peptide (Agrawal et al., 2004), monoclonal antibodies (Trkola et al., 2001) are few molecules reported for targeting this receptor. Tuftsin, a tetrapeptide derived from IgG, targets its receptors in macrophages and dendritic cells, activating phagocytosis to attenuate HIV reservoirs in these cells (Jain et al., 2010; Tzehoval et al., 1978). Numerous glycan residues like mannose are exposed on the surface of HIV glycoprotein gp120 which makes carbohydrate binding agents congruent targeting ligands for these receptors (Ji et al., 2005; Pollicita et al., 2008; Ramana et al., 2014). Actinohivin found in Actinomycetes is one such agent reported (Hoorelbeke et al., 2010). Human leukocyte antigen receptor (HLA-DR) (Gray et al., 1999), C-type lectin DC-SIGN (DC specific ICAM-3-grabbing nonintegrin) (W. Jin et al., 2014) and LFA-1 (leukocyte function-associated antigen) (Hioe et al., 2011; Kim et al., 2010) are promising targets identified for HIV therapy.

1.5.4.3.3 Tuberculosis

Tuberculosis is a notoriously recurring, tenacious infectious disease with alarming rise in resistant Mycobacterium strains infecting macrophages. Scientists are taking concerted efforts in developing targeted chemotherapy to overcome this hurdle. Lemmer Y and his team have maneuvered nanoparticles with mycolic acid (MA), a mycobacteial lipid which interacts with anti-MA antibodies, to be taken up by infected macrophages (Lemmer et al., 2015). Isoniazid loaded PLGA- MA targeted nanoparticles on assessment revealed higher uptake into *Mycobacterium* infected macrophages. Deol et al. (1997) have designed O-Stearyl amylopectin coated stealth liposomes for lung specific tuberculosis therapy. These systems loaded with isoniazid, and rifampin significantly reduced colony forming units in lung, liver and spleen compared to free drug. Novel mannosylated dendrimers reported by Kumar et al. displayed selectively higher uptake by alveolar macrophages compared to drug solution (Kumar et al., 2006).

1.5.4.4   Localization of Drug at Infection Site

Localization of drug at the site of infection allows achievement of therapeutically effective concentrations for longer duration. This approach is advantageous for antibiotic therapy since it reduces incidence of drug resistance by lowering drug dose, dosing frequency and improving patient compliance (Gao et al., 2011). Polymeric carriers are preferred for such applications as exemplified by hydrogels and mucoadhesive systems discussed below. Hydrogels are a class of swelling hydrophilic matrix systems which inherently possess antimicrobial activity owing to the polymeric material and/or antibiotics housed within, finding wide application in wound dressings (Ng et al., 2014). These swellable gels can also stabilize nanoparticles like liposomes within their matrix (Gao et al., 2011). Such dual action is incredibly efficient in breaking down biofilms and overcoming resistance (Li et al., 2013; Ng et al., 2014; Wu et al., 2014). Mucoadhesive systems localize drug along the mucosal epithelia and are applied to combat a variety of infections like, corneal infections, periodontitis, oropharyngeal candidiasis, infection of oral cavity (Nguyen and Hiorth, 2015). These systems have been extensively studied for antibiotic delivery to stomach and eradication of *H. Pylori* infections (Adebisi and Conway, 2015; Parreira et al., 2014; Siddalingam and Chidambaram, 2014). Combination of mucoadhesion and pH responsiveness of polymer present a dual attack against stomach ulcers of *H. Pylori* (Du et al., 2015).

1.6   ROLE OF DRUG TARGETING IN ONCOLOGY

1.6.1   INTRODUCTION

Cancer, a genetic disease, results from germline somatic and epigenetic alterations leading to uncontrolled growth of cells (Zhao and Adjei, 2014). These cellular aberrations are associated with constant proliferative signaling, perpetual replication and ability to withstand cell apoptosis. The interplay of proto oncogenes and tumor suppressor genes is responsible for the aforementioned peculiarities of cancerous cells.

1.6.2 SHORTCOMINGS OF CONVENTIONAL THERAPY

Classic chemotherapy, entails the use of cytotoxic/ cytostatic drugs or molecularly targeted therapies limited in their action by the physiology of the tumorous tissue and/or physicochemical and biological properties of anticancer agents being used per se. For example, 5-fluorouracil used in the treatment of colorectal cancer generates an overall response rate of only 10–15% (Zhang et al., 2008). Physiological factors associated with the tumor leading to chemotherapy failure include heterogeneous blood supply to tumors and heterogeneities in cells varying from tumor to tumor, high interstitial pressure owing to absence of lymphatics causing extravasation and dilution in the surrounding tissue, interstitial fibrosis leading to increased diffusional distance between the interstitium and tumor cells and resistance to chemotherapeutic agents by means of genetic alterations. These elements of tumor tissue make drug diffusion a formidable challenge thereby abating their chemotherapeutic response in treatment of cancer. Besides, drug properties like undesirable hydrophobicity, poor tumor specificity and off target toxicity, rapid elimination from circulation owing to a short half-life, metabolism to inactive compounds and unfavorable pharmacokinetic and biodistribution profiles further dwindle cancer treatment (Arias, 2011; Ruoslahti et al., 2010).

1.6.3 STRATEGIES FOR TUMOR TARGETING: THE CHEMOTHERAPEUTIC APPROACH

The confrontations posed by both the aforementioned aspects of chemotherapy failure have been exploited by scientists and contained within the armamentarium of cancer treatment. Development of personalized nanomedicine is the need of the hour for treating cancer. The tumor microenvironment and defective vasculature have been constructively used to deliver drugs via passive targeting by the EPR effect (explained in Section 1.4.2.2) while receptor overexpression on tumorous tissue have been capitalized employing surface engineered nanosystems by active targeting.

A successful tumor targeting technology relies on ability of nano-carrier to retain the cargo, evade the RES, target the tumor and then release the cargo only at the intended site of action (Bae and Park, 2011). Besides, maintaining high blood levels of the nanocarrier insures diffusion of the nanosystems towards the tumor without causing emanation in the backward direction, since diffusion is a concentration gradient dependent process.

The microenvironment presents parenchyma containing neoplastic cells, stroma containing host cells, extracellular matrix (containing collagen, glycosaminoglycans, proteins, debris and tumor associated macrophages and dendritic cells) and interstitial fluid barriers to the nanosystems extravasating from the neovasculature surrounding the tumor. The dissemination of colloidal nanosystems to the target tumor cells, therefore, depends on their concentration in the blood, extent of extravasation into the tumor interstitium, diffusion through the latter and interaction with the components of the interstitium. Each of these components play an elusive, yet important role in targeting the tumor cells by the EPR effect. While pharmacological approaches such as the use of enzymes like heparinases, hyaluronidases and collagenases might be able to overcome the extracellular matrix barrier in certain tumors (Alexandrakis et al., 2004; Lieleg et al., 2009), pharmaceutical technological approaches focus on controlling the physicochemical properties of nanosystems. The latter involve size, charge, surface hydrophilicity and shape of the nanosystems. While a hydrophilic coat of PEG on the surface of nanosystems confers a longer circulation half-life, a size of less than 200 nm helps evade RES effectively (Li et al., 2001; Owens and Peppas, 2006). Controlling each of the aforementioned factors is vital to accomplish tumor targeting by the EPR effect. The roles of these factors on targeting have been extensively reviewed in literature (Arias, 2011; Bertrand et al., 2014) and a few examples concerning their influence on passive targeting have been summarized in Table 1.4.

Active targeting to tumors entails the presence of a ligand on the nano-carrier surface, possessing avidity towards a surface molecule or specific receptor which may be overexpressed or exclusively expressed on diseased

TABLE 1.4 Examples of Active Targeting in Cancer Therapy

Cellular internalization mechanism	Ligand/receptor involved	Type of cancer: few examples	References
Phagocytosis (tumor associated macrophages)	Mannose, fructose, scavenger receptors	Breast cancer	(Aderem and Underhill, 1999) (Rabinovitch, 1995) (Zhu et al., 2013) (Williams et al., 2016)
Clathrin-dependent endocytosis (CME)	Polymers like PLA, PLL	Lung cancer, breast cancer, leukemia	(Vasir and Labhasetwar, 2008) (Harush-Frenkel et al., 2007)
	Aminosilane functionalized silica nanotubes		(Harush-Frenkel et al., 2008)
	Ligands like mannose-6-phosphate, transferrin, nicotinic acid		(Ng et al., 2015) (Kim et al., 2013) (Vizoso et al., 2015) (Kratz et al., 1998)
Caveolae mediated endocytosis	Pegylated nanoparticles like DOXIL, galactose residues, poly acrylic acid, cyclic RGD peptide	Liver cancer, Breast cancer, pancreatic tumor, glioblastoma, prostate cancer	(Sahay et al., 2010) (Nishikawa et al., 2009) (Zhang and Monteiro-Riviere, 2009) (Oba et al., 2008) (Furger et al., 2003) (Sheldrake and Patterson, 2009) (Hosotani et al., 2002)

TABLE 1.4 (Continued)

Cellular internalization mechanism	Ligand/receptor involved	Type of cancer: few examples	References
Clathrin and caveolae independent mechanisms	Interleukin-2, growth hormones, glycosylphosphatidylinositol	Metastatic renal cell carcinoma, malignant melanoma	(Doherty and McMahon, 2009) (Antony and Dudek, 2010)
Macropinocytosis	Nonspecific entry of nanomaterials, both small and large which cannot undergo phagocytosis		(Sahay et al., 2010)
Multiple pathways	PEG, Polyethylene imine, TAT peptide	Breast cancer, Ovarian cancer	(Rejman et al., 2005) (Patel et al., 2007) (Aroui et al., 2010)

target (Cheng et al., 2012; Koshkaryev et al., 2013; Peer et al., 2007; Shi et al., 2011; Tabernero et al., 2013). The ligand used for active targeting to tumor cells may be carbohydrate, protein or lipid in nature (Rieux et al., 2013; Yu et al., 2010) and can be appended to the surface of the nanocarrier either prior to or post formulation. Upon in vivo administration, such ligand decorated nanocarriers extravasate into the interstitial space of the tumors by the EPR effect, thus bringing them in close proximity of target cell receptors. This is followed by receptor-ligand interaction leading to internalization of the nanocarriers and subsequent intracellular release of cargo. Both passive (EPR effect) and ligand mediated active targeting, thus, work in concordance to accomplish targeted drug delivery to tumors.

1.6.3.1 Designing Nanocarriers for Active Targeting: Basic Considerations

The factors imperative in designing surface modified nanosystems vary, depending on the latter's proximity to the target and can be classified for convenience as follows:

1. Factors affecting extravasation of nanocarrier within tumor interstitium
   a. Nanocarrier size and shape
   b. Nanocarrier surface and charge
2. Factors affecting receptor-ligand interaction
   a. Ligand density and multivalency
   b. Nanocarrier size and ligand charge
   c. Surface hydrophobicity
3. Factors affecting nanocarrier internalization and intracellular trafficking
   a. Nanocarrier size, shape and charge
   b. Type of cell internalization mechanism determined by choice of ligand (described in Table 1.3)

1.6.3.1.1 Nanocarrier Size and Shape

Size and shape of the system dictate its extravasation into the tumor interstitium facilitated by the EPR effect in addition to opsonization by the RES in blood, as previously mentioned. Besides, size is governed by the ligand present on the surface of nanocarrier. Both size and shape also play a role in receptor interaction and cellular uptake of systems. Jiang et al. (2008) have demonstrated the importance of nanocarrier size at different stages in vivo. With the aid of Herceptin receptors attached to the surface of gold nanoparticles (Her-GNPs), the authors have deduced a higher avidity for 40 nm Her-GNPs vis-à-vis 2 nm particles, thereby allowing prolonged binding to ErbB2 receptors overexpressed in ovarian and breast cancers. Based on their findings, they have reported a size range of 40 to 50 nm to be vital to maximum receptor mediated intracellular uptake, attributing it to a balance between cross-linking with surface receptors and membrane wrapping of nanocarriers. Gratton et al. (2008) have demonstrated the role of micropinocytosis in the cellular uptake of large, micron sized particles (1 to 3 µ), while caveolae and clathrin mediated endocytosis were shown to facilitate internalization of nanoparticles in the range of 100 to 200 nm.

Shape of certain nanocarriers like nanotubes affect their transvascular transport, regardless of size constraints placed by the EPR effect. Rod shaped nanoparticles with dimensions of 54 nm length and 15 nm diameter revealed 4.1 times deeper tumor penetration vis-à-vis nanospheres of 35 nm diameter, when tested in female SCID mice bearing orthotopic E0771 mammary tumors (Chauhan et al., 2011). Besides, Barua et al. (2013) inferred a higher specific cellular uptake for nanorods compared to spheres, thereby disclosing the role of particle shape in intracellular trafficking of nanocarriers.

1.6.3.1.2 Nanocarrier Hydrophobicity, Ligand Density and Charge

Surface hydrophobicity and charge on the nanocarriers can be modified by the physicochemical properties of ligand present. The extent of ligand display (ligand density) on the surface of nanocarriers further determine ligand receptor interaction and consequent internalization. Insufficient surface revelation of ligand owing to increased hydrophobicity and reduced surface ligand density culminated into decreased binding to membrane receptors as shown by investigations carried out by Valencia et al. (2011). In the same study, higher surface hydrophobicity was also seen to be associated with increased macrophage uptake.

Besides, an overall cationic nanoparticle charge was shown to improve nonspecific cellular uptake (Zhao et al., 2011), while negatively charged nanocarriers were found to undergo decelerated uptake as compared to their positively charged counterparts (Sahay et al., 2010).

In order to devise an active targeting strategy with ligand decorated nanocarriers, the nature of ligand, its density and the overall nanoparticle architecture upon incorporation of ligand needs to be pre-investigated. Pragmatic choice of ligand, should, therefore, be the primary leitmotif of any investigator studying nanoparticle fabrication for oncology.

1.7   PITFALLS OF PRESENT DAY DRUG TARGETING

For over a century now, many targeting strategies have been evaluated and developed with aim of finding newer magic bullets for the wide-ranging sufferings of mankind. Inspite of these efforts, majority of them have failed to translate into clinical use. The reason for the same may be diverse like (Bae and Park, 2011; Lammers et al., 2012)

a) Over interpretation and/or misunderstanding of basic concepts used in targeting:
   i. EPR effect is highly heterogeneous phenomenon varying substantially from one tumor model to other and from patient to patient. There may be huge differences among vasculature within a single tumor.
   ii. Overexpression is a misnomer. It is a term that describes the relative abundance of specific marker on target tissue in comparison to nontarget ones. The true expression levels of target markers need to be reviewed (Bae and Park, 2011).
   iii. Inappropriate animal models in preclinical studies: The tumor growth in rodent models is much faster than that in humans, resulting in leakier vasculature in rodents than humans (Alley et al., 2010).

b) Systems are too complex to be up scaled for large scale production.

c) Lack of image guided insight to personalize therapeutic interventions.

d) Toxicity of nanosystems: Increased access of small sized nano-
    carrier systems with cellular and subcellular components esca-
    lates toxicity manifestation such as oxidative stress, inflammatory
    responses (Jain et al., 2015), platelet aggregation, mitochondrial
    damage, cardiovascular effects, hemolysis and blood clotting
    (De Jong and Borm, 2008). Accumulation of non-biodegradable
    material within the body induces chronic inflammatory response.

1.8   CONCLUSION

With over a century of research in drug delivery and targeting, today,
we are standing on a mountain of information that can be judiciously
employed to appease human suffering. The ultimate goal of targeted
drug delivery is to develop clinically useful formulations by improving
the safety and efficacy of chemotherapy. From the brief discussions of
few of the applications of targeted systems, it is evident that targeting has
great potentials in resolving important clinical problems. The different
targeting modalities are not independent but are interrelated with each
other. With increasing incidence of resistance, the future of targeted systems now moves towards development of cocktail of multiple strategies
within a single system. Whilst developing targeted systems one must bear
in mind the following considerations: (1) targeted systems alone hold the
key in overcoming existing problems of chemotherapy; (2) targeting must
result in preferential accumulation of material at target site. Hence, careful selection of target is vital; (3) though some basic concepts of targeting
are well understood and established, true targeted system is far from reality; and (4) simultaneous monitoring of biodistribution of nanoparticles
and assessing of therapeutic efficacy of system can help personalization
of nanomedicine. This can be achieved by co-incorporation of drug and
imaging agent within single formulation. Whilst advancing research in
development of nanosystems, it must be borne in mind that these systems pose significant toxicity issues. These observations clearly warrant
new ways of handling toxicity of nanocarriers wherein safety evaluation
and risk benefit analysis be performed on case to case basis and not rely
on toxicological data of bulk material alone. There is a need for a new

framework of regulatory guidelines along with laws, ethics and testing protocols.

KEYWORDS

- active targeting
- cancer
- infectious disease
- nanocarriers
- passive targeting
- targeting

REFERENCES

Abraham, E. P.; Chain, E. An enzyme from bacteria able to destroy penicillin. *Nature* **1940**, *146*(3713), 837.

Adebisi, A. O.; Conway, B. R. Modification of drug delivery to improve antibiotic targeting to the stomach. *Therapeutic delivery* **2015**, *6*(6), 741–762.

Aderem, A.; Underhill, D. M. Mechanisms of phagocytosis in macrophages. *Annu. Rev. Immunol.* **1999**, *17*(1), 593–623.

Agrawal, A. K.; Singhal, A.; Gupta, C. Functional drug targeting to erythrocytes in vivo using antibody bearing liposomes as drug vehicles. *Biochem. Biophys. Res. Commun.* **1987**, *148*(1), 357–361.

Agrawal, L.; VanHorn-Ali, Z.; Berger, E. A.; Alkhatib, G. Specific inhibition of HIV-1 coreceptor activity by synthetic peptides corresponding to the predicted extracellular loops of CCR5. *Blood* **2004**, *103*(4), 1211–1217.

Alexandrakis, G.; Brown, E. B.; Tong, R. T.; McKee, T. D.; Campbell, R. B.; Boucher, Y.; Jain, R. K. Two-photon fluorescence correlation microscopy reveals the two-phase nature of transport in tumors. *Nat. Med.* **2004**, *10*(2), 203–207.

Alley, S. C.; Okeley, N. M.; Senter, P. D. Antibody–drug conjugates: targeted drug delivery for cancer. *Curr. Opin. Chem. Bio.* **2010**, *14*(4), 529–537.

Andrews, G. P.; Laverty, T. P.; Jones, D. S. Mucoadhesive polymeric platforms for controlled drug delivery. *Eur. J. Biopharm.* **2009**, *71*(3), 505–518.

Antony, K. G.; Dudek, Z. A. Interleukin 2 in cancer therapy. *Curr. Med. Chem.* **2010**, *17*(29), 3297–3302.

Arias, J. L. Drug targeting strategies in cancer treatment: an overview. *Mini Rev. Med. Chem.* **2011**, *11*(1), 1–17.

Aroui, S.; Brahim, S.; De Waard, M.; Kenani, A. Cytotoxicity, intracellular distribution and uptake of doxorubicin and doxorubicin coupled to cell-penetrating peptides in different cell lines: a comparative study. *Biochem. Biophys. Res. Commun.* **2010**, *391*(1), 419–425.

Azzopardi, E. A.; Ferguson, E. L.; Thomas, D. W. The enhanced permeability retention effect: a new paradigm for drug targeting in infection. *J. Antimicrob. Chemother.* **2012**, 379.

Bae, Y. H.; Park, K. Targeted drug delivery to tumors: myths, reality and possibility. *J. Controll. Release* **2011**, *153*(3), 198.

Bakliwal, S.; Pawar, S. In-situ gel: new trends in controlled and sustained drug delivery system. *Int. J. PharmTech. Res.* **2010**, *2*, 1398–1408.

Bareford, L. M.; Swaan, P. W. Endocytic mechanisms for targeted drug delivery. *Adv. Drug Deliv. Rev.* **2007**, *59*(8), 748–758.

Barthelemy, P.; Autissier, D.; Gerbaud, G.; Courvalin, P. Enzymic hydrolysis of erythromycin by a strain of Escherichia coli. A new mechanism of resistance. *J. Antibiot.* **1984**, *37*(12), 1692–1696.

Barua, S.; Yoo, J.-W.; Kolhar, P.; Wakankar, A.; Gokarn, Y. R.; Mitragotri, S. Particle shape enhances specificity of antibody-displaying nanoparticles. *Proc. Natl. Acad. Sci.* **2013**, *110*(9), 3270–3275.

Bawa, P.; Pillay, V.; Choonara, Y. E.; du Toit, L. C. Stimuli-responsive polymers and their applications in drug delivery. *Biomed. Mater.* **2009**, *4*(2), 022001.

Bengoechea, J.-A.; Lindner, B.; Seydel, U.; Ramón, D.; Ignacio, M. Yersinia pseudotuberculosis and Yersinia pestis are more resistant to bactericidal cationic peptides than Yersinia enterocolitica. *Microbiology* **1998**, *144*(6), 1509–1515.

Bertrand, N.; Wu, J.; Xu, X.; Kamaly, N.; Farokhzad, O. C. Cancer nanotechnology: the impact of passive and active targeting in the era of modern cancer biology. *Adv. Drug Deliv. Rev.* **2014**, *66*, 2–25.

Blecher, K.; Nasir, A.; Friedman, A. The growing role of nanotechnology in combating infectious disease. *Virulence* **2011**, *2*(5), 395–401.

Brody, E. N.; Gold, L. Aptamers as therapeutic and diagnostic agents. *Reviews in Mol. Biotechnol.* **2000**, *74*(1), 5–13.

Brown, M. H.; Paulsen, I. T.; Skurray, R. A. The multidrug efflux protein NorM is a prototype of a new family of transporters. *Mol. Microbiol.* **1999**, *31*(1), 394–395.

Byrne, J. D.; Betancourt, T.; Brannon-Peppas, L. Active targeting schemes for nanoparticle systems in cancer therapeutics. *Adv. Drug Deliv. Rev.* **2008**, *60*(15), 1615–1626.

Cabral, H.; Matsumoto, Y.; Mizuno, K.; Chen, Q.; Murakami, M.; Kimura, M.; Terada, Y.; Kano, M.; Miyazono, K.; Uesaka, M. Accumulation of sub-100 nm polymeric micelles in poorly permeable tumors depends on size. *Nat. Nanotechnol.* **2011**, *6*(12), 815–823.

Chang, W.-J.; Rothberg, K. G.; Kamen, B. A.; Anderson, R. Lowering the cholesterol content of MA104 cells inhibits receptor-mediated transport of folate. *J. Cell Biol.* **1992**, *118*(1), 63–69.

Chaubey, P.; Patel, R. R.; Mishra, B. Development and optimization of curcumin-loaded mannosylated chitosan nanoparticles using response surface methodology in the treatment of visceral leishmaniasis. *Expert Opin. Drug Deliv.* **2014**, *11*(8), 1163–1181.

Chauhan, V. P.; Popović, Z.; Chen, O.; Cui, J.; Fukumura, D.; Bawendi, M. G.; Jain, R. K. Fluorescent Nanorods and Nanospheres for Real-Time In Vivo Probing of Nanoparticle Shape-Dependent Tumor Penetration. *Angewandte Chemie International Edition* **2011**, *50*(48), 11417–11420.

Chen, X.; Zenger, K.; Lupp, A.; Kling, B.; Heilmann, J. R.; Fleck, C.; Kraus, B.; Decker, M. Tacrine-silibinin codrug shows neuro-and hepatoprotective effects in vitro and procognitive and hepatoprotective effects in vivo. *J. Med. Chem.* **2012**, *55*(11), 5231–5242.

Cheng, H.; Tumpey, T. M.; Staats, H. F.; van Rooijen, N.; Oakes, J. E.; Lausch, R. N. Role of macrophages in restricting herpes simplex virus type 1 growth after ocular infection. *Invest. Ophthalmol. Vis. Sci.* **2000**, *41*(6), 1402–1409.

Cheng, Z.; Al Zaki, A.; Hui, J. Z.; Muzykantov, V. R.; Tsourkas, A. Multifunctional nanoparticles: cost versus benefit of adding targeting and imaging capabilities. *Science* **2012**, *338*(6109), 903–910.

Cho, K.; Wang, X.; Nie, S.; Shin, D. M. Therapeutic nanoparticles for drug delivery in cancer. *Clin. Cancer Res.* **2008**, *14*(5), 1310–1316.

Clayton, R.; Ohagen, A.; Nicol, F.; Del Vecchio, A. M.; Jonckers, T. H.; Goethals, O.; Van Loock, M.; Michiels, L.; Grigsby, J.; Xu, Z. Sustained and specific in vitro inhibition of HIV-1 replication by a protease inhibitor encapsulated in gp120-targeted liposomes. *Antiviral Res.* **2009**, *84*(2), 142–149.

Dahan, A.; Zimmermann, E. M.; Ben-Shabat, S. Modern prodrug design for targeted oral drug delivery. *Molecules* **2014**, *19*(10), 16489–16505.

De Jong, W. H.; Borm, P. J. Drug delivery and nanoparticles: applications and hazards. *Int. J. Nanomedicine* **2008**, *3*(2), 133.

Decuzzi, P.; Pasqualini, R.; Arap, W.; Ferrari, M. Intravascular delivery of particulate systems: does geometry really matter? *Pharm. Res.* **2009**, *26*(1), 235–243.

Deol, P.; Khuller, G.; Joshi, K. Therapeutic efficacies of isoniazid and rifampin encapsulated in lung-specific stealth liposomes against Mycobacterium tuberculosis infection induced in mice. *Antimicrob. Agents Chemother.* **1997**, *41*(6), 1211–1214.

des Rieux, A.; Pourcelle, V.; Cani, P. D.; Marchand-Brynaert, J.; Préat, V. Targeted nanoparticles with novel nonpeptidic ligands for oral delivery. *Adv. Drug Deliv. Rev.* **2013**, *65*(6), 833–844.

Devarajan, P. V.; Jain, S. (2015). *Targeted Drug Delivery: Concepts and Design*: Springer.

Doherty, G. J.; McMahon, H. T. Mechanisms of endocytosis. *Annu. Rev. Biochem.* **2009**, *78*, 857–902.

Dreher, M. R.; Liu, W.; Michelich, C. R.; Dewhirst, M. W.; Yuan, F.; Chilkoti, A.; Tumor vascular permeability, accumulation, and penetration of macromolecular drug carriers. *J. Natl. Cancer Inst.* **2006**, *98*(5), 335–344.

Du, H.; Liu, M.; Yang, X.; Zhai, G. The design of pH-sensitive chitosan-based formulations for gastrointestinal delivery. *Drug Discov. Today* **2015**, *20*(8), 1004–1011.

Duzgunes, N.; Pretzer, E.; Simoes, S.; Slepushkin, V.; Konopka, K.; Flasher, D.; Pedroso de Lima, M. C. Liposome-mediated delivery of antiviral agents to human immunodeficiency virus-infected cells. *Mol. Membr. Biol.* **1999**, *16*(1), 111–118.

Ehrlich, P. Address in pathology, on chemiotherapy: delivered before the Seventeenth International Congress of Medicine. *BMJ* **1913**, *2*(2746), 353.

Ezekovvitz, R.; Williamsi, D.; Kozieltil, H. Uptake of Pneumocystis carinii mediated by the macrophage mannose receptor. *Nature* **1991**, *351*(6322), 155–158.

Ferrara, K. W. Driving delivery vehicles with ultrasound. *Adv. Drug Deliv. Rev.* **2008**, *60*(10), 1097–1102.

Furger, K. A.; Allan, A. L.; Wilson, S. M.; Hota, C.; Vantyghem, S. A.; Postenka, C. O.; Al-Katib, W.; Chambers, A. F.; Tuck, A. B. β3 Integrin Expression Increases Breast Carcinoma Cell Responsiveness to the Malignancy-Enhancing Effects of Osteopontin. *Mol. Cancer Res.* **2003**, *1*(11), 810–819.

Gagné, J.-F.; Désormeaux, A.; Perron, S.; Tremblay, M. J.; Bergeron, M. G. Targeted delivery of indinavir to HIV-1 primary reservoirs with immunoliposomes. *BBA-Biomemb.* **2002**, *1558*(2), 198–210.

Gan, C. W.; Feng, S.-S. Transferrin-conjugated nanoparticles of poly (lactide)-D-α-tocopheryl polyethylene glycol succinate diblock copolymer for targeted drug delivery across the blood–brain barrier. *Biomaterials* **2010**, *31*(30), 7748–7757.

Ganta, S.; Devalapally, H.; Shahiwala, A.; Amiji, M. A review of stimuli-responsive nanocarriers for drug and gene delivery. *J. Control. Release* **2008**, *126*(3), 187–204.

Gao, P.; Nie, X.; Zou, M.; Shi, Y.; Cheng, G. Recent advances in materials for extended-release antibiotic delivery system. *J. Antibiot.* **2011**, *64*(9), 625–634.

Garg, M.; Asthana, A.; Agashe, H. B.; Agrawal, G. P.; Jain, N. K. Stavudine-loaded mannosylated liposomes: in-vitro anti-HIV-I activity, tissue distribution and pharmacokinetics. *J. Pharm. Pharmacol.* **2006**, *58*(5), 605–616.

Gaumet, M.; Vargas, A.; Gurny, R.; Delie, F. Nanoparticles for drug delivery: the need for precision in reporting particle size parameters. *Eur. J. Pharm. Biopharm.* **2008**, *69*(1), 1–9.

Goodman, L.; Wintrobe, M.; Dameshek, W.; Goodman, M.; Gilman, A.; McLennan, M. Use of methyl-bis (beta-chloroethyl) amine hydrochloride and tris (beta-chloroethyl) amine hydrochloride for Hodgkin's disease, lymphosarcoma, leukemia and certain allied and miscellaneous disorders. *JAMA* **1946**, *132*, 126–132.

Gratton, S. E.; Ropp, P. A.; Pohlhaus, P. D.; Luft, J. C.; Madden, V. J.; Napier, M. E.; DeSimone, J. M. The effect of particle design on cellular internalization pathways. *Proc. Natl. Acad. Sci.* **2008**, *105*(33), 11613–11618.

Gray, C. M.; Lawrence, J.; Schapiro, J. M.; Altman, J. D.; Winters, M. A.; Crompton, M.; Loi, M.; Kundu, S. K.; Davis, M. M.; Merigan, T. C. Frequency of class I HLA-restricted anti-HIV CD8+ T cells in individuals receiving highly active antiretroviral therapy (HAART). *J. Immunol.* **1999**, *162*(3), 1780–1788.

Greupink, R.; Bakker, H. I.; Reker-Smit, C.; Kok, R.-J.; Meijer, D. K.; Beljaars, L.; Poelstra, K. Studies on the targeted delivery of the antifibrogenic compound mycophenolic acid to the hepatic stellate cell. *J. Hepatol.* **2005**, *43*(5), 884–892.

Gupta, Y.; Jain, A.; Jain, S. K. Transferrin-conjugated solid lipid nanoparticles for enhanced delivery of quinine dihydrochloride to the brain. *J. Pharm. Pharmacol.* **2007**, *59*(7), 935–940.

Harush-Frenkel, O.; Debotton, N.; Benita, S.; Altschuler, Y. Targeting of nanoparticles to the clathrin-mediated endocytic pathway. *Biochem. Biophys. Res. Commun.* **2007**, *353*(1), 26–32.

Harush-Frenkel, O.; Rozentur, E.; Benita, S.; Altschuler, Y. Surface charge of nanoparticles determines their endocytic and transcytotic pathway in polarized MDCK cells. *Biomacromolecules* **2008**, *9*(2), 435–443.

He, C.; Hu, Y.; Yin, L.; Tang, C.; Yin, C. Effects of particle size and surface charge on cellular uptake and biodistribution of polymeric nanoparticles. *Biomaterials* **2010**, *31*(13), 3657–3666.

He, J.; Chen, Y.; Farzan, M.; Choe, H.; Ohagen, A.; Gartner, S.; Busciglio, J.; Yang, X.; Hofmann, W.; Newman, W. CCR3 and CCR5 are coreceptors for HIV-1 infection of microglia. *Nature* **1997**, *385*, 645–649.

Hioe, C. E.; Tuen, M.; Vasiliver-Shamis, G.; Alvarez, Y.; Prins, K. C.; Banerjee, S.; Nadas, A.; Cho, M. W.; Dustin, M. L.; Kachlany, S. C. HIV envelope gp120 activates LFA-1 on CD4 T-lymphocytes and increases cell susceptibility to LFA-1-targeting leukotoxin (LtxA). *PLoS One* **2011**, *6*(8), e23202.

Hoffman, A. S. The origins and evolution of "controlled" drug delivery systems. *J. Control. Release* **2008**, *132*(3), 153–163.

Hoorelbeke, B.; Huskens, D.; Férir, G.; François, K. O.; Takahashi, A.; Van Laethem, K.; Schols, D.; Tanaka, H.; Balzarini, J. Actinohivin, a broadly neutralizing prokaryotic lectin, inhibits HIV-1 infection by specifically targeting high-mannose-type glycans on the gp120 envelope. *Antimicrob. Agents Chemother.* **2010**, *54*(8), 3287–3301.

Hosotani, R.; Kawaguchi, M.; Masui, T.; Koshiba, T.; Ida, J.; Fujimoto, K.; Wada, M.; Doi, R.; Imamura, M. Expression of integrin αvβ3 in pancreatic carcinoma: relation to MMP-2 activation and lymph node metastasis. *Pancreas* **2002**, *25*(2), e30-e35.

Huang, P.; Wang, D.; Su, Y.; Huang, W.; Zhou, Y.; Cui, D.; Zhu, X.; Yan, D. Combination of small molecule prodrug and nanodrug delivery: amphiphilic drug–drug conjugate for cancer therapy. *J. Am. Chem. Soc.* **2014**, *136*(33), 11748–11756.

Huo, M.; Yuan, J.; Tao, L.; Wei, Y. Redox-responsive polymers for drug delivery: from molecular design to applications. *Polym. Chem.* **2014**, *5*(5), 1519–1528.

Jain, K.; Kesharwani, P.; Gupta, U.; Jain, N. Dendrimer toxicity: Let's meet the challenge. *Int. J. Pharm.* **2010**, *394*(1), 122–142.

Jain, K.; Kumar Mehra, N.; K Jain, N. Nanotechnology in drug delivery: safety and toxicity issues. *Cur. Pharm. Des.* **2015**, *21*(29), 4252–4261.

Ji, X.; Olinger, G. G.; Aris, S.; Chen, Y.; Gewurz, H.; Spear, G. T. Mannose-binding lectin binds to Ebola and Marburg envelope glycoproteins, resulting in blocking of virus interaction with DC-SIGN and complement-mediated virus neutralization. *J. Gen. Virol.* **2005**, *86*(9), 2535–2542.

Jiang, W.; Kim, B. Y.; Rutka, J. T.; Chan, W. C. Nanoparticle-mediated cellular response is size-dependent. *Nat. Nanotechnol.* **2008**, *3*(3), 145–150.

Jin, W.; Li, C.; Du, T.; Hu, K.; Huang, X.; Hu, Q. DC-SIGN plays a stronger role than DCIR in mediating HIV-1 capture and transfer. *Virology* **2014**, *458*, 83–92.

Jin, Y.; Song, L.; Su, Y.; Zhu, L.; Pang, Y.; Qiu, F.; Tong, G.; Yan, D.; Zhu, B.; Zhu, X. Oxime linkage: a robust tool for the design of pH-sensitive polymeric drug carriers. *Biomacromolecules* **2011**, *12*(10), 3460–3468.

Johal, H. S.; Garg, T.; Rath, G.; Goyal, A. K. Advanced topical drug delivery system for the management of vaginal candidiasis. *Drug Deliv.* **2016**, *23*(2), 550–563.

Joshi, V. M.; Devarajan, P. V. Receptor-mediated hepatocyte-targeted delivery of primaquine phosphate nanocarboplex using a carbohydrate ligand. *Drug Deliv. Ttransl. Res.* **2014**, *4*(4), 353–364.

Juillerat-Jeanneret, L.; Schmitt, F. Chemical modification of therapeutic drugs or drug vector systems to achieve targeted therapy: looking for the grail. *Med. Res. Rev.* **2007**, *27*(4), 574–590.

Kaur, A.; Jain, S.; Tiwary, A. Mannan-coated gelatin nanoparticles for sustained and targeted delivery of didanosine: in vitro and in vivo evaluation. *Acta Pharm.* **2008**, *58*(1), 61–74.

Khandare, J.; Minko, T. Polymer–drug conjugates: progress in polymeric prodrugs. *Prog. Polym. Sci.* **2006**, *31*(4), 359–397.

Kim, S.-H.; Choe, C.; Shin, Y.-S.; Jeon, M.-J.; Choi, S.-J.; Lee, J.; Bae, G.-Y.; Cha, H.-J.; Kim, J. Human lung cancer-associated fibroblasts enhance motility of nonsmall cell lung cancer cells in coculture. *Anticancer Res.* **2013**, *33*(5), 2001–2009.

Kim, S.-S.; Peer, D.; Kumar, P.; Subramanya, S.; Wu, H.; Asthana, D.; Habiro, K.; Yang, Y.-G.; Manjunath, N.; Shimaoka, M. RNAi-mediated CCR5 silencing by LFA-1-targeted nanoparticles prevents HIV infection in BLT mice. *Mol. Ther.* **2010**, *18*(2), 370–376.

Kinman, L.; Bui, T.; Larsen, K.; Tsai, C.-C.; Anderson, D.; Morton, W. R.; Hu, S.-L.; Ho, R. J. Optimization of lipid-indinavir complexes for localization in lymphoid tissues of HIV-infected macaques. *J Acquir Immune Defic Syndr.* **2006**, *42*(2), 155–161.

Koshkaryev, A.; Sawant, R.; Deshpande, M.; Torchilin, V. Immunoconjugates and long circulating systems: origins, current state-of-the-art and future directions. *Adv. Drug Deliv. Rev.* **2013**, *65*(1), 24–35.

Kosowska, K.; Jacobs, M.; Bajaksouzian, S.; Koeth, L.; Appelbaum, P. Alterations of penicillin-binding proteins 1A, 2X, and 2B in Streptococcus pneumoniae isolates for which amoxicillin MICs are higher than penicillin MICs. *Antimicrob. Agents Chemother.* **2004**, *48*(10), 4020–4022.

Kratz, F.; Beyer, U.; Roth, T.; Tarasova, N.; Collery, P.; Lechenault, F.; Cazabat, A.; Schumacher, P.; Unger, C.; Falken, U. Transferrin conjugates of doxorubicin: synthesis, characterization, cellular uptake, and in vitro efficacy. *J. Pharm. Sci.* **1998**, *87*(3), 338–346.

Kreuter, J. Nanoparticles—a historical perspective. *Int. J. Pharm.* **2007**, *331*(1), 1–10.

Kumar, A.; Schweizer, H. P. Bacterial resistance to antibiotics: active efflux and reduced uptake. *Adv. Drug Deliv. Rev.* **2005**, *57*(10), 1486–1513.

Kumar, P. V.; Asthana, A.; Dutta, T.; Jain, N. K. Intracellular macrophage uptake of rifampicin loaded mannosylated dendrimers. *J. Drug Target* **2006**, *14*(8), 546–556.

Lammers, T.; Kiessling, F.; Hennink, W. E.; Storm, G. Drug targeting to tumors: principles, pitfalls and (pre) clinical progress. *J. Control. Release* **2012**, *161*(2), 175–187.

Lang, B.-C.; Yang, J.; Wang, Y.; Luo, Y.; Kang, Y.; Liu, J.; Zhang, W.-S. An improved design of water-soluble propofol prodrugs characterized by rapid onset of action. *Anesth. Analg.* **2014**, *118*(4), 745–754.

Laverman, P.; Boerman, O. C.; Oyen, W. J.; Corstens, F. H.; Storm, G. In vivo applications of PEG liposomes: unexpected observations. *Crit Rev Ther Drug Carrier* **2001**, *18*(6).

Laverman, P.; Dams, E. T. M.; Storm, G.; Hafmans, T. G.; Croes, H. J.; Oyen, W. J.; Corstens, F. H.; Boerman, O. C. Microscopic localization of PEG-liposomes in a rat model of focal infection. *J. Control. Release* **2001**, *75*(3), 347–355.

Lee, B.; Sharron, M.; Montaner, L. J.; Weissman, D.; Doms, R. W. Quantification of CD4, CCR5, and CXCR4 levels on lymphocyte subsets, dendritic cells, and differentially conditioned monocyte-derived macrophages. *Proc. Nat.l Acad. Sci.* **1999**, *96*(9), 5215–5220.

Lemmer, Y.; Kalombo, L.; Pietersen, R.-D.; Jones, A. T.; Semete-Makokotlela, B.; Van Wyngaardt, S.; Ramalapa, B.; Stoltz, A. C.; Baker, B.; Verschoor, J. A. Mycolic acids, a promising mycobacterial ligand for targeting of nanoencapsulated drugs in tuberculosis. *J. Control. Release* **2015**, *211*, 94–104.

Leslie, A.; Moody, P.; Shaw, W. V. Structure of chloramphenicol acetyltransferase at 1.75-A resolution. *Proc. Nat.l Acad. Sci* **1988**, *85*(12), 4133–4137.

Li, M. C.; Hertz, R.; Bergenstal, D. M. Therapy of choriocarcinoma and related trophoblastic tumors with folic acid and purine antagonists. *N. Engl. J. Med* **1958**, *259*(2), 66–74.

Li, Y.-P.; Pei, Y.-Y.; Zhang, X.-Y.; Gu, Z.-H.; Zhou, Z.-H.; Yuan, W.-F.; Zhou, J.-J.; Zhu, J.-H.; Gao, X.-J. PEGylated PLGA nanoparticles as protein carriers: synthesis, preparation and biodistribution in rats. *J. Control. Release* **2001**, *71*(2), 203–211.

Li, Y.; Fukushima, K.; Coady, D. J.; Engler, A. C.; Liu, S.; Huang, Y.; Cho, J. S.; Guo, Y.; Miller, L. S.; Tan, J. P. Broad-Spectrum Antimicrobial and Biofilm-Disrupting Hydrogels: Stereocomplex-Driven Supramolecular Assemblies. *Angewandte Chemie International Edition* **2013**, *52*(2), 674–678.

Lieleg, O.; Baumgärtel, R. M.; Bausch, A. R. Selective filtering of particles by the extracellular matrix: an electrostatic bandpass. *Biophys J.* **2009**, *97*(6), 1569–1577.

Longmuir, K. J.; Robertson, R. T.; Haynes, S. M.; Baratta, J. L.; Waring, A. J. Effective targeting of liposomes to liver and hepatocytes in vivo by incorporation of a Plasmodium amino acid sequence. *Pharm Res.* **2006**, *23*(4), 759–769.

Mansuri, S.; Kesharwani, P.; Jain, K.; Tekade, R. K.; Jain, N. Mucoadhesion: A promising approach in drug delivery system. *Reactive and Functional Polymers* **2016**, *100*, 151–172.

Marger, M. D.; Saier, M. H. A major superfamily of transmembrane facilitators that catalyze uniport, symport and antiport. *Trends Biochem. Sci.* **1993**, *18*(1), 13–20.

Mariam, D. H.; Mengistu, Y.; Hoffner, S. E.; Andersson, D. I. Effect of rpoB mutations conferring rifampin resistance on fitness of Mycobacterium tuberculosis. *Antimicrob. Agents Chemother.* **2004**, *48*(4), 1289–1294.

Mishra, V.; Mahor, S.; Rawat, A.; Gupta, P. N.; Dubey, P.; Khatri, K.; Vyas, S. P. Targeted brain delivery of AZT via transferrin anchored pegylated albumin nanoparticles. *J. Drug Target.* **2006**, *14*(1), 45–53.

Moriyón, I.; López-Goñi, I. Structure and properties of the outer membranes of Brucella abortus and Brucella melitensis. *Int. Microbiol.* **2010**, *1*(1), 19–26.

Muller, R. H.; Keck, C. M. Challenges and solutions for the delivery of biotech drugs–a review of drug nanocrystal technology and lipid nanoparticles. *J. Biotechnol.* **2004**, *113*(1), 151–170.

Mura, S.; Nicolas, J.; Couvreur, P. Stimuli-responsive nanocarriers for drug delivery. *Nat. mater.* **2013**, *12*(11), 991–1003.

Neyfakh, A. A. The multidrug efflux transporter of Bacillus subtilis is a structural and functional homolog of the Staphylococcus NorA protein. *Antimicrob. Agents Chemother* **1992**, *36*(2), 484–485.

Ng, C. T.; Tang, F. M. A.; Li, J. J. E.; Ong, C.; Yung, L. L. Y.; Bay, B. H. Clathrin-Mediated Endocytosis of Gold Nanoparticles In Vitro. *Anat. Rec.* **2015**, *298*(2), 418–427.

Ng, V. W.; Chan, J. M.; Sardon, H.; Ono, R. J.; García, J. M.; Yang, Y. Y.; Hedrick, J. L. Antimicrobial hydrogels: A new weapon in the arsenal against multidrug-resistant infections. *Adv. Drug Deliv. Rev.* **2014**, *78*, 46–62.

Nguyen, S.; Hiorth, M. Advanced drug delivery systems for local treatment of the oral cavity. *Ther. Deliv.* **2015**, *6*(5), 595–608.

Nie, S.; Understanding and overcoming major barriers in cancer nanomedicine. *Nanomed.* **2010**, *5*(4), 523–528.

Nimje, N.; Agarwal, A.; Saraogi, G. K.; Lariya, N.; Rai, G.; Agrawal, H.; Agrawal, G. Mannosylated nanoparticulate carriers of rifabutin for alveolar targeting. *J. Drug Target.* **2009**, *17*(10), 777–787.

Nishikawa, T.; Iwakiri, N.; Kaneko, Y.; Taguchi, A.; Fukushima, K.; Mori, H.; Morone, N.; Kadokawa, J.-I. Nitric oxide release in human aortic endothelial cells mediated by delivery of amphiphilic polysiloxane nanoparticles to caveolae. *Biomacromolecules* **2009**, *10*(8), 2074–2085.

Oba, M.; Aoyagi, K.; Miyata, K.; Matsumoto, Y.; Itaka, K.; Nishiyama, N.; Yamasaki, Y.; Koyama, H.; Kataoka, K. Polyplex micelles with cyclic RGD peptide ligands and disulfide cross-links directing to the enhanced transfection via controlled intracellular trafficking. *Mol. Pharm.* **2008**, *5*(6), 1080–1092.

Owais, M.; Ahmed, I.; Krishnakumar, B.; Jaina, R.; Bachhawat, B.; Gupta, C. Tuftsin-bearing liposomes as drug vehicles in the treatment of experimental aspergillosis. *FEBS letters* **1993**, *326*(1–3), 56–58.

Owais, M.; Varshney, G. C.; Choudhury, A.; Chandra, S.; Gupta, C. M. Chloroquine encapsulated in malaria-infected erythrocyte-specific antibody-bearing liposomes effectively controls chloroquine-resistant Plasmodium berghei infections in mice. *Antimicrob. Agents Chemother.* **1995**, *39*(1), 180–184.

Owens, D. E.; Peppas, N. A. Opsonization, biodistribution, and pharmacokinetics of polymeric nanoparticles. *Int J. Pharm.* **2006**, *307*(1), 93–102.

Pang, X.; Du, H.-L.; Zhang, H.-Q.; Zhai, Y.-J.; Zhai, G.-X. Polymer–drug conjugates: present state of play and future perspectives. *Drug Discov. Today* **2013**, *18*(23), 1316–1322.

Park, E.-J.; Zhang, Y.-Z.; Vykhodtseva, N.; McDannold, N. Ultrasound-mediated blood-brain/blood-tumor barrier disruption improves outcomes with trastuzumab in a breast cancer brain metastasis model. *J. Control. Release* **2012**, *163*(3), 277–284.

Parreira, P.; Fátima Duarte, M.; Reis, C. A.; Martins, M. C. L. Helicobacter pylori infection: A brief overview on alternative natural treatments to conventional therapy. *Crit. Rev. Microbiol.* **2014**, 1–12.

Patel, L. N.; Zaro, J. L.; Shen, W.-C. Cell penetrating peptides: intracellular pathways and pharmaceutical perspectives. *Pharm. Res.* **2007**, *24*(11), 1977–1992.

Patel, M. D.; Majee, S. B.; Samad, A.; Devarajan, P. V. Ionic complexation as a noncovalent approach for the design of folate anchored rifampicin gantrez nanoparticles. *J. Biomed. Naotechnol.* **2013**, *9*(5), 765–775.

Paulsen, I. T.; Skurray, R. A.; Tam, R.; Saier, M. H.; Turner, R. J.; Weiner, J. H.; Goldberg, E. B.; Grinius, L. L. The SMR family: a novel family of multidrug efflux proteins involved with the efflux of lipophilic drugs. *Mol. Microbiol* **1996**, *19*(6), 1167–1175.

Pawar, V. K.; Kansal, S.; Garg, G.; Awasthi, R.; Singodia, D.; Kulkarni, G. T. Gastroretentive dosage forms: a review with special emphasis on floating drug delivery systems. *Drug Deliv.* **2011**, *18*(2), 97–110.

Peer, D.; Karp, J. M.; Hong, S.; Farokhzad, O. C.; Margalit, R.; Langer, R. Nanocarriers as an emerging platform for cancer therapy. *Nat. Nanotechnol.* **2007**, *2*(12), 751–760.

Petros, R. A.; DeSimone, J. M. Strategies in the design of nanoparticles for therapeutic applications. *Nat Rev Drug Discov* **2010**, *9*(8), 615–627.

Pollicita, M.; Schols, D.; Aquaro, S.; Peumans, W.; Van Damme, E.; Perno, C.; Balzarini, J. Carbohydrate-binding agents (CBAs) inhibit HIV-1 infection in human primary monocyte-derived macrophages (MDMs) and efficiently prevent MDM-directed viral capture and subsequent transmission to CD4+ T lymphocytes. *Virology* **2008**, *370*(2), 382–391.

Pruthi, J.; Mehra, N. K.; Jain, N. K. Macrophages targeting of amphotericin B through mannosylated multiwalled carbon nanotubes. *J. Drug Target* **2012**, *20*(7), 593–604.

Rabinovitch, M. Professional and nonprofessional phagocytes: an introduction. *Trends Cell Biol* **1995**, *5*(3), 85–87.

Radovic-Moreno, A. F.; Lu, T. K.; Puscasu, V. A.; Yoon, C. J.; Langer, R.; Farokhzad, O. C. Surface charge-switching polymeric nanoparticles for bacterial cell wall-targeted delivery of antibiotics. *ACS nano* **2012**, *6*(5), 4279–4287.

Ramana, L. N.; Anand, A. R.; Sethuraman, S.; Krishnan, U. M. Targeting strategies for delivery of anti-HIV drugs. *J. Control. Release* **2014**, *192*, 271–283.

Rao, K. S.; Reddy, M. K.; Horning, J. L.; Labhasetwar, V.; TAT-conjugated nanoparticles for the CNS delivery of anti-HIV drugs. *Biomaterials* **2008**, *29*(33), 4429–4438.

Rautio, J.; Laine, K.; Gynther, M.; Savolainen, J. Prodrug approaches for CNS delivery. *The AAPS Journal* **2008**, *10*(1), 92–102.

Rejman, J.; Bragonzi, A.; Conese, M. Role of clathrin-and caveolae-mediated endocytosis in gene transfer mediated by lipo- and polyplexes. *Mol. Ther.* **2005**, *12*(3), 468–474.

Ruoslahti, E.; Bhatia, S. N.; Sailor, M. J. Targeting drugs and nanoparticles to tumors. *J. Cell Biol* **2010**, *188*(6), 759–768.

Sahay, G.; Alakhova, D. Y.; Kabanov, A. V. Endocytosis of nanomedicines. *J. Control. Release* **2010**, *145*(3), 182–195.

Sahay, G.; Kim, J. O.; Kabanov, A. V.; Bronich, T. K. The exploitation of differential endocytic pathways in normal and tumor cells in the selective targeting of nanoparticulate chemotherapeutic agents. *Biomaterials* **2010**, *31*(5), 923–933.

Saier, M.; Tam, R.; Reizer, A.; Reizer, J. Two novel families of bacterial membrane proteins concerned with nodulation, cell division and transport. *Mol. Microbiol.* **1994**, *11*(5), 841–847.

Santos-Magalhães, N. S.; Mosqueira, V. C. F. Nanotechnology applied to the treatment of malaria. *Adv. Drug Deliv. Rev.* **2010**, *62*(4), 560–575.

Schmitt-Sody, M.; Strieth, S.; Krasnici, S.; Sauer, B.; Schulze, B.; Teifel, M.; Michaelis, U.; Naujoks, K.; Dellian, M. Neovascular targeting therapy paclitaxel encapsulated in cationic liposomes improves antitumoral efficacy. *Clin Cancer Res* **2003**, *9*(6), 2335–2341.

Schnitzer, J. E.; Caveolae: from basic trafficking mechanisms to targeting transcytosis for tissue-specific drug and gene delivery in vivo. *Adv. Drug Deliv. Rev.* **2001**, *49*(3), 265–280.

Sheldrake, H. M.; Patterson, L. H. Function and Antagonism of β. *Curr. Cancer Drug Target* **2009**, *9*(4), 519–540.

Sheng, Y.; Yang, X.; Pal, D.; Mitra, A. K. Prodrug approach to improve absorption of prednisolone. *Int. J. Pharm.* **2015**, *487*(1), 242–249.

Shi, J.; Xiao, Z.; Kamaly, N.; Farokhzad, O. C. Self-assembled targeted nanoparticles: evolution of technologies and bench to bedside translation. *Acc. Chem Res.* **2011**, *44*(10), 1123–1134.

Shiratsuchi, A.; Kaido, M.; Takizawa, T.; Nakanishi, Y. Phosphatidylserine-mediated phagocytosis of influenza A virus-infected cells by mouse peritoneal macrophages. *J. Clin. Virol.* **2000**, *74*(19), 9240–9244.

Shukla, S.; Ablack, A. L.; Wen, A. M.; Lee, K. L.; Lewis, J. D.; Steinmetz, N. F. Increased tumor homing and tissue penetration of the filamentous plant viral nanoparticle Potato virus X. *Mol. Pharm.* **2012**, *10*(1), 33–42.

Siddalingam, R.; Chidambaram, K. Helicobacter pylori—current therapy and future therapeutic strategies. *Trends in Helicobacter Pylori Infection. Intech* **2014**.

Sievers, E. L.; Senter, P. D. Antibody-drug conjugates in cancer therapy. *Annu. Rev. Med.* **2013**, *64*, 15–29.

Singhal, A.; Gupta, C. Antibody-mediated targeting of liposomes to red cells in vivo. *FEBS letters* **1986**, *201*(2), 321–326.

Smith, A. W. Biofilms and antibiotic therapy: is there a role for combating bacterial resistance by the use of novel drug delivery systems? *Adv. Drug Deliv. Rev.* **2005**, *57*(10), 1539–1550.

Strebhardt, K.; Ullrich, A. Paul Ehrlich's magic bullet concept: 100 years of progress. *Nat. Rev. Cancer* **2008**, *8*(6), 473–480.

Strieth, S.; Eichhorn, M. E.; Werner, A.; Sauer, B.; Teifel, M.; Michaelis, U.; Berghaus, A.; Dellian, M. Paclitaxel encapsulated in cationic liposomes increases tumor microvessel leakiness and improves therapeutic efficacy in combination with Cisplatin. *Clin. Cancer Res.* **2008**, *14*(14), 4603–4611.

Sudhakar, A.; History of cancer, ancient and modern treatment methods. *J. Cancer Sci. Ther.* **2009**, 1(2), 1.

Tabernero, J.; Shapiro, G. I.; LoRusso, P. M.; Cervantes, A.; Schwartz, G. K.; Weiss, G. J.; Paz-Ares, L.; Cho, D. C.; Infante, J. R.; Alsina, M. First-in-humans trial of an RNA interference therapeutic targeting VEGF and KSP in cancer patients with liver involvement. *Cancer Discov.* **2013**, *3*(4), 406–417.

Tait-Kamradt, A.; Clancy, J.; Cronan, M.; Dib-Hajj, F.; Wondrack, L.; Yuan, W.; Sutcliffe, J. mefE is necessary for the erythromycin-resistant M phenotype in Streptococcus pneumoniae. *Antimicrob. Agents Chemother.* **1997**, *41*(10), 2251–2255.

Thamphiwatana, S.; Gao, W.; Pornpattananangkul, D.; Zhang, Q.; Fu, V.; Li, J.; Obonyo, M.; Zhang, L. Phospholipase A2-responsive antibiotic delivery via nanoparticle-stabilized liposomes for the treatment of bacterial infection. *J. Mater. Chem. B* **2014**, *2*(46), 8201–8207.

Torchilin, V. P. Passive and active drug targeting: drug delivery to tumors as an example. In *Drug Delivery*, Springer: **2010**, pp. 3–53.

Torchilin, V.; Tumor delivery of macromolecular drugs-based on the EPR effect. *Adv. Drug Deliv. Rev.* **2011**, *63*(3), 131–135.

Trkola, A.; Ketas, T. J.; Nagashima, K. A.; Zhao, L.; Cilliers, T.; Morris, L.; Moore, J. P.; Maddon, P. J.; Olson, W. C. Potent, broad-spectrum inhibition of human immunodeficiency virus type 1 by the CCR5 monoclonal antibody PRO 140. *J. Virol* **2001**, *75*(2), 579–588.

Tzehoval, E.; Segal, S.; Stabinsky, Y.; Fridkin, M.; Spirer, Z.; Feldman, M. Tuftsin (an Ig-associated tetrapeptide) triggers the immunogenic function of macrophages: implications for activation of programmed cells. *Proc. Natl. Acad. Sci.* **1978**, *75*(7), 3400–3404.

Valencia, P. M.; Hanewich-Hollatz, M. H.; Gao, W.; Karim, F.; Langer, R.; Karnik, R.; Farokhzad, O. C. Effects of ligands with different water solubilities on self-assembly and properties of targeted nanoparticles. *Biomaterials* **2011**, *32*(26), 6226–6233.

van Tilburg Bernardes, E.; Lewenza, S.; Reckseidler-Zenteno, S. Current Research Approaches to Target Biofilm Infections. *Journal of Postdoctoral Research* **2015**, *36*, 49.

Van Veen, H. W.; Konings, W. N. The ABC family of multidrug transporters in microorganisms. *BBA-Bioenergetics* **1998**, *1365*(1), 31–36.

Vasir, J. K.; Labhasetwar, V. Quantification of the force of nanoparticle-cell membrane interactions and its influence on intracellular trafficking of nanoparticles. *Biomaterials* **2008**, *29*(31), 4244–4252.

Vig, B. S.; Huttunen, K. M.; Laine, K.; Rautio, J. Amino acids as promoieties in prodrug design and development. *Adv. Drug Deliv. Rev.* **2013**, *65*(10), 1370–1385.

Vizoso, M.; Puig, M.; Carmona, F. J.; Maqueda, M.; Velásquez, A.; Gómez, A.; Labernadie, A.; Lugo, R.; Gabasa, M.; Rigat-Brugarolas, L. G. Aberrant DNA methylation in nonsmall cell lung cancer-associated fibroblasts. *Carcinogenesis* **2015**, *36*(12), 1453–1463.

Vyas, S. P.; Quraishi, S.; Gupta, S.; Jaganathan, K. Aerosolized liposome-based delivery of amphotericin B to alveolar macrophages. *Int. J. Pharm.* **2005**, *296*(1), 12–25.

Weiser, J. R.; Saltzman, W. M. Controlled release for local delivery of drugs: barriers and models. *J. Control. Release* **2014**, *190*, 664–673.

Williams, C. B.; Yeh, E. S.; Soloff, A. C. Tumor-associated macrophages: unwitting accomplices in breast cancer malignancy. *Breast Cancer* **2016**, *2*, 15025.

Willmott, C.; Maxwell, A. A single point mutation in the DNA gyrase A protein greatly reduces binding of fluoroquinolones to the gyrase-DNA complex. *Antimicrob. Agents Chemother.* **1993**, *37*(1), 126–127.

Winau, F.; Westphal, O.; Winau, R. Paul Ehrlich—in search of the magic bullet. *Microbes and Infection* **2004**, *6*(8), 786–789.

Wu, F.; Meng, G.; He, J.; Wu, Y.; Wu, F.; Gu, Z. Antibiotic-loaded chitosan hydrogel with superior dual functions: antibacterial efficacy and osteoblastic cell responses. *ACS applied materials & interfaces* **2014**, 6(13), 10005–10013.

Xiao, K.; Li, Y.; Luo, J.; Lee, J. S.; Xiao, W.; Gonik, A. M.; Agarwal, R. G.; Lam, K. S. The effect of surface charge on in vivo biodistribution of PEG-oligocholic acid-based micellar nanoparticles. *Biomaterials* **2011**, *32*(13), 3435–3446.

Yang, H.-W.; Hua, M.-Y.; Liu, H.-L.; Huang, C.-Y.; Tsai, R.-Y.; Lu, Y.-J.; Chen, J.-Y.; Tang, H.-J.; Hsien, H.-Y.; Chang, Y.-S. Self-protecting core-shell magnetic nanoparticles for targeted, traceable, long half-life delivery of BCNU to gliomas. *Biomaterials* **2011**, *32*(27), 6523–6532.

Yang, W.; Moore, I. F.; Koteva, K. P.; Bareich, D. C.; Hughes, D. W.; Wright, G. D. TetX is a flavin-dependent monooxygenase conferring resistance to tetracycline antibiotics. *J. Biol. Chem.* **2004**, *279*(50), 52346–52352.

Yoshida, H.; Bogaki, M.; Nakamura, S.; Ubukata, K.; Konno, M. Nucleotide sequence and characterization of the Staphylococcus aureus norA gene, which confers resistance to quinolones. *J. Bacteriol* **1990**, *172*(12), 6942–6949.

You, Y. C.; Dong, L. Y.; Dong, K.; Xu, W.; Yan, Y.; Zhang, L.; Wang, K.; Xing, F. J. In vitro and in vivo application of pH-sensitive colon-targeting polysaccharide hydrogel used for ulcerative colitis therapy. *Carb Polym* **2015**, *130*, 243–253.

Yu, B.; Tai, H. C.; Xue, W.; Lee, L. J.; Lee, R. J. Receptor-targeted nanocarriers for therapeutic delivery to cancer. *Mol. Membrane Biol.* **2010**, *27*(7), 286–298.

Yuan, Q.; Shah, J.; Hein, S.; Misra, R. Controlled and extended drug release behavior of chitosan-based nanoparticle carrier. *Acta biomaterialia* **2010**, *6*(3), 1140–1148.

Zhang, D.-Y.; Shen, X.-Z.; Wang, J.-Y.; Dong, L.; Zheng, Y.-L.; Wu, L.-L. Preparation of chitosan-polyaspartic acid-5-fluorouracil nanoparticles and its anticarcinoma effect on tumor growth in nude mice. *World J Gastroenterol* **2008**, *14*(22), 3554–3562.

Zhang, L. W.; Monteiro-Riviere, N. A.; Mechanisms of quantum dot nanoparticle cellular uptake. *Toxicol Sci* **2009**, *110*(1), 138–155.

Zhang, Y.; Thomas, T. P.; Desai, A.; Zong, H.; Leroueil, P. R.; Majoros, I. J.; Baker, Jr. J. R. Targeted dendrimeric anticancer prodrug: a methotrexate-folic acid-poly (amidoamine) conjugate and a novel, rapid, "one pot" synthetic approach. *Bioconjugate Chem* **2010**, *21*(3), 489–495.

Zhao, F.; Zhao, Y.; Liu, Y.; Chang, X.; Chen, C.; Zhao, Y. Cellular uptake, intracellular trafficking, and cytotoxicity of nanomaterials. *Small* **2011**, *7*(10), 1322–1337.

Zhao, Y.; Adjei, A. A. Targeting oncogenic drivers. **2014**.

Zhu, S.; Niu, M.; O'Mary, H.; Cui, Z. Targeting of tumor-associated macrophages made possible by PEG-sheddable, mannose-modified nanoparticles. *Mol. Pharm.* **2013**, *10*(9), 3525–3530.

CHAPTER 2

NANO-ONCOTARGETS AND INNOVATIVE THERAPIES

YAMUNA MOHANRAM and LAKSHMI KIRAN CHELLURI*

*Department of Transplant Biology, Immunology and Stem Cell Unit, Global Hospitals, Lakdi-ka-Pul, Hyderabad–500 004 (TS), India, Tel.: 0091-40-30244501; Fax: 0091-40-23244455, *E-mail: lkiran@globalhospitalsindia.com*

CONTENTS

ABSTRACT

Nano-oncotargets applied to novel therapeutic interventions are vastly explored in cancer. Targeted drug delivery is the key to successful and sustained release drug formulations. The chemistry is evolving with technological advancements. It is not only restricted to chemical moieties, but also has been focused on novel contrast agents and process development tools in cell-based therapies. The presentation is focused on the impact of the nano size, shape, charge and surface modification for a successful targeted drug delivery. A few illustrations have been explored, and the subject progress has been dwelt at length. The multidisciplinary approach has been elaborated for innovative approaches in treating the cancer. The role of nano-particles in current applications is elucidated. Biomaterials that make nano-particles compatible for human applications have been described in forwarding our understanding of this highly exploratory science. Looking forward, nano materials hold a pivotal role in resolving issues that are highly beneficial in reducing the costs in treating cancer.

2.1   INTRODUCTION

Nanotechnology is a multi-interdisciplinary field of science that was developed due to the cumulative advancements in the research and development fields of material sciences, physics, chemistry, biology and electronics. This technology aims at resizing/restructuring or recreating existing materials in nanometer (nm) scales in order to bring out its maximum effective potential (Aitken et al., 2004; Schleich et al., 2014). The etiology to the term 'nm' traces its origin to the Greek work "nanos" referring to dwarf while scientifically it implies to the unit of length that equates to 1 billionth of a meter or 1×10^{-9} m and represented as nm in SI units (De Jong and Borm, 2008). The concept of particle existence at nano scale level has long existed in nature. To get a better perception to this unit of length, the biological moieties such as nucleic acid, proteins, cyto-skeletal polysaccharides and cell membrane channels are considered as a scale of reference (Aitken et al., 2004). For example, the blood cells – neutrophils, eosinophil, basophils and lymphocytes are present within a size range of 10–15 nm, while the nucleotides are of 0.33 nm and are aligned to form the helically coiled DNA spanning a width of 2–3 nm. These examples clearly demonstrate that the "nano" ideology has not been created defying the laws of nature.

2.1.1   NANOPARTICLES

The British Standard of Institution, defines nanoparticles as an engineered material that measures below 100 nm in two or more of its dimensions (Aitken et al., 2004; De Jong and Borm, 2008; Schleich et al., 2014; Taqaddas,

2014) and are constructed by either the "bottom–up" or "top–down" approach. In theory, the nanoscale lies intermediary to the atomic/molecular/ micro and the bulk/macro scale, and hence the formulation of these particles is approached with two different methodologies. The "bottom up" approach strategizes by building the entire nanoparticle from an atom with either an aqueous or gaseous base. On the contrary, the "top down" approach aims at breaking or etching macro structures through wet/dry grinding system (Horikosh and Serpone, 2013; Jortner and Rao, 2002). The nanoparticles constructed through these methodologies can exist in more than one form such as the quantum dots, nanocluster, nanocrystal, nanowires or even as nanotubes (Jortner and Rao, 2002; Lin et al., 2014).

Till date, the construction of nanoparticles have successfully been carried out using various materials and few of them are, albumin (Chithrani et al., 2006), collagen (El-Samaligy and Rohdewald, 1983; Papi et al., 2011), gelatin (Balthasar et al., 2005; Coester et al., 2000; Leo et al., 1997), silk fibroin (Kundu et al., 2010; Park et al., 2004; Yan et al., 2009), gold (Asadishad et al., 2010; El-Sayed et al., 2005; Hwu et al., 2008; Kang and Wang, 2014; Prabaharan et al., 2009; Sokolov et al., 2003; Tomuleasa, 2012; Wang et al., 2011), iron (Chertok et al., 2008; Hwu et al., 2008; Kayal and Ramanujan, 2010; Schleich et al., 2013; Verma et al., 2015), lipids (Dong et al., 2009; Liu et al., 2011;), polymers (Gref et al., 2000; Patlolla et al., 2010; Tao et al., 2014), ceramics (Shenoy et al., 2005) and organic carbon (Peter et al., 2010; Ray et al., 2009).

Nanoparticles, despite being constructed with materials of varied chemical composition, exhibit uniformity in their unique characteristic feature that is unlike their micro and macro counterparts. Interestingly, their distinctive properties have been documented not only in their size, but also in the structure, assembly and more specifically with their spectral, electrical and magnetic properties (Buzea et al., 2007; Jortner and Rao, 2002; Mahmoudi et al., 2011; Wilczewska et al., 2012). In addition, the nanoparticles also differ from their micro and macro counterpart by displaying a larger surface area for its corresponding smaller mass. According to Hochella Jr. et al. (2008) the differences in the physical and physicochemical properties of these particles remain unexplainable by the classical laws of physics but nevertheless, it has been attributed to the larger surface area consisting of higher population of the reactive atoms and its existences in the transitional state that is sandwiched between the atomic/ molecular and bulk state (Hochella et al., 2008; Morais et al., 2014).

These interesting behavioral patterns of the particles have attracted immense interest of research over the years, from multidisciplinary areas of science and technology. The research outcome from the various studies has significantly created awareness and has increased the potential interest in its application in the commercial market for the development of devices and also in the overall improvement of the existing technologies (Jortner and Rao, 2002).

2.1.2 APPLICATIONS OF NANOPARTICLES

Over the years, there has been a progressive increase in the commercial interest of the application of nanoparticles in various fields including optoelectronics (Zhang et al., 2013), cosmetics (Benson, 2005; Kokura et al., 2010; Nguyen et al., 2011), economic solar cells (Yang et al., 2010), high performance batteries (Chan et al., 2008; Magasinski et al., 2010), generators (Jung et al., 2015; Liu et al., 2015), single atom transistors (Jung et al., 2015), foods (Duncan, 2011), constructions (Van Broekhuizen et al., 2011), textiles (Gittard et al., 2010; Ravindra et al., 2010; Shim et al., 2008), waste water treatment (Kaegi et al., 2013) and even in medical applications (Des Rieux et al., 2006; Gelperina et al., 2005; Jin et al., 2009; Mahmoudi et al., 2011; Salari et al., 2015).

The commercialization of the nanoparticle-based products is often a result of an extensive basic and application specific laboratory characterization. Experimental tests investigating the molecular structure, chemical composition, melting point, vapor pressure, flash point, pH, size distribution, shape, charge and application specific toxicity studies are crucial determining factors in the translation of the material for commercial use (Jortner and Rao, 2002).

2.2 NANOTECHNOLOGY IN DRUG DELIVERY

Amongst the various applications of nanotechnology, its utilization in medicine has been highly beneficial in revolutionizing the current methods of disease diagnosis and treatment (Bamrungsap et al., 2012; Betty et al., 2010; Mahmoudi et al., 2011). Nanomedicine is a generic terminology used for the application of nanotechnology in medicine and is a vast field having extensive relevance.

Nanomaterials, finds its application as a biomimetic agent mimicking the biological moieties, upon conjugation with functional molecules to its surface (Bamrungsap et al., 2012; Betty et al., 2010; Moghimi et al., 2005). Besides, it can be used as a biosensor that detects the molecular changes in the body for disease diagnosis and also as a self assembling biomaterial for damaged tissue repair providing a temporary support while healing. In addition, it also finds its use as a contrast agent in magnetic resonance imaging (MRI) (Bamrungsap et al., 2012; Betty et al., 2010; Moghimi et al., 2005). However, amidst the various applications, the use of nanoparticles as drug delivery agent has been leading the field of nanomedicine (Moghimi et al., 2005).

Since the inception of this technology, there has been a tremendous growth in the translation of the bench side research of these nano-particle-based drug carriers. Successful bench side formulations have moved to drug toxicity testing in animals and in to clinical trials. Nano-based forays in the pharma industry have also led to Food and Drug Administration (FDA) approved drugs that are commercially available for clinical application in humans. The very first nano-based drug formulation to receive the approval was the liposome encapsulated doxorubicin, DOXIL for the treatment of AIDS-related Kaposi's sarcoma (Teske et al., 2015). Till date various drugs have been commercialized, while many are still in the pipeline for the treatment of cancer, HIV associated Kaposi's syndrome, hepatitis, infectious disease, anesthetics, cardiac/vascular disorders, inflammatory/immune disorders, endocrine/exocrine disorders even degenerative disorders (Etheridge et al., 2013). Few of such products are listed in Table 2.1.

The utilization of nano-particles as a drug delivery agent has been highly advantageous due to the feasibility of engineering particles with a biocompatible material and customizing it-based on size, shape and charge. The leverage of engineering and customizing material helps in the production of nano sized particles with a larger surface area, convenient for conjugation of drugs (Bamrungsap et al., 2012; Betty et al., 2010; Buzea et al., 2007; Mahmoudi et al., 2011). The nano biocompatible materials

TABLE 2.1 List of Pharmaceutical Companies and Their Nanomedicinal Products

S. No.	Company	Product Name	Nanoparticle	Size	Administration	Target Tissue
1	Astellas	AmBisome®	Amphotericin B encapsulated in a single bilayer liposomes	<100 nm	Intravenous	Fungal infections
2	Janssen Pharmaceuticals	Doxil®	Doxorubicin HCl liposeomes	85 nm	Intravenous	Ovarian cancer, breast cancer and AIDS-related Kaposi's syndrome
3	NeXstar Pharmaceuticals	Daunoxome®	Daunorubicin citrate liposome	45 nm	Intravenous	Kaposi's Sarcoma
4	Taiwan Liposome Company	Lipo-Dox®	Doxorubicin encapsulated liposomes		Intravenous	Ovarian cancer, breast cancer and AIDS-related Kaposi's syndrome
5	Taiwan Liposome Company	AmBiL®	Amphotericin B encapsulated liposomes		Intravenous	Fungal infections
6	Sigma-Tau Pharmaceuticals, Inc.	DepoCyt®	Cytarabine encapsulated lipid vesicles	35 nm	Intrathecal	Lymphomatous menigitis
7	Taiwan Liposome Company	ProFlow®	Lipid-based emulsion of Prostaglandin E1	100–200 nm	Intravenous	Diabetic neuropathy and ulcers
8	Seqqus Pharmaceuticals	Amphotec®	Lipid colloid dispersion of Amphotericin B	~130 nm	Intravenous	Fungal infections
9	Cristal Therapeutics	CriPec® Docetaxal	Polymer	30–100 nm	Intravenous or Sub cutaneous	Solid tumors
10	CytImmune Sciences	Aurimune	Gold nanoparticles coated with polyethylene glycol and tumor necrosis factor alpha	27 nm	Intravenous	Advanced stage cancer

TABLE 2.1 (Continued)

S. No.	Company	Product Name	Nanoparticle	Size	Administration	Target Tissue
11	Nanobiotix	NanoXray – NBTXR3, NBTX-IV, NBTX-TOPO	Crystallized hafnium oxide	50 nm	Intravenous, intratumor or direct application	primary liver cancer, glioblastoma, rectal cancer, prostate cancer, breast cancer, head and neck cancer
12	Parvus Therapeutics Inc	Navacims™	Iron oxide nanoparticles coated with major histocompatibility protein and a very short antigenic peptide	1–100 nm		Type 1 diabetes, multiple sclerosis, rheumatoid arthritis, allergic asthma, primary biliary cirhosis and in cancer
13	MagForce Nanotechnologies	NanoTHERM	Iron oxide coated with aminosilane coating	15 nm	Intravenous	Tumor
14	Genetic Immunity	Dermavir	Synthetic polymer – Mannosylated linear polyethyleneimine PEIm coated with pDNA	100 nm	Topical through DermaPrep	HIV/AIDS
15	Intezyne	IVECT™ Method	Triblock copolymer micelle with GRP78 suppressors	30–80 nm		Neuroendocrine tumors and noon-small cell lung cancers
16	Smith & Nephew	ACTICOAT	Silver coated polyurethane layer on a polyurathane foam pad and an adhesive water proof polyurathane film layer		Wound dressing material	Antibacterial
17	Celgene	ABRAXANE	Paclitaxel bound to albumin	130 nm	Injectable	Metastatic breast cancers

are not only similar in scale to the biological moieties but are also beneficial in navigating across the various nano size physiological barriers that have been hard to reach (Bamrungsap et al., 2012; Betty et al., 2010). Nano particle-based drug formulations are highly dependent on the core material of choice, which determines the mode of drug loading such as encapsulation or adsorption or even conjugation to outer surface. Similarly, the core material can also determine the mode of drug targeting to the tissue. Drug targeting to specific tissue can be achieved by surface modification of the material with the conjugation of ligands such as tissue specific antibodies. This mode of drug targeting is active targeting (Figure 2.1).

The thriving interest of big pharmaceutical industries in the development of nanomedicinal drugs and its huge financial investments in these molecules has been primarily due to its potential to directly address the shortcoming of the existing pharmaceutical agents (Mahmoudi et al., 2011; Moghimi et al., 2005). Despite the establishment of beneficial therapeutic values by many pharmacological agents in the preclinical setting, they

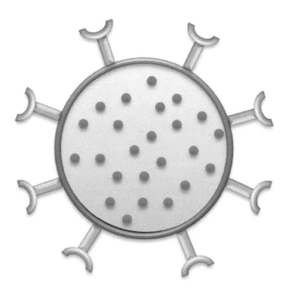

FIGURE 2.1 Diagrammatic representation of active tissue targeting nanoparticle drug carrier. The drug encapsulated and surface modified nanoparticles selectively target tissues with the ligands attached to its surface. The ligands recognize and bind to its corresponding cell surface epitopes (figure adapted).

do not all necessarily produce satisfactory evidences in the clinical trials. This however is not always due to the poor pharmacological kinetics of the drug but is also heavily dependent on the drug delivery agent. Drug delivery agents play a key role in the release of the drug which influences the treatment outcome. Particularly, in the treatment of cancer, it is essential that there is slow and sustained release of the high drug dosage to the site of cancer for an effective treatment. However, with poor pharmacological formulations, there is an early release of the drug at a high concentration, which could potentially lead to drug associated side effects. In addition, the fluctuation of the drug concentration results in insufficient therapeutic drug dosage required for action. These factors thus drastically alter the treatment efficacy. Additionally, the poor water solubility of the carriers' results in an unequivocal distribution of the compounds in the system resulting in diluting the cancer-toxic molecules. Moreover, the concern with the administration of the existing pharmaceutical agents is its inability to target against specific tissue. In the case of cancer, the chemotherapeutic agents lack of tissue specific targeting. This could lead to drug related side effects such as cellular cytotoxicity in non-cancerous cells, requiring further medical assistance and prolonged treatment plan for obtaining satisfactory recovery rate (Mahmoudi et al., 2011; Moghimi et al., 2005).

On the other hand, the feasibility of engineering nanoparticles can modulate the drug release rate, the degradation kinetics and also its clearance from the system. Regulation of these factors can thus assure a sustained drug release from its cavity or surface (Bamrungsap et al., 2012; Moghimi et al., 2005). Besides, the surface manipulation of the nano particle-based drug carrier with ligands can potentially reduce systemic cytotoxicity, thus reducing the drug associated side effects (Bamrungsap et al., 2012). With a robust formulation, the nano drug carrier-based medicinal products could enhance the treatment efficiency, reduce the drug dosage, treatment length and systemic cytoxicity compared to the traditional pharmacology formulations (Bamrungsap et al., 2012; Mahmoudi et al., 2011; Moghimi et al., 2005). Besides, a potent formulation could revolutionize the cancer treatment in a cost effective manner (Bamrungsap et al., 2012; Mahmoudi et al., 2011; Nigam et al., 2011). This however, requires extensive characterization-based on size, shape, charge and surface property of the material, as explained in the following subsections.

2.2.1 CRITICAL FACTORS FOR SUCCESSFUL DRUG DELIVERY

2.2.1.1 Size

The size of the nanoparticle has been recognized as the key factor for its utilization in the pharmacological formulation as it can navigate through the various nano sized physiological barriers. However, the size of these nanoparticles also determines the mobility, circulation, navigation/migration of the nanoparticles upon administration. In addition, the particle size also regulates the fate of the drug carriers-based on their degradation kinetics and clearance rate from the system (Jiang et al., 2008; Moghimi et al., 2005).

While engineering the nanoparticle-based drug delivery agent, it is essential that the particles are neither too large as they undergo a rapid clearance from the system nor are they too small to drift through the pores of the vasculature. Hence, to attain higher particle retention within the system the particles have to be large to avoid vascular leakage yet small enough to prevent rapid clearance by the liver and spleen and also be able to navigate into the target tissues. Thus, the ideal particle size range should be between 50–100 nm which permits the particles to migrate into the cancerous cells and yet be retained within them (Alexander-Bryant et al., 2013; Bamrungsap et al., 2012).

2.2.1.2 Surface Charge

The total surface charge of the nanoparticle determines the behavioral patterns of the material-based on its electrostatic repulsion. The electrostatic repulsive pattern of the nanoparticles highly influences its cellular uptake, physiological translocation, aggregation and also its stability (Bruge et al., 2015; Honary and Zahir, 2013). The charge on the particle is highly affected by the alteration in the pH or ionic strength. It is measured through dynamic light scatter technology that calculates the zeta potential or the total electric charge of all the ions in the solution. The results are presented within the range of 0 ± 60 mV (Berg et al., 2009).

Nanoparticles display a behavioral difference with the alteration in their zeta potential. Particles with higher than 30 mV demonstrate an increased electrostatic repulsive force that leads to altering its stability. However, with lower zeta potential reading there is a reduced electrostatic repulsive force resulting in the coagulation of the particles (Honary and Zahir, 2013). The positively charged nanoparticles have a higher potential for cellular and lysosomal uptake, than negatively charged particles. In addition it can also evade the process of early clearance from the system by resisting the opsonization (Alexander-Bryant et al., 2013; Honary and Zahir, 2013). Thus the use of positively charged nanoparticles and the positively charged substrates such as polymers have been of increased interest of research.

2.2.1.3 Shape

Besides, the size, charge and surface property, the shape of the nanoparticles also play an equally important role in determining the fate of the nanoparticles. The shape of the nanoparticle influences in the migration, circulation, cellular uptake, translocation across the physiological barriers and also the intra cellular retention time and clearance rate (Bamrungsap et al., 2012; Moghimi et al., 2005). Spherical nanoparticles are comparatively lesser toxic that any other shaped nanoparticles. Silver nanoparticles shaped as a nano plate exhibited higher detrimental effect in zebra fishes than as a sphere or wire (George et al., 2009). Similarly spherical nickel nanoparticles exhibit less toxicity than the dendrimeric structure (Ispas et al., 2009).

2.2.1.4 Surface Properties

For a nanoparticle-based drug delivery system it is highly essential to determine its surface properties in terms of its hydrophobicity and hydrophilicity. It is this factor that determinates the clearance rate of the circulating drug loaded nanocarrier by the mononuclear phagocytic system (MPS) (Gref et al., 2000). The MPS is a highly efficient method of foreign body clearance within the circulatory system. It functions by coating the foreign bodies with the opsonin proteins of the blood serum. This process is referred to as opsonization. The labeling or coating of the foreign bodies with opsonin acts as a signaling mechanism for the macrophage and the mononuclear phagocytic system for a rapid identification and clearance from the blood stream (Gref et al., 2000). However, the MPS clearance varies-based on the hydrophobic and hydrophilic nature of the nanoparticles. Nanoparticles with hydrophobic surfaces are highly prone to opsonization due to the strong interaction mediated by its surface property, leading to its immediate removal by the reticulo-endothelial system (RES). The hydrophilic surface nanoparticles, however evade this process (Ahmadi et al., 2015; Gref et al., 2000; Nigam et al., 2011).

Iron nanoparticles are highly hydrophobic and are highly prone to opsonization and immediate clearance from the system. In order to enhance its circulatory retention time in the system, nanoparticles are often surface masked with hydrophilic components (Ahmadi et al., 2015; Gref et al., 2000; Nigam et al., 2011). This thus helps in the increased stabilization and prolonged circulatory/retention time. In addition, it increases the probability of drug delivery at the target site (Nigam et al., 2011; Schleich et al., 2013). Polyethylene glycol (Ahmadi et al., 2015; Garcia-Fuentes et al., 2005; Gref et al., 2000; Schleich et al., 2013; Shenoy et al., 2005; Zhao et al., 2015) and poly vinyl alcohol (Teske and Detweiler, 2015) are two of the well-characterized biocompatible hydrophilic polymer that are employed as masking agents.

2.2.2 TISSUE TARGETING

Nano particle-based drug delivery systems have been strategized to deliver drugs to the cells through a passive or active targeting mechanisms. The current pharmacological formulation has adopted either of the targeting protocol for the delivery of the drugs to the cancerous cells. A diagrammatic representation of the two processes has been illustrated in Figure 2.2.

2.2.2.1 Passive Tissue Targeting

The passive tissue targeting strategy involves a straightforward approach. Intravenous administration of a well-formulated nano particle-based drug carrier, circulates in the system and migrates through the fenestrations of the endothelial tissues with a pore size ranging between 10–1000 nm. These fenestrations are commonly observed during disease manifestation such as cancers. The presence large fenestration aids in the easier transportation/migration of nanoparticles with an optimal size range of 20–400 nm

(Chertok et al., 2008; El-Sayed et al., 2005; Schleich et al., 2014), permitting their selective drifting or pinocytosis into the cancerous tissues for targeted accumulation (Alexander-Bryant et al., 2013; Bamrungsap et al., 2012; El-Sayed et al., 2005; Etheridge et al., 2013; Martins et al., 2013; Schleich et al., 2014; Sokolov et al., 2003; Tewes et al., 2007). The cancerous cells exhibit a characteristic feature of enhanced permeability and retention effect (EPR) (Alexander-Bryant et al., 2013; Bamrungsap et al., 2012; El-Sayed et al., 2005; Martins et al., 2013; Sokolov et al., 2003; Tewes et al., 2007). The EPR effect is beneficial in the permeability of the nanoparticles in to the tissues and in its retention due to the poor lymphatic system. This thus helps in retaining the nanoparticle within the cells resulting in the release of the drug in to the cancerous cells (Alexander-Bryant et al., 2013; El-Sayed et al., 2005; Schleich et al., 2014).

2.2.2.2 Active Tissue Targeting

Active tissue targeting is the second approach that has been adopted for pharmacological formulations of the nano particle-based drug carriers. This approach is very similar to the passive tissue targeting strategy yet differs in its potential to selectively target tissues (Alexander-Bryant

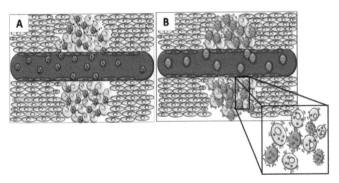

FIGURE 2.2 Passive (A) and Active Cell (B) Targeting Strategies Employed by Nano-Drug Carriers. In passive cellular targeting (A) the drug loaded nanoparticles infiltrate through the leaky vasculature, bind and nonspecifically deliver the drug to the underlying tissues. Active targeting (B) of tissues is carried out with nanoparticular infiltration through the vasculature and selective deliver drugs by binding (ligand) to cells expressing the epitope (figure adapted from Mahmoudi et al., 2011).

et al., 2013; Bamrungsap et al., 2012; El-Sayed et al., 2005; Martins et al., 2013). The selective tissue targeting has been approached by engineering the nanoparticles with tissue specific ligands. The conjugation of ligands such as antibodies and lipoproteins to the surface of the nanoparticles, increases its specificity in binding to the cells where the epitopes reside (Alexander-Bryant et al., 2013; Bamrungsap et al., 2012; Etheridge et al., 2013; Martins et al., 2013; Schleich et al., 2014). Similar to passive targeting, up on intravenous administration, these nanoparticle-based drug carriers circulate and migrate through the endothelial fenestrations. However, they differ in the cellular binding by having a higher specific to the selective epitopes.

These, surface bound nanoparticles are then internalized through endocytosis (Martins et al., 2013; Schleich et al., 2014). Due to the EPR effect of the tumor tissue, the nanoparticles are retained within the target tissue and gradually break down to release the drug (Schleich et al., 2014). Comparison of the two tissue targeting procedures, active targeting has been found to be beneficial as it has a higher specificity in its method of delivery (Alexander-Bryant et al., 2013; Schleich et al., 2014).

Till date various formulations have been laboratory tested and few of which have also been formulated for drug delivery in human. Coating of an integrin specific binding receptor, arginyl-glycyl-aspartic acid (RGD), to the nanoparticles demonstrates a preferential binding to the integrin overexpressing tumor endothelial surfaces. Transferrin coating on gold nanoparticles ensures internalization through the clathrin mediated endocytsis pathway. While, albumin coating reduces the inflammatory response and has increases the nanoparticles internalization into the tumor cells (Chithrani et al., 2006).

Likewise, nanoparticles have also been coated with tumor tissue specific antibodies, which aid in the enhancing the tumoricidal properties of the drug (Jiang et al., 2008). Rituximab and trastuzumab are the two monoclonal antibody-based drugs that have received USFDA approval for clinical treatment while there are various other ongoing studies that are in the pipeline that are formulated for cancer tissue targeting (Etheridge et al., 2013). Immunoliposome coated with monoclonal HER2 antibody and encapsulated with doxorubicin have a higher tumor targeting efficiency in mice models. Demonstrating an improved pharmacological property of the drug with selective drug delivery to the tumor cells (Park et al., 2002). Besides, investigating the use of a single ligand for tissue targeting, studies have been carried out with the combination use of two different antibodies. The dual ligand targeting all the more increases the tissue identification and binding sensitivity. Targeting B-chronic lymphocytic leukemic cells with monoclonal CD 37 with antiCD 19/antiCD 20 improves the tumor cell cytotoxicity over the use of a CD 37 on its own (Yu et al., 2013).

As a continuous development to the use of these antibody coated nanoparticles for attaining a more satisfactory result in terms of tumoricidal properties, the use of laser irradiation have also been tested. Gold nanoparticles coated with antiepidermal growth factor (EGF) upon excitation of its surface plasma resonance with argon ion laser exposure at 514 nm demonstrate to have a higher specificity for targeting cancerous cells, leading to a photo thermal destruction (El-Sayed et al., 2005; Park et al., 2004; Prabaharan et al., 2009; Sokolov et al., 2003). Similarly, laser irradiation on anti-CD 30 targeted L-428 Hodgkin cells had an increased cell death (Qu et al., 2012).

2.2.3 NANO DRUG DELIVERY AGENTS

2.2.3.1 Polymer Based Nanoparticles

Biocompatible polymers of both natural and synthetic origin have been extensively researched as a drug delivery agent for its stability, sustained drug release and cost effective industrial fabrication. Few of the commonly used natural polymers are chitosan, gelatin, albumin, dextran and sodium alginate (Alexander-Bryant et al., 2013; Bruge et al., 2015) while synthetic polymers include poly lactic acid (PLA), poly ethylene glycol (PEG) (Alexander-Bryant et al., 2013; Bruge et al., 2015; Etheridge et al., 2013; Gref et al., 2000), poly vinyl alcohol (Sokolov et al.., 2003) or a combination of one or more of these such as poly(lactide-coglycolide) (PLGA) (Chertok et al., 2008; Schleich et al., 2014). The use of the synthetic and natural polymer as a drug delivery agent is highly determined-based on its physical properties such as size, surface chemistry, drug loading potential, biocompatibility, sustained drug release potential and its degradation rate and also the toxicity of the by-products upon degradation (Bamrungsap et al., 2012; Sokolov et al., 2003; Zhao et al., 2015). Till date, based on the above mention characteristics, various polymer-based drugs carriers have been successfully formulated within the size range of 10–100 nm that either encapsulate or surface adsorb the drugs to the polymeric structures (Alexander-Bryant et al., 2013; Bamrungsap et al., 2012).

2.2.3.1.1 *Pacitaxel*

Paclitaxel is an approved anticancer drug used for the treatment against breast, colon, primary ovarian and non-small cell lung cancers (Chertok et al., 2008; Hwu et al., 2008; Schleich et al., 2013; Sokolov et al., 2003; Wang et al., 2011). The mode of action of the drug has been attributed to its potential to alter the microtubular structures of the cells in their late G2 and M phases, resulting in arresting the process of cellular division (Chertok et al., 2008; Schleich et al., 2013; Wang et al., 2011). Despite it having clinical significance in the treatment of tumor, it has also been known to have poor water solubility and also lack tumor targeting ability (Chertok et al., 2008; Schleich et al., 2013; Wang et al., 2011). However, these issues have been addressed in more than one way with use of various types of polymer with/without ligands to aid in tissue targeting and enhancing its pharmacological properties (Wang et al., 2011). Drug loading with polymer materials have been commonly carried by encapsulation within its hollow cavity and is useful for a slow and sustained release of the drug by either diffusion or desorption from its surface. Poly (lactic-coglycolic acid) is one such polymer that has been employed in drug formulation and is constructed with the combination of two polymers: PLA and PGA. Drug release kinetic studies carried out on paclitaxel encapsulated PLGA with d-alpha-tocopheryl polyethylene glycol 100 succinate (Vitamin E TPGS) has been found to be controlled with 100% encapsulation efficiency (Mu and Feng, 2003). Developmental advances with the use of polymer have also lead to the construction of pH sensitive polymer, that exhibit a differential response to the change of pH in micro environment. Poly (β-amino ester) (PbAE) engineered with the similar technology has been documented to have a higher cellular uptake and also an enhanced cytotoxic potential up on the release of the encapsulated paclitaxel drug (Sokolov et al., 2003).

2.2.3.1.2   9-Nitrocamptothecin

9-Nitrocamptothecin (9-NC) is another example of a water insoluble drug which when administered acts up on the topo isomerase I enzyme of the nucleus. Targeting the nuclear enzyme potential leads to cellular death and hence finds its use for treating cancers. The treatment efficiency of this drug however has been restricted in many ways however, encapsulation with polymers have demonstrated significant difference in the drug release kinetics and its pharmacological properties. In addition, studies also have been carried out to improve the pharmacokinetic properties of the drug with the use of combinational polymers such as PLGA and PEG. Utilization of the polymers by altering the composition, has been beneficial in the attaining the advantages of both polymers in the sustained drug release for the induction of apoptosis in the ovarian carcinoma cells (Ahmadi et al., 2015; Derakhshandeh et al., 2010).

2.2.3.2   Lipid Based Nanoparticles

Lipid-based nanoparticle constructions have been studied since 1991. These nanoparticles were an outcome of advancement of then well researched materials such as emulsion, liposomes and polymeric nanoparticles (Battaglia et al., 2014). Lipid nanoparticles are well studied due its biocompatibility, biodegradability, drug solubility, drug release kinetics and drug bio-distribution (Muller et al., 2000). Moreover, its cost effective industrial fabrication and similarity to the physiological lipid molecules has increased its pharmacological implications (Battaglia et al., 2014; Etheridge et al., 2013).

2.2.3.2.1   Doxorubicin

Doxorubicin was the first nanoparticle-based anti cancer drug that received the USFDA approval and are commercially available as DOXIL, formulated with a lipid polymer combination (Bamrungsap et al., 2012). The presence of PEG- liposome combinations have shown to alter the biodistribution and the drug release kinetics. It also aids prolonging the retention period in circulation. The combination encapsulation of the drug successfully evades the opsonization, clearance by the RES and passively targets the cancer cells where the doxorubicin are released, over time (Ahmadi et al., 2015; Judson et al., 2001; Lyass et al., 2000). Clinical trials carried with DOXIL have established a beneficial effect with the use of the drug in advanced soft tissue sarcomas (Judson et al., 2001; Lyass et al., 2000).

2.2.3.2.2 Amphotericin B

Amphotericin B, is a water insoluble drug administered for the treatment of systemic fungal infections that have issues of poor bio distribution. However, the encapsulation of the drug with lipids has been reported to have beneficial effect in the improvement of the pharmacokinetics of the drug when tested in patients with systemic fungal infections. The lipid complexes are believed to help in the sustained release and improve the overall stability of the drug (Adedoyin et al., 2000).

2.2.4 METAL NANOPARTICLES

Till date various metal and metal oxides have been investigated as drug carrier and drug delivering agent, including iron, gold, silver, zinc and manganese, of which gold and iron are well studied.

2.2.4.1 Gold Nanoparticles

Engineering gold nanoparticles has been of interest due to the effortless manipulation in the size, shape and the surface chemistry. The chemical stability, bio-compatibility and surface plasma resonances have been an added advantage in its choice as a drug delivery agent (Schleich et al., 2013). Enhancement of the pharmacokinetics has also been possible with the surface modification of the material using conjugated antibodies on polymer masked or unmasked materials. The advancements in the engineering of gold nanoparticles have employed its potential to absorb high surface plasma resonance and its resonance light scattering potential for the use as a nonphotobleachable optical contrast agent in the endoscopic compatible optical imaging for cancer detections (Asadishad et al., 2010;

Kang et al., 2014). Unlike the other contrast agents, the gold nanoparticles excited with single laser scatter light emitting multicolors of high intensity and brightness (Asadishad et al., 2010) which can be controlled with the manipulation of the shape and size of the particles (Prabaharan et al., 2009).

2.2.4.1.1 Doxorubicin

The increased pharmacokinetics of doxorubicin, a water insoluble anticancer drug has also been tested with the gold nanoparticles and is documented to have enhanced performance in its tumoricidal effect. Efficient passive cellular targeting and internalization process in melanoma cell lines have been achieved with 2.8 nm gold nanoparticles that deliver the doxorubicin drug (Wang et al., 2011). In the continuous search for attaining a better delivery there have been various approaches that are adopted with the use of stabilizers and pH sensitive polymers. The use of stabilizers such as folate on the surface of the nanoparticles aids in the folate mediated internalization of the drug (Wang et al., 2011) and in combination with pH sensitive polymer coated nanoparticles assures the delivery of the drug in the acidic microenvironments (Judson et al., 2001; Tomuleasa et al., 2012).

2.2.4.2   Iron Oxide Nanoparticles

Iron oxide nanoparticles, unlike the other materials display a super paramagnetic property under the influence of strong magnetic field, otherwise present in a nonmagnetic state. This property of the biocompatible super paramagnetic iron oxide nanoparticles (SPION) has found its application as a contrast agent in magnetic resonance imaging (MRI) (Arbab et al., 2003; Mahmoudi et al., 2011; Schleich et al., 2014; Sokolov et al., 2003). MRI is a non-invasive diagnostic tool that employs the use of contrast agents and strong magnetic field to delineate tissues to understand its anatomical and physiological properties (Arbab et al., 2003). In comparison to the other contrast agents, SPIONs have been a preferred choice not only for its biocompatibility and magnetic property but also for in its safe degradation by the internal iron metabolic pathway (Arbab et al., 2003;

Judson et al., 2001). In addition, its higher retention system provides a longer time period for imaging the tissues (Arbab et al., 2003; Mahmoudi et al., 2011). Besides, SPIONs are also used as drug carriers. These multifunctional SPION are an excellent theranostic agent, intergrating its dual functionality of the drug delivery for therapy and in diagnostics (Arbab et al., 2003; Chertok et al., 2008; El-Sayed et al., 2005; Schleich et al., 2014). Theranostic agents are highly beneficial in providing a real time nanoparticle tracking of its bio-distribution, tissue targeting and also the drug response within the tissues (Arbab et al., 2004; Chertok et al., 2008; Schleich et al., 2013).

Selective drug targeting with SPIONs is achieved with the use of external magnetic field, which is higher than the blood flow rate in the system (Arbab et al., 2004; El-Sayed et al., 2005). The strong magnetic fields are particularly useful in driving the nanoparticles to the target site for its accumulation and passive tissue uptake (Arbab et al., 2004; Mahmoudi et al., 2011). Chertok et al., through their study were able to demonstrate a higher of the iron oxide nanoparticles accumulation in the gliosarcomas in rat tumor models. Nanoparticles of 100 nm size were directed to the tumor site under the influence of magnetic field 0.4 T for a 30 min period. Through this study, they were able to demonstrate a significant difference in the accumulation of the iron nanoparticles in the tumor specific sites of the brain, to the non-cancerous region (Chertok et al., 2008).

Alternatively, the iron nanoparticles can also be used for active targeting under magnetic field, following surface modification (El-Sayed et al., 2005; Schleich et al., 2014). As a result of the unification of the two popularly approached strategies, a more accurate targeting can be achieved. In a comparative study carried out by Schleich et al. in 2014, the authors have demonstrated a higher specificity in paclitaxel drug delivery by the PLGA coated RGD labeled nanoparticles in the tumor tissues. Through this method of drug targeting, the magnetic field secured the nanoparticles at the target site which the RGD coating influenced in the selective cancer cell binding. The clinical relevance of this strategy for drug targeting was document to be more efficient that drug targeting in the absence of ligand conjugation and also with the traditional targeting approach (Schleich et al., 2014).

The very first cancer drug formulation with iron oxide nanoparticles was investigated in 1996 with epirubicin loaded ferro fluids, in patients with advanced solid tumors. Since then, there have been various studies and formulations developed for the treatment of infection, cardiovascular, inflammation, arthritis, and also for stem cell tracking (Mahmoudi et al., 2011). However, prior to the application of SPIONs as a theranostic agent, extensive characterization of the material is crucial. In poorly formulated drugs SPIONs have been known to particle aggregation in blood vessels and reactive oxygen species (ROS) production in cells which result in toxicity. Thus extensive characterization-based on its size, shape, charge, drug loading and drug dosage is mandatory (Schleich et al., 2014), as these are key factors in determining the ROS formation. Hence, a careful manipulation can prevent its production (Schleich et al., 2014). Similarly, the alteration of the nanoparticular surface property with polymers is essential to hinder particle agglomeration within the blood vessels (Mahmoudi et al., 2011).

2.2.4.2.1 Doxorubicin

Iron oxide nanoparticles have a higher cellular uptake with materials that are masked with synthetic/natural polymers (Arbab et al., 2005; Chertok et al., 2010; Javid et al., 2013; Kievit et al., 2011; Schleich et al., 2013; Sokolov et al., 2003). Thus, emphasizing the need for stabilization of the particles prior to use and till date various methods have been adopted. One such method is the use of chitosan, a natural polymer. The drug efficiency with chitosan coated and doxorubicin loaded SPION have higher drug efficiency against ovarian cancerous cells than the drug load SPIONs (Javid et al., 2013). Likewise, synthetic polymers such poly vinyl alcohol (PVA), PEG and PLGA, have been studied for its ability to stabilize the material (Chertok et al., 2010; Kievit et al., 2011; Schleich et al., 2013; Sokolov et al., 2003). Despite, the difference in the chemical composition it has been demonstrated that the stabilization of the iron nanoparticles is essential for its use as a drug carrier.

An iron oxide nanoparticle-based drug carrier development study, has shown PVA coating of the material surface to have a slow and sustained release of the doxorubicin drug (Kayal and Ramanujan, 2010). Stabilization studies with polyethylenimine on iron nanoparticles loaded doxorubicin on pH sensitive spacer were observed to be beneficial in improving the drug pharmacokinetic in multidrug resistant cancerous cells (Kievit et al., 2011). Similarly, iron oxide nanoparticle coated with a phospholipid-PEG complex and loaded with doxorubicin could induce cytotoxicity in cancerous cells. In addition, this novel construction could induce hyperthermia in tumor cells with the SPION core, thus together it was able to target the tumor specific cells (Quinto, et al., 2015).

2.2.5 CELLULAR LABELING

Clinical applications of stem cells have been extensively researched, for its potential to repair and restore damaged tissues. The success of the treatment modality has been staggered over the years, due to the lack of understanding of in the cellular migration, bio-distribution, engraftment and clearance within the system. The combinational use of iron oxide nanoparticles and MRI could circumvent this issue. Labeling stem cells with iron oxide nanoparticles and its real time tracking with MRI will be beneficial in determining the efficiency of the treatment and help in the further standardization.

Stabilization of iron nanoparticles was determined essential with either transfecting agents or cationic polymeric materials to increase the cellular uptake (Arbab et al., 2003, 2005, 2006; Frank et al., 2003; Verma et al., 2015). Masking the iron oxide nanoparticles with these agents such as Poly L-Lysine, PEG, dextran and heparin sulfate did not alter the multilineage differentiation potential, cellular viability and proliferation. Longer exposure of these materials to the cells neither imparted cytotoxicity (Arbab et al., 2003, 2005, 2006; Verma et al., 2015).

Effective MR signals can be obtained with low concentration (12.5 μg) of iron oxide nanoparticles and were demonstrated to be safe for cell labeling (Verma et al., 2015). Low intensity MRI was used in the tracking of endothelial progenitor cell migration to tumor sites in mice (Arbab et al., 2006). Similarly, the navigation of the neuronal progenitor cells was tracked to ischemic specific regions of the brain in a rat stroke model. Its successful cellular therapy was established with the detection of vasculature formation at the ischemic borders in the brain (Jiang et al., 2005).

Iron oxide nanoparticles have also been modified to perform a multifunction role in cell sorting and labeling. Cell specific sorters are constructed with the conjugation of tissue specific antibodies or peptides to the surface of the nanoparticles. Cardiac precursor cell sorting was developed with the use of cardiac specific antibodies. The functionality of the construct was confirmed by sorting of the precursor cells and by MRI (Rashi and Yamuna et al., 2016). The conjugation of iron oxide nanoparticles with human immuno deficiency virus-transactivator of transcription (HIV-TAT) peptide were used for cellular labeling in stem cells. The MR tracking of the labeled cells were tracked to the bone marrow of immune-compromised mice (Lewin et al., 2000). The use of iron oxide nanoparticles in complementation with MRI can be advantageous in cellular tracking. The presence of magnetic field is beneficial in targeting the cells to the site for regeneration. However, a comprehensive characterization of the material, the cell and its combination is required to be investigated for achieving a successful MR guided cell therapy.

2.3   CONCLUSION

The application of nanotechnology in medicine promises to bring a change by revolutionizing the way the current pharmacological agents are delivered for cancer treatment. Formulations of drug with the nanoparticles are highly advantageous. (1) Size – easier transportation/movement through the cellular barriers. (2) Engineering – modification of the size, shape, charge and surface properties help in the evading early clearance from the system and also have a large surface area for drug loading. (3) Tissue specific targeting – delivery of drug to cancerous cells can be achieved with the use of antibodies that have tissue specific epitopes for binding. In summary, with a well-formulated drug, the nano particle-based drug carrier can potentially enhance the cancer treatment modality as a result can also reduce the drug dosage and treatment period over time. In order to have more pharmacological agents that are available for treatment, it is highly essential to address various drawbacks that are associated with its use. Firstly, development of newer drug requires hefty financial investments from big pharmaceutical companies and more multifunctional nanoparticle-based drugs should be developed. During formulation of these new drugs, their shortest and safest route of drug of administration should be studied. Careful examination of the toxicity imparted by the bye products up on degradation must be evaluated. Long-term side effects of with the use of the materials should be studied. Advanced diagnostic tool and techniques should also be used for tracking drug bio-distribution. Furthermore, this application of nano parti-cle-based drug delivery should also be developed for other diseases such as cardiovascular and diabetes. In conclusion, formulation of drugs that can address the various issues shall be highly beneficial in treating cancer in a cost effective and highly efficient manner.

KEYWORDS

- bio-materials
- contrast agents
- innovative cancer therapies
- nanomaterials
- non-invasive imaging
- oncotargets
- targeted drug delivery

REFERENCES

Adedoyin, A.; Swenson, C. E.; Bolcsak, L. E.; Hellmann, A.; Radowska, D.; Horwith, G.; Janoff, S. A.; Branch, R. A. A pharmacokinetic study of amphotericin B lipid com-plex injection (Abelcet) in patients with definite or probable systemic fungal infec-tions. *Antimicrob Agents and Chemother.* **2000**, *44*(10), 2900–2902.

Ahmadi, K. D.; Jalalizadeh, A.; Mostafaie, A.; L Hosseinzadeh, F. Encapsulation in PLGA-PEG enhances 9-nitro-camptothecin cytotoxicity to human ovarian carcinoma cell line through apoptosis pathway. *Res Pharm Sci.* **2015**, *10*(2), 161–168.

Aitkens, R. J.; Creely, S. K.; Tran, C. L. Nanoparticles: An Occupational Hygiene Review. *HSE books*, **2004**.

Alexander-Bryant, A. A.; Vanden Berg-Foels, W. S.; & Wen, X. Bioengineering strategies for designing targeted cancer therapies. *Adv Cancer Res*, **2013**, *118*, 1–59.

Arbab, A. S.; Bashaw, L. A.; Miller, B. R.; Jordan, E. K.; Lewis, B. K.; Kalish, H.; Frank, J. A. Characterization of biophysical and metabolic properties of cells labeled with superparamagnetic iron oxide nanoparticles and transfection agent for cellular MR imaging. *Radiology.* **2003**, *229*(3), 838–846.

Arbab, A. S.; Frenkel, V.; Pandit, S. D.; Anderson, S. A.; Yocum, G. T.; Bur, M.; Khuu, M. H.; Read, J. E.; Frank, J. A. Magnetic resonance imaging and confocal microscopy studies of magnetically labeled endothelial progenitor cells trafficking to sites of tumor angiogenesis. *Stem Cells.* **2006**, *24*(3), 671–678.

Arbab, A. S.; Jordan, E. K.; Wilson, L. B.; Yocum, G. T.; Lewis, B. K.; Frank, J. A. In vivo trafficking and targeted delivery of magnetically labeled stem cells. *Human gene therapy.* **2004**, *15*(4), 351–360.

Arbab, A. S.; Yocum, G. T.; Rad, A. M.; Khakoo, A. Y.; Fellowes, V.; Read, E. J.; Frank, J. A. Labeling of cells with ferumoxides-protamine sulfate complexes does not inhibit function or differentiation capacity of hematopoietic or mesenchymal stem cells. *NMR in Biomedicine.* **2005**, *18*(8), 553–559.

Asadishad, B.; Vossoughi, M.; Alemzadeh, I. Folate-receptor-targeted delivery of doxorubicin using polyethylene glycol-functionalized gold nanoparticles. *Ind Eng Chem Res.* **2010**, *49*(4), 1958–1963.

Balthasar, S.; Michaelis, K.; Dinauer, N.; von Briesen, H.; Kreuter, J.; Langer, K. Preparation and characterization of antibody modified gelatin nanoparticles as drug carrier system for uptake in lymphocytes. *Biomaterials.* **2005**, *26*(15), 2723–2732.

Bamrungsap, S.; Zhao, Z.; Wang, L.; Li, C.; Fu, T.; Tang, W. A focus on nanoparticles as a drug delivery system. *Nanomed.* **2012**, *7*(8), 1253–1271

Battaglia, L.; Gallarate, M.; Panciani, P. P.; Ugazio, E.; Sapino, S.; Peira, E.; Chirio, D. Techniques for the preparation of solid lipid nano and microparticles. **2014**, Application of nanotechnology in drug delivery, Ali Demir Sezer (Ed.), InTech.

Benson, H. A. Transdermal drug delivery: penetration enhancement techniques. *Curr Drug Deliv.* **2005**, *2*(1), 23–33.

Berg, J. M.; Romoser, A.; Banerjee, N.; Zebda, R.; Sayes, C. M. The relationship between pH and zetapotential of ~30 nm metal oxide nanoparticle suspensions relevant to in vitro toxicological evaluations. *Nanotoxicology.* **2009**, *3*(4).

Betty, Y. S.; Kim, M. D.; Rutka, J. T.; Chan, W. C. W. Nanomedicine. *N Engl J Med.* **2010**, *363*(25), 2435–2443

Brugè, F.; Damiani, E.; Marcheggiani, F.; Offerta, A.; Puglia, C.; Tiano, L. A comparative study on the possible cytotoxic effects of different nanostructured lipid carrier (NLC) compositions in human dermal fibroblasts. *Int J Pharm.* **2015**, *495*(2), 879–885.

Buzea, C.; Pacheco, I. I.; Robbie, K. Nanomaterials and nanoparticles: sources and toxicity. *Biointerphases.* **2007**, *2*(4), MR17–MR71.

Chan, C. K.; Peng, H.; Liu, G.; McIlwrath, K.; Zhang, X. F.; Huggins, R. A.; Cui, Y. High-performance lithium battery anodes using silicon nanowires. *Nat. Nanotechnol.* **2008**, *3*(1), 31–35.

Chertok, B.; David, A. E.; Yang, V. C. Polyethyleneimine-modified iron oxide nanoparticles for brain tumor drug delivery using magnetic targeting and intracarotid administration. *Biomaterials.* **2010**, *31*(24), 6317–6324.

Chertok, B.; Moffat, B. A.; David, A. E.; Yu, F.; Bergemann, C.; Ross, B. D.; Yang, V. C. Iron oxide nanoparticles as a drug delivery vehicle for MRI monitored magnetic targeting of brain tumors. *Biomaterials.* **2008**, *29*(4), 487–496.

Chithrani, B. D.; Ghazani, A. A.; Chan, W. C. Determining the size and shape dependence of gold nanoparticle uptake into mammalian cells. *Nano Letters.* **2006**, *6*(4), 662–668.

Coester, C. J.; Langer, K.; Von Briesen, H.; Kreuter, J. Gelatin nanoparticles by two step desolvation a new preparation method, surface modifications and cell uptake. *J Microencapsul.* **2000**, *17*(2), 187–193.

De Jong.W. H.; Borm, A. J. P. Drug delivery and nanoparticles: applications and hazards. *Int J Nanomed.* **2008**, *3*(2) 133–149.

Derakhshandeh, K.; Soheili, M.; Dadashzadeh, S.; Saghiri, R. Preparation and in vitro characterization of 9-nitrocamptothecin-loaded long circulating nanoparticles for delivery in cancer patients. *Int J Nanomed.* **2010**, *5*, 463.

Des Rieux, A.; Fievez, V.; Garinot, M.; Schneider, Y. J.; Préat, V. Nanoparticles as potential oral delivery systems of proteins and vaccines: a mechanistic approach. *J. Control. Release.* **2006**, *116*, 1–27.

Dong, X.; Mattingly, C. A.; Tseng, M.; Cho, M.; Adams, V. R.; Mumper, R. J. Development of new lipid-based paclitaxel nanoparticles using sequential simplex optimization. *Eur J Pharm Biopharm.* **2009**, *72*(1), 9–17.

Duncan, T. V. Applications of nanotechnology in food packaging and food safety: barrier materials, antimicrobials and sensors. *J Colloid Interf Sci.* **2011**, *363*(1), 1–24.

El-Sayed, I. H.; Huang, X.; El-Sayed, M. A. Surface plasmon resonance scattering and absorption of anti-EGFR antibody conjugated gold nanoparticles in cancer diagnostics: applications in oral cancer. *Nano letters.* **2005**, *5*(5), 829–834.

El-Samaligy, M. S.; Rohdewald, P. Reconstituted collagen nanoparticles, a novel drug carrier delivery system. *J Pharm Pharmacol.* **1983**, *35*(8), 537–539.

Etheridge, M. L.; Campbell, S. A.; Erdman, A. G.; Haynes, C. L.; Wolf, S. M.; McCullough, J. The big picture on nanomedicine: the state of investigational and approved nanomedicine products. *Nanomedicine.* **2013**, *9*(1), 1–14.64.

Frank, J. A.; Miller, B. R.; Arbab, A. S.; Zywicke, H. A.; Jordan, E. K.; Lewis, B. K.; Henry, L.; Bryant, Jr.; Bulte, J. W. Clinically applicable labeling of mammalian and stem cells by combining superparamagnetic iron oxides and transfection agents. *Radiology.* **2003**, *228*(2), 480–487.

Garcia-Fuentes, M.; Prego, C.; Torres, D.; Alonso, M. J. A comparative study of the potential of solid triglyceride nanostructures coated with chitosan or poly (ethylene glycol) as carriers for oral calcitonin delivery. *Eur J Pharm Sci.* **2005**, *25*(1), 133–143.

Gelperina, S.; Kisich, K.; Iseman, M. D.; Heifets, L.; The potential advantages of nanoparticle drug delivery systems in chemotherapy of tuberculosis. *Am. J. Respir. Crit. Care. Med.* **2005**, *172*, 1487–1490.

George, S.; Lin, S.; Ji, Z.; Thomas, C. R.; Li, L. J.; Mecklenburg, M.; Meng, H.; Wang, X.; Zhang, H.; Xia, T.; Hohman, N.; Lin, S.; Zink, J. I.; Weiss, P. S.; Nel, A. E. Surface defects on plate-shaped silver nanoparticles contribute to its hazard potential in a fish gill cell line and zebrafish embryos. *ACS Nano.* **2012**, *6*(5), 3745–3759

Gittard, S. D.; Hojo, D.; Hyde, G. K.; Scarel, G.; Narayan, R. J.; Parsons, G. N. Antifungal textiles formed using silver deposition in supercritical carbon dioxide. *J. Mater. Eng. Perform.* **2010**, *19*(3), 368–373.

Gref, R.; Lück, M.; Quellec, P.; Marchand, M.; Dellacherie, E.; Harnisch, S.; Blunk. T.; Müller, R. H. 'Stealth' corona-core nanoparticles surface modified by polyethylene glycol (PEG): influences of the corona (PEG chain length and surface density) and of the core composition on phagocytic uptake and plasma protein adsorption. *Colloids Surfaces B.* **2000**, *18*(3), 301–313.

Hochella, M. F.; Lower, S. K.; Maurice, P. A.; Penn, R. L.; Sahai, N.; Sparks, D. L.; Twining, B. S. Nanominerals, mineral nanoparticles, and earth systems. *Science.* **2008**, *319*(5870), 1631–1635.

Honary, S.; Zahir, F. Effect of zeta potential on the properties of nano-drug delivery systems-a review. *Trop J Pharm Res.* **2013**, *12*(2), 255–264

Horikoshi, S and Serpone, N. Intrduction to nanoparticles. Microwave in nanoparticle synthesis; Horikoshi, S.; Serpone, N. W.; Ed; Wiley: New York, **2013**.

Hwu, J. R.; Lin, Y. S.; Josephrajan, T.; Hsu, M. H.; Cheng, F. Y.; Yeh, C. S.; Su, C. W.; Shieh, D. B. Targeted paclitaxel by conjugation to iron oxide and gold nanoparticles. *J Am Chem Soc.* **2008**, *131*(1), 66–68.

Ispas, C.; Andreescu, D.; Patel, A.; Goia, D. V.; Andreescu, S.; Wallace, K. N. Toxicity and developmental defect of different sizes and shape nickel nanoparticles in zebrafish. *Environ. Sci. Technol.* **2009**, *43*(16), 6349–6356

Javid, A.; Ahmadian, S.; Saboury, A. A.; Kalantar, S. M.; Rezaei-Zarchi, S. Chitosan-coated superparamagnetic iron oxide nanoparticles for doxorubicin delivery: synthesis and anticancer effect against human ovarian cancer cells. *Chem Bio Drug Des.* **2013**, *82*(3), 296–306.

Jiang, Q.; Zhang, Z. G.; Ding, G. L.; Zhang, L.; Ewing, J. R.; Wang, L.; Zhang, R.; Li, L.; Lu, M.; Meng.h.; Arbab, S. A.; Hu, J.; Li, J. Q.; Pourabdollah, S.; Nejad, D.; Athiraman, H.; Chopp, M. Investigation of neural progenitor cell induced angiogenesis after embolic stroke in rat using MRI. *Neuroimage.* **2005**, *28*(3), 698–707

Jiang, W.; Kim, B. Y.; Rutka, J. T.; Chan, W. C. Nanoparticle-mediated cellular response is size-dependent. *Nat Nanotechnol.* **2008**, *3*(3), 145–150.

Jortner, J.; Rao, C. N. R.; Nanostructured advanced materials- perspective and directions. *Pure Appl.Chem.* **2002**, *74*, 1489–1783

Judson, I.; Radford, J. A.; Harris, M.; Blay, J. Y.; Van Hoesel, Q.; Le Cesne, A.; Oosterom, V. T. A.; Clemons, M. J.; Kamby, C.; Herman, C.; Whittaker, J.; Donato di Paola, E.; Verweij, J.; Nielsen, S. Randomised phase II trial of PEGylated liposomal doxorubicin (DOXIL®/CAELYX®) versus doxorubicin in the treatment of advanced or metastatic soft tissue sarcoma: a study by the EORTC Soft Tissue and Bone Sarcoma Group. *Eur J Cancer.* **2001**, *37*(7), 870–877.

Jung, W. S.; Kang, M. G.; Moon, H. G.; Baek, S. H.; Yoon, S. J.; Wang, Z. L.; Kim, S. W.; Kang, C. Y. High output piezo/triboelectric hybrid generator. *Sci Rep.* **2015**, 5.

Kaegi, R.; Voegelin, A.; Ort, C.; Sinnet, B.; Thalmann, B.; Krismer, J.; Hagendorfer, H.; Elumelu, M.; Mueller, E. Fate and transformation of silver nanoparticles in urban wastewater systems. *Water Res.* **2013**, *47*(12), 3866–3877.

Kang, K. A.; Wang, J. Conditionally activating optical contrast agent with enhanced sensitivity via gold nanoparticle plasmon energy transfer: feasibility study. *J Nanobiotechnology.* **2014**, *12*(1), 1–9.

Kayal, S.; Ramanujan, R. V. Doxorubicin loaded PVA coated iron oxide nanoparticles for targeted drug delivery. *Mat Sci Eng: C.* **2010**, *30*(3), 484–490.

Kievit, F. M.; Wang, F. Y.; Fang, C.; Mok, H.; Wang, K.; Silber, J. R.; Ellenbogen, R. G.; Zhang, M. Doxorubicin loaded iron oxide nanoparticles overcome multidrug resistance in cancer in vitro. *J Controlled Release.* **2011**, *152*(1), 76–83.

Kokura, S.; Handa, O.; Takagi, T.; Ishikawa, T.; Naito, Y.; Yoshikawa, T. Silver nanoparticles as a safe preservative for use in cosmetics. *Nanomed Nanotech Biol and Med.* **2010**, *6*(4), 570–574.

Kundu, J.; Chung, Y. I.; Kim, Y. H.; Tae, G.; Kundu, S. C. Silk fibroin nanoparticles for cellular uptake and control release. *Int J Pharm*. **2010**, *388*(1), 242–250.

Leo, E.; Vandelli, M. A.; Cameroni, R.; Forni, F. Doxorubicin-loaded gelatin nanoparticles stabilized by glutaraldehyde: involvement of the drug in the cross-linking process. *Int J Pharm*. **1997**, *155*(1), 75–82.

Lewin, M.; Carlesso, N.; Tung, C.; Tang, X.; Cory, D.; Scadden, D. T.; Weissleder, R. Tat peptide-derivatized magnetic nanoparticles allow in vivo tracking and recovery of progenitor cells. *Nat Biotech*. **2000**, *18*, 410–414.

Lin, C. P.; Lin, S.; Wang. C. P.; Sridhar, R. Techniques for physiochemical characterization of nanomaterials. *Biotechnol Adv*. **2014**, *32*(4), 711–726

Liu, G.; Xu, W.; Xia, X.; Shi, H.; Hu, C. Newton's cradle motion-like triboelectric nano-generator to enhance energy recycle efficiency by using elastic deformation. *J. Mater. Chem. A*. **2015**, *3*, 21133–21139.

Liu, H.; Ye, T.; Mao, C. Fluorescent carbon nanoparticles derived from candle soot. *Angew Chem Int Ed*. **2007**, *46*(34), 6473–6475.

Liu, Y.; Zhang, B.; & Yan, B. Enabling anticancer therapeutics by nanoparticle carriers: the delivery of Paclitaxel. *Int J Mol Sci*. **2011**, *12*(7), 4395–4413.

Lyass, O.; Uziely, B.; Ben-Yosef, R.; Tzemach, D.; Heshing, N. I.; Lotem, M.; Brufman, G.; Gabizon, A. Correlation of toxicity with pharmacokinetics of PEGylated liposomal doxorubicin (Doxil) in metastatic breast carcinoma. *Cancer*. **2000**, *89*(5), 1037–1047.

Magasinski, A.; Dixon, P.; Hertzberg, B.; Kvit, A.; Ayala, J.; Yushin, G. High-performance lithium-ion anodes using a hierarchical bottom-up approach. *Nat. Mater*. **2010**, *9*(4), 353–358.

Mahmoudi, M.; Sant, S.; Wang, B.; Laurent, S.; Sen, T. Superparamagnetic iron oxide nanoparticles (SPIONs): development, surface modification and applications in chemotherapy. *Adv Drug Delivery Rev*. **2011**, *63*(1), 24–46.

Martins, P.; Rosa, D.; R Fernandes, A.; Baptista, P. V. Nanoparticle drug delivery systems: recent patents and applications in nanomedicine. *Recent Patents on Nanomedicine*. **2013**, *3*(2), 105.

Moghimi, S. M.; Hunter, A. C.; Murray, J. C. Nanomedicine: current status and future prospects. *FASEB J*. **2005**, *19*(3), 311–330

Morais, M. G. D.; Martins, V. G.; Steffens, D.; Pranke, P.; da Costa, J. A. V. Biological applications of nanobiotechnology. *J Nanosci Nanotechnol*, **2014**, *14*(1), 1007–1017.

Mu, L. and Feng, S. S. A novel controlled release formulation for anticancer drug paclitaxel (Taxol): PLGA nanoparticles containing vitamin E TPGS. *J Control Release*, **2003**, *86*(1), 33–48

Muller, R. H.; Mader, K.; Gohla, S. Solid lipid nanoparticles (SLN) for controlled drug delivery-a review of the state-of-the-art. *Eur J Pharm Biopharm*. **2000**, *50*(1), 161–177.

Nguyen, T. H.; Kim, Y. H.; Song, H. Y.; Lee, B. T. Nano Ag loaded PVA nano-fibrous mats for skin applications. *J Biomed Mater Res B Appl Biomater*. **2011**; *96*(2), 225–233.

Nigam, S.; Barick, K. C.; Bahadur, D. Development of citrate-stabilized Fe_3O_4 nanoparticles: conjugation and release of doxorubicin for therapeutic applications. *J Magn Magn Mater*. **2011**, *323*(2), 237–243.

Papi, M.; Palmieri, V.; Maulucci, G.; Arcovito, G.; Greco, E.; Quintiliani, G.; Fraziano, M.; De Spirito, M. Controlled self assembly of collagen nanoparticle. *J Nanopart Res.* **2011**, *13*(11), 6141–6147.

Park, J. W.; Hong, K.; Kirpotin, D. B.; Colbern, G.; Shalaby, R.; Baselga, J.; Shao, Y.; Nielsen.B. U.; Marks, D. J.; Moore, D.; Papahadjopoulous, D.; Benz, C. C. Anti-HER2 immunoliposomes enhanced efficacy attributable to targeted delivery. *Clin Cancer Res.* **2002**, *8*(4), 1172–1181.

Park, W. H.; Jeong, L.; Yoo, D. I.; Hudson, S. Effect of chitosan on morphology and conformation of electrospun silk fibroin nanofibers. *Polymer.* **2004**, *45*(21), 7151–7157.

Patlolla, R. R.; Chougule, M.; Patel, A. R.; Jackson, T.; Tata, P. N.; Singh, M. Formulation, characterization and pulmonary deposition of nebulized celecoxib encapsulated nanostructured lipid carriers. *J Controlled Release.* **2010**, *144*(2), 233–241.

Peter, M.; Binulal, N. S.; Soumya, S.; Nair, S. V.; Furuike, T.; Tamura, H.; Jayakumar, R. Nanocomposite scaffolds of bioactive glass ceramic nanoparticles disseminated chitosan matrix for tissue engineering applications. *Carbohyd Polym.* **2010**, *79*(2), 284–289.

Prabaharan, M.; Grailer, J. J.; Pilla, S.; Steeber, D. A.; Gong, S. Gold nanoparticles with a monolayer of doxorubicin-conjugated amphiphilic block copolymer for tumor-targeted drug delivery. *Biomaterials.* **2009**, *30*(30), 6065–6075.

Qu, X.; Yao, C.; Wang, J.; Li, Z.; Zhang, Z. Anti-CD30-targeted gold nanoparticles for photothermal therapy of L-428 Hodgkin's cell. *Int. J. Nanomed.* **2012**, *7*, 6095–6103.

Quinto, C. A.; Mohindra, P.; Tong, S.; Bao, G. Multifunctional superparamagnetic iron oxide nanopaticles for combined chemotherapy and hyperthermia cancer treatment. *Nanoscale,* **2015**, *7*(29), 12728–12736.

Rashi Jain, Yamuna Mohanram, Lakshmi Kiran Chelluri, Deepak Kumar, Nataliya Smith, Debra Saunders, Shubmitha Bhatnagar, Venkat Vamsi Venuganti, Ravindranath Kancherla, Ratnakar Kamaraju, Rao, V. L. Papineni, Rheal A. Towner, Partha Ghosal. A modified approach to image guided cell-based therapy for cardiovascular diseases using cardiac precursor nanoprobe – GloTrack. *Nature; Prot Exch;* **2016**; doi: 10.1038/protex.2016.004.

Ravindra, S.; Mohan, Y. M.; Reddy, N. N.; Raju, K. M. Fabrication of antibacterial cotton fibers loaded with silver nanoparticles via "Green Approach." *Colloids and Surf A Physicochem Eng Asp.* **2010**, *367*(1), 31–40.

Ray, S. C.; Saha, A.; Jana, N. R.; Sarkar, R. Fluorescent carbon nanoparticles: synthesis, characterization, and bioimaging application. *J Phys Chem C.* **2009**, *113*(43), 18546–18551.

Salari, F.; Varasteh, A. R.; Vahedi, F.; Hashemi, M.; Sankian, M. Down-regulation of Th2 immune responses by sublingual administration of poly (lactic-coglycolic) acid (PLGA)-encapsulated allergen in BALB/c mice. *Int Immunopharmacol.* **2015**.

Schleich, N.; Po, C.; Jacobs, D.; Ucakar, B.; Gallez, B.; Danhier, F.; Préat, V. Comparison of active, passive and magnetic targeting to tumors of multifunctional paclitaxel/SPIO-loaded nanoparticles for tumor imaging and therapy. *J Control Release.* **2014**, *194*, 82–91.

Schleich, N.; Sibret, P.; Danhier, P.; Ucakar, B.; Laurent, S.; Muller, R. N.; Jerome, C.; Gallez, B.; Preat, V.; Danhier, F. Dual anticancer drug/superparamagnetic iron oxide-loaded

PLGA-based nanoparticles for cancer therapy and magnetic resonance imaging. *Int J Pharma.* **2013**, *447*(1), 94–101.

Shenoy, D.; Little, S.; Langer, R.; Amiji, M. Poly(ethylene oxide)-modified poly(β-amino ester) nanoparticles as a pH-sensitive system for tumor-targeted delivery of hydrophobic drugs: part 2. In vivo distribution and tumor localization studies. *Pharmaceut Res* **2005**, *22*(12), 2107–2114.

Shim, B. S.; Chen, W.; Doty, C.; Xu, C.; Kotov, N. A. Smart electronic yarns and wearable fabrics for human biomonitoring made by carbon nanotube coating with polyelectrolytes. *Nano Lett.* **2008**, *8*(12), 4151–4157.

Sokolov, K.; Follen, M.; Aaron, J.; Pavlova, I.; Malpica, A.; Lotan, R.; Richards-Kortum, R.. Real-time vital optical imaging of precancer using antiepidermal growth factor receptor antibodies conjugated to gold nanoparticles. *Cancer Res.* **2003**, *63*(9), 1999–2004.

Tao, Y.; He, J.; Zhang, M.; Hao, Y.; Liu, J.; Ni, P. Galactosylated biodegradable poly (ε-caprolactone-cophosphoester) random copolymer nanoparticles for potent hepatoma-targeting delivery of doxorubicin. *Polym. Chem.* **2014**, *5*(10), 3443–3452.

Taqaddas, A.; Use of Magnetic Nanoparticles in Cancer Detection with MRI. *World Academy of Science, Engineering and Technology, Int J Medical, Health, Biomedical, Bioengineering and Pharmaceutical Engineering* **2014**, *8*(9).

Teske, S. S.; Detweiler, S. C. The biomechanisms of metal and metal-oxide nanoparticles' interactions with cells. *Int J Eniron Res Public Health.* **2015**, *12*, 1112–1134.

Tewes, F.; Munnier, E.; Antoon, B.; Okassa, L. N.; Cohen-Jonathan, S.; Marchais, H.; Eyrolles, L. D.; Souce, M.; Dubois, P.; Chourpa, I. Comparative study of doxorubicin-loaded poly (lactide-coglycolide) nanoparticles prepared by single and double emulsion methods. *Eur J Pharma Biopharma.* **2007**, *66*(3), 488–492.

Tomuleasa, C.; Soritau, O.; Orza, A.; Dudea, M.; Petrushev, B.; Mosteanu, O.; Susman, S.; Florea, A.; Pall, E.; Aldea, M.; Kacso, G.; Cristea, V.; Neagoe, I. B.; Irimie, A. Gold nanoparticles conjugated with cisplatin/doxorubicin/capecitabine lower the chemoresistance of hepatocellular carcinoma-derived cancer cells. *J Gastrointestin Liver Dis.* **2012**, *21*(2), 187–196

Van Broekhuizen, P.; van Broekhuizen, F.; Cornelissen, R.; Reijnders, L. Use of nanomaterials in the European construction industry and some occupational health aspects thereof. *J Nanopart Res.* **2011**, *13*(2), 447–462.

Verma, V. K.; Beevi, S. S.; Debnath, T.; Shalini, U.; Kamaraju, S. R.; Kona, L. K.; Mohanram, Y.; Chelluri, L. K. Signal regulatory protein alpha (SIRPA) and kinase domain receptor (KDR) are key expression markers in cardiac specific precursor selection from hADSCs. *New Horizons in Translational Medicine,* **2015**, *2*(4), 93–101.

Verma, V. K.; Kamaraju, S. R.; Kancherla, R.; Kona, L. K.; Beevi, S. S.; Debnath, T.; Usha, P. S.; Vadapalli, R.; Arbab, S. A.; Chelluri, L. K. Fluorescent magnetic iron oxide nanoparticles for cardiac precursor cell selection from stromal vascular fraction and optimization for magnetic resonance imaging. *Int J Nanomed.* **2015**, *10*(1), 711–726.

Wang, F.; Wang, Y. C.; Dou, S.; Xiong, M. H.; Sun, T. M.; Wang, J. Doxorubicin-tethered responsive gold nanoparticles facilitate intracellular drug delivery for overcoming multidrug resistance in cancer cells. *Acs Nano.* **2011**, *5*(5), 3679–3692

Wang, Y.; Li, X.; Wang, L.; Xu, Y.; Cheng, X.; Wei, P. Formulation and pharmacokinetic evaluation of paclitaxel nanosuspension for intravenous delivery. *Int J Nanomed.* **2011**, *6*, 1497–1507

Wilczewska, A. Z.; Niemirowicz, K.; Markiewicz, K. H.; Car, H. Nanoparticles as drug delivery systems. *Pharmacol Rep.* **2012**, *64*, 1020–1037.

Yan, H. B.; Zhang, Y. Q.; Ma, Y. L.; Zhou, L. X. Biosynthesis of insulin-silk fibroin nanoparticles conjugates and in vitro evaluation of a drug delivery system. *J Nanoparticle Res.* **2009**, *11*(8), 1937–1946.

Yang, Z.; Chen, C. Y.; Liu, C. W.; Chang, H. T. Electrocatalytic sulfur electrodes for CdS/CdSe quantum dot-sensitized solar cells. *Chem Comm.* **2010**, *46*(30), 5485–5487.

Yu, B.; Mao, Y.; Yuan, Y.; Yue, C.; Wang, X.; Mo, X.; Jarjoura, D.; Paulaitis, E. M.; Lee, J. R.; Byrd, C. J.; Lee, J. L.; Muthusamy, N. Targeted drug delivery and cross-linking induced apoptosis with anti-CD37 based dual-ligand immune-liposomes in B chronic lymphocytic leukemia cells. *Biomaterials*, **2013**, *34*(26), 6185–6193.

Zhang, S.; Lu, Z.; Gu, L.; Cai, L.; Cao, X. Deterministic growth of AgTCNQ and CuTCNQ nanowires on large-area reduced graphene oxide films for flexible optoelectronics. *Nanotechnology.* **2013**,*24*(46),465202 doi.org/10.1088/0957-4484/24/46/465202.

Zhao, Z.; Li, Y.; Zhang, Y. Preparation and Characterization of Paclitaxel Loaded SF/PLLA-PEG-PLLA Nanoparticles via Solution-Enhanced Dispersion by Supercritical CO_2. *J Nanomater.* **2015**, *501*, 913254.

CHAPTER 3

DENDRIMER CONJUGATES AS NOVEL DNA AND siRNA CARRIERS

YUYA HAYASHI,[1,2] KEIICHI MOTOYAMA,[1] TAISHI HIGASHI,[1] HIROFUMI JONO,[3] YUKIO ANDO,[4] and HIDETOSHI ARIMA[1,5,*]

[1]Department of Physical Pharmaceutics, Graduate School of Pharmaceutical Sciences, Kumamoto University, 5-1 Oe-honmachi, Chuo-ku, Kumamoto 862-0973, Japan

[2]Research Fellow of Japan Society for the Promotion of Science, Graduate School of Pharmaceutical Sciences, Kumamoto University, 1-1-1 Honjo, Chuo-ku, Kumamoto 860-8556, Japan

[3]Department of Clinical Pharmaceutical Sciences, Graduate School of Pharmaceutical Sciences, Kumamoto University, 1-1-1 Honjo, Chuo-ku, Kumamoto 860-8556, Japan

[4]Department of Neurology, Graduate School of Medical Sciences, Kumamoto University, 1-1-1 Honjo, Chuo-ku, Kumamoto 860-8556, Japan

[5]Program for Leading Graduate Schools "HIGO (Health life science: Interdisciplinary and Glocal Oriented) Program," Kumamoto University, Chuo-ku, Kumamoto 860-8556, Japan, *E-mail: arimah@gpo.kumamoto-u.ac.jp

CONTENTS

ABSTRACT

Gene and small interfering RNA (siRNA) delivery can be particularly used for the treatment of diseases by the introduction of genetic materials mammalian to cells either to express new proteins or to suppress the expression of proteins, respectively. The hepatocyte may be a potentially important target for gene and RNA interference (RNAi), because various crucial diseases occur in this organ. The potential use of various polyamidoamine (PAMAM) starburst™ dendrimer conjugates with α-cyclodextrin (α-CDE) as a novel DNA and siRNA carriers (vectors) has been demonstrated. This review provides the recent findings on hepatocyte-targeted α-CDE for DNA and siRNA delivery.

3.1 INTRODUCTION

The liver is the largest internal organ in the body and associated with metabolism, detoxification, synthesis, and secretion of major plasma proteins, and iron homeostasis. In addition, the liver is one of the crucial target organs for gene and RNA interference (RNAi) therapy, because many diseases such as liver tumors, hepatic cirrhosis, hereditary metabolic diseases, chronic viral hepatitis, lysosome disease, etc. (Raper and Wilson, 1995). The liver-directed gene and RNAi therapy has been developed as an alternative to orthotropic liver transplantation, which is the only actual therapy for liver diseases, and could modernize the treatment of many inherited hematologic and metabolic diseases (Nguyen and Ferry, 2004). Recently, Mipomersen, a second-generation antisense oligonucleotide (ODN) developed against the coding region of human apolipoprotein B (apoB) mRNA, have been approved by the Food and Drug Administration (FDA) for use in patients with homozygous familial hypercholesterolemia (Gotto and Moon, 2013; Sahebkar and Watts, 2013). Additionally, ND-L02-s, 0201, vitamin A-coupled lipid nanoparticles to deliver small interfering RNA (siRNA) against heat shock protein (HSP) 47, has been developed for the cure of liver fibrosis (Sato et al., 2008). Also, Nitto Denko has been performing a placebo-controlled phase I study, which investigates the safety and tolerability of single, ascending intravenous doses of ND-L02-s0201 in healthy volunteers. In vivo delivery of DNA and RNA to hepatocytes has been achieved by using hydrodynamic injection (Zhang et al., 1999), hepatocyte-direct injection (Kawakami et al., 2002), mechanical massage (Liu and Huang, 2002), site-specific tissue suction device (Shimizu et al., 2012), etc. Thereby, many drug carriers have been established to increase cellular uptake and optimize tissue distribution of gene and ODNs. Unfortunately, the delivery systems hardly accomplish the criteria of an ideal delivery system lacking of hepatocyte-targeting ability and safety profile. Therefore, novel-underlying carriers to address them must be developed. In this review, we focus on the hepatocyte-targeted DNA and RNA delivery systems using cyclodextrin (CyD), polyamidoamine (PAMAM) starburst™ dendrimer (dendrimer) and these conjugates.

3.2 CYCLODEXTRIN-BASED DELIVERY SYSTEM FOR GENE AND OLIGONUCLEOTIDES

Cyclodextrins (CyDs) are cyclic (α-1,4)-linked oligosaccharides of α-D-glucopyranose containing a hydrophobic interior cavity and hydrophilic exterior surface, and are known to have inclusion ability (Figure 3.1) (Uekama et al., 1998). The α-, β-, and γ-CyDs are parent CyDs, consisting of six, seven, and eight glucose units, respectively. CyDs can enhance the

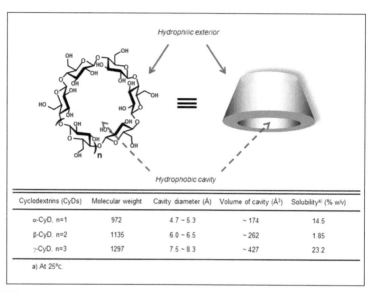

Cyclodextrins (CyDs)	Molecular weight	Cavity diameter (Å)	Volume of cavity (Å³)	Solubility[a] (% w/v)
α-CyD, n=1	972	4.7 ~ 5.3	~ 174	14.5
β-CyD, n=2	1135	6.0 ~ 6.5	~ 262	1.85
γ-CyD, n=3	1297	7.5 ~ 8.3	~ 427	23.2

a) At 25°C.

FIGURE 3.1   Structure and properties of natural CyDs. (Reprinted from Uekama, K.; Hirayama, F.; Irie, T. Cyclodextrin drug carrier systems. *Chem. Rev.*, **1998**, *98*, 2045–2076. © 1998 American Chemical Society. With permission.)

solubility, dissolution rate and bioavailability of drugs, and so the widespread use of CyDs is acknowledged in the pharmaceutical field (Szente and Szejtli, 1999; Uekama, 2004). CyDs have been reported to interact with lipid membrane components of cells, leading to the induction of hemolysis (Fauvelle et al., 1997; Irie et al., 1982; Ohtani et al., 1989), although CyDs enter cells only very slightly owing to their high molecular weight (ca. 1,000) and hydrophilicity (Uekama et al., 1998). In terms of the delivery of genes and ODNs using CyDs, CyDs unable to interact with the anionic DNA/RNA polymers (Arima and Motoyama, 2009). Thereby, the combination of CyDs with some cell-penetrating carriers was required to uptake into the cells.

Recently, CyDs have been used to as carries for gene and ODNs (Abdou et al., 1997; Croyle et al., 1998; Davis et al., 2004; Davis et al., 2010; Ortiz Mellet et al., 2011; Roessler et al., 2001). Davis and co-workers demonstrated a novel CyD polymer (CDP) for the delivery of DNA and siRNA (Davis et al., 2004; Davis, 2009; Gonzalez et al., 1999). This CDP system is comprised of three delivery components and the M2 subunit of ribonucleotide reductase (RRM2) against siRNA (Heidel et al., 2007). That is the three delivery components; CDP, adamantine-PEG (AD-PEG) and AD-PEG-transferrin (AD-PEG-Tf), each plays a specific but complimentary role in overcoming numerous in vivo barriers. This delivery system, CALAA-01, was the first targeted polycation for siRNA delivery to come in the clinical trial for cancer (Davis et al., 2010), although the clinical trial was discontinued.

Wu et al. demonstrated that an in vitro targeting system of pDNA to hepatocytes using a soluble DNA carrier that takes advantage of asialoglycoprotein receptor 1 (ASGPR1)-mediated endocytosis to achieve internalization has been investigated (Wu and Wu, 1998; Wu and Wu, 1988). Additionally, Hara et al. reported ASGPR1-mediated transfer of pDNA to murine hepatocyte using asialofetuin (AF)-labeled liposomes (Hara et al., 1995). ASGPR1 is known to be a high-capacity galactose-binding receptor expressed on hepatocytes, which binds to its native substrates such as AF with low affinity (Westerlind et al., 2004). Therefore, various ASGPR1-mediated delivery systems have been reported, e.g., PEG-graft-poly L-lysine (PLL) (Hu et al., 2012), galactosylated N-2-hydroxypropyl methacrylamide-b-N-3-guanidinopropyl methacrylamide block copolymers (Qin et al., 2011), galactosylated PEG-graft-polyethyleneimine (PEI) (Nie et al., 2011), etc. Thus, a galactosylation of carriers is one of the effective methods to deliver gene and siRNA to hepatocytes and/or to increase the transfer activity. Also, Pun et al. demonstrated that the potential of pDNA complexes with both β-CyD-containing polycations (βCDPs) and modified the polyplex with galactosylated adamantine thorough host-guest interaction as a hepatocyte-selective polyplexes. The galactosylated polyplex showed 10-fold higher gene transfect efficiency than glucozylated polyplex in HepG2 cells (ASGPR1 (+)), and predictably did not show preferential transfection of galactosylated polyplexes to HeLa cells (ASGPR1 (−)). Therefore, this carrier may be useful as the gene delivery system to hepatocytes (Pun and Davis, 2002). Recently, McMahon et al. prepared galactosylated amphiphilic CyDs as a hepatocyte-targeted gene delivery system (McMahon et al., 2012). These CyDs had lipophilic groups (C6 or C16) and hydrophilic groups (cationic primary amine groups or galactose targeting ligands) in a CyD molecule, and form nanoparticles with pDNA. In this system, gene expression of the nanoparticles having a galactose moiety was higher than that of lacking galactose moiety nanoparticles in HepG2 cells. In addition, its gene expression was significantly increased in the presence of the dioleoylphosphatidylethanolamine (DOPE), probably due to an increase in endosomal escaping ability (McMahon et al., 2012; O'Neill et al., 2011). In addition, Symens et al. developed a multivalent D-galactose ligand modified polycationic amphiphilic CyD (pacds) (Symens et al., 2012), and demonstrated that nucleic acid/galactose modified paCD complexes (CDplexes) efficiently entered the cells. Meanwhile, when galactose-modified CDplexes containing mRNA encoding the green fluorescent protein were transfected to HepG2 cells, the fractions of GFP-positive cells of upto 31% were observed. These results suggest that the galactose-modified CDplexes are specifically internalized via the ASGPR1 and that the galactose-modified paCDs may serve as a valuable mRNA delivery platform for the transient transfection of hepatocytes (Symens et al., 2012).

3.3   DENDRIMER-BASED DELIVERY SYSTEM FOR GENE AND OLIGONUCLEOTIDES

Dendrimers, which are developed by Tomalia and co-workers are biocompatible, nonimmunogenic and water-soluble (Tomalia et al., 1985). In addition, dendrimers are a unique class of synthetic macromolecules having highly branched, three-dimensional, nanoscale architectures with very low polydispersity and high functionality. These properties have allowed their application in pharmaceutical, medicinal chemistry and nanotechnology fields to be attractive (Challa and Stefanovic, 2011). Since polyamidoamine dendrimers (dendrimers) possess terminal modifiable amine functional group, they form complexes with pDNA (Dufes et al., 2005, Dutta et al., 2010), short hairpin RNA (shRNA) (Kim et al., 2006; Liu et al., 2011; Yuan et al., 2010) and siRNA (Troiber and Wagner, 2011; Wang et al., 2010; Yuan et al., 2011) through the electrostatic interaction as well as the binding to cellular surface glycosaminoglycans (Ruponen et al., 2003). As a result, dendrimers are known to offer efficient transfer activity for gene and ODN drugs. Additionally, the high transfection efficiency of dendrimers can be due to their well-defined shape and the proton-sponge effect. Herein, the proton-sponge effect is believed to be caused by cationic polymers that promote endosome osmotic swelling, disruption of the endosomal membrane and intracellular release of gene and ODN drugs (Klajnert and Bryszewska, 2001). Clearly, the properties of dendrimers as nonviral vectors significantly are associated with their generation (G). Regarding pDNA delivery, gene transfer activity of dendrimers with high generations is likely to be superior to that of low generation (Braun et al., 2005; Kukowska-Latallo et al., 1996). However, high generation dendrimers induced cytotoxicity in a generation-dependent manner. Recently, dendrimers having low generation and asymmetric structure are reported to be useful to reduce its cytotoxicity (Shah et al., 2011). As recently reported, double-stranded RNA is less flexible than pDNA, which can lead to the insufficient encapsulation or the formation of undesirably large complexes (Merkel et al., 2010). Additionally, dendrimer-mediated siRNA delivery has typically focused on the use of high generations, such as G6 or G7, since the use of low-generation dendrimers (e.g., G1–3) has not consistently lead to the formation of uniformly small complexes, . Meanwhile, Perez et al. reported that the siRNA complexes with lower generation dendrimers (G4–7) prepared in low ionic strength media were efficiently taken up by cells (Perez et al., 2009). Therefore, there has been a growing interest in developing dendrimers with low generation (<G4) because of their extremely low cytotoxicity (Morgan et al., 1989). Interestingly, the potential use of CyD conjugates with PAMAM dendrimer using low generation as nonviral vectors will be described below. The various types of dendrimers for gene and ODN have been reported such as dendritic poly L-lysine (Ohsaki et al., 2002), carbosilane dendrimers (Bermejo et al., 2007), triazine dendrimers (Merkel et al., 2010), etc. Several excellent books and reviews on this subject have appeared in recent years (Dutta et al., 2010; Yuan et al., 2011; Shcharbin et al., 2013; Singha et al., 2011). In the field of ODN delivery, Liu et al. reported that the siRNA delivery systems are-based on neutral cross-linked dendrimers for hepatocyte-targeted gene silencing (Liu et al., 2012). By replacing the terminal amines with hydrazide groups and N-acetylgalactosamine (GalNAc) ligands, cationic PAMAM dendrimers were transformed into neutral glycosylated carriers for siRNA delivery. In vitro cellular uptake and RNAi experiments showed that the

GalNAc-modified cross-linked dendrimers effectively associated with HepG2 cells and suppressed the expression of a luciferase reporter gene. These results suggest that GalNAc-modified cross-linked dendritic systems provide a new paradigm for designing siRNA delivery systems with biocompatibility and targeting capability.

3.4 CYCLODEXTRIN/DENDRIMER CONJUGATES (CDES)-BASED DELIVERY SYSTEM FOR GENE AND OLIGONUCLEOTIDES

Arima et al. (2001) previously reported the utility of dendrimer conjugates with CyD (CDEs), i.e., of three CDEs with α-, β- or γ-CyD, dendrimers (G2) functionalized with α-CyD (α-CDE) showed luciferase gene expression approximately 100 times higher than the unmodified dendrimer or physical mixtures of the dendrimer and α-CyD. Additionally, in various generations of α-CDEs, α-CDE (G3) with the degree of substitution (DS) of 2 had the highest transfection efficiency in vitro and in vivo with negligible cytotoxicity (Kihara et al., 2002). Moreover, α-CDE (G3) could apply the various types of gene and ODNs such as pDNA (Kihara et al., 2002, 2003), siRNA (Arima et al., 2011; Tsutsumi et al., 2007), shRNA (Tsutsumi et al., 2008) and microRNA (miRNA). The potential use of α-CDE (G3) for gene and ODN carriers was reported as excellent review articles (Arima et al., 2012, 2012, 2013). The CyD conjugates with dendrimer using low generation will be described below.

3.4.1 GALACTOSYLATED DENDRIMER/α-CYCLODEXTRIN CONJUGATE (GAL-α-CDE)

To achieve hepatocyte-specific gene transfer carries-based on α-CDE (G2), Wada et al. prepared galactose-appended α-CDE (G2) (Gal-α-CDE (G2)) with the various degrees of substitution of the galactose moiety (DSG) (Figure 3.2) (Wada et al., 2005). In the case of Gal-α-CDEs (G2), galactose residues were attached to primary amino residues of α-CDE (G2) using α-D-galactopyranosylphenyl isothiocyanate. Herein, Gal-α-CDEs (G2) formed complexes with pDNA, protected the degradation of pDNA by DNase I and exhibited no cytotoxicity. Then, to investigate whether Gal-α-CDEs (G2) have hepatocyte-specific gene transfer activity, Renilla

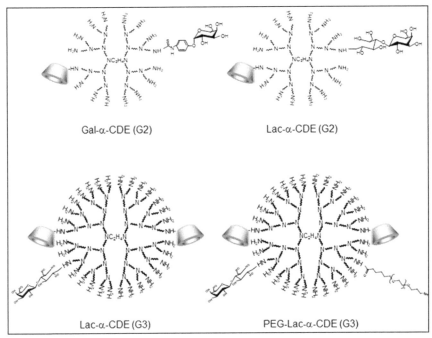

Gal-α-CDE (G2)   Lac-α-CDE (G2)

Lac-α-CDE (G3)   PEG-Lac-α-CDE (G3)

FIGURE 3.2 Chemical structures of various hepatocyte-targeted α-CDEs. (Reprinted from Wada, K.; Arima, H.; Tsutsumi, T.; Chihara, Y.; Hattori, K.; Hirayama, F.; Uekama, K. Improvement of gene delivery mediated by mannosylated dendrimer/α-cyclodextrin conjugates. *J. Control. Release,* **2005,** *104,* 397–413. © 2005 with permission from Elsevier.)

luciferase activity after transfection of pDNA complexes with dendrimer (G2), α-CDE (G2) and Gal-α-CDEs (DSG4, 8, 15) were determined in various cells. It should be noted that Gal-α-CDE (G2, DSG4) showed much higher gene transfer activity than dendrimer (G2), α-CDE (G2) and Gal-α-CDEs (DSG8, 15) in HepG2 cells (ASGPR1 (+)). Similarly, gene transfer activity of Gal-α-CDE (G2, DSG4) was higher than that of dendrimer (G2) and α-CDE (G2) in both NIH3T3 and A549 cells (ASGPR1 (-)). Therefore, these results suggest that Gal-α-CDE (G2, DSG4) augments gene transfer activity in an ASGPR1-independent manner. To confirm whether gene transfer activity of Gal-α-CDE (G2, DSG4) is associated with the galactose-binding lectin recognition, the competitive studies using AF and galactose were performed in HepG2 cells. However, no competitive effects of AF and galactose on gene transfer activity of Gal-α-CDE (G2, DSG4) were observed in HepG2 cells. Thus, it is evident that the enhancing effect of Gal-α-CDE (G2, DSG4) on gene transfer activity may be in an ASGPR1-independent manner. Interestingly, gene transfer activity of Gal-α-CDE (G2, DSG4) was not lowered by the addition of 10% serum, although that of dendrimer tended to be suppressed. Therefore, gene transfer activity of Gal-α-CDE (G2, DSG4) may be considerably resistant to serum under the experimental conditions. The detail mechanisms for the enhancing effect of Gal-α-CDE (G2, DSG4) on gene transfer activity are still unknown. Several possibilities for the enhancing mechanism of Gal-α-CDE (G2, DSG4) in cells could be proposed: (1) the feasible intracellular trafficking and nuclear translocation of the pDNA complex with Gal-α-CDE (G2, DSG4), which may be due to the interaction with intracellular galactose-binding lectins such as galectins (Liu et al., 2002; Wang et al., 2004) and CBP35, a nuclear galactose-binding lectin (Cowles et al., 1989; Wang et al., 1991) in a direct or indirect manner, but not the cell surface ASGPR1; (2) the resistance to methylation of pDNA; and/or (3) the feasible release of pDNA from the complexes with Gal-α-CDE (G2, DSG4) in cytoplasm and/or nucleus, because of the lack of pDNA condensation in the Gal-α-CDE (G2, DSG4) system. Consequently, these results suggest the potential use of Gal-α-CDE (G2, DSG4) as a nonviral vector in various cells, but not hepatocyte-specific vector.

A similar trend was also observed in the other type of sugar-appended α-CDE. To develop antigen presenting cell (APC)-specific gene delivery of α-CDE (G2), Wada et al. prepared Man-α-CDEs (G2) with the various DS of the mannose moiety (DSM) and examined their gene transfer activity in a variety of cells (Wada et al., 2005), because APC express mannose receptors. In the case of Man-α-CDEs (G2), the mannose residues were attached to primary amino residues of α-CDE (G2) using α-D-mannopyranosylphenyl isothiocyanate. Man-α-CDEs (G2, DSM3, 5) showed much higher gene transfer activity than dendrimer (G2), α-CDE (G2) and Man-α-CDE (G2, DSM1, 8) in various cells, which are independent of the expression of cell surface mannose receptors. The surface plasmon resonance (SPR) study revealed that the specific binding activity of Man-α-CDE (G2, DSM3) to concanavalin A, a mannose lectin, was not very strong. It should be noted that Man-α-CDE (G2, DSM3) provided gene transfer activity higher than dendrimer (G2) and α-CDE (G2) in kidney 12 h after intravenous administration in mice. These results suggest the potential use of Man-α-CDE (G2, DSM3) as a nonviral vector, although Man-α-CDE (G2, DSM3) did not show APC-specific gene delivery. In addition, to improve APC-specific gene transfer activity of Man-α-CDE (G2), Arima et al. newly prepared Man-α-CDEs (G3) with various DSM (5, 10, 13 and 20) and compared their gene transfer activity and cytotoxicity, and elucidated the enhancing mechanism for the gene transfer activity (Arima et al., 2006). In the various DSM carriers, Man-α-CDE (G3, DSM10) provided the highest gene transfer activity in NR8383, A549, NIH3T3 and HepG2 cells, and the activity of Man-α-CDE (G3, DSM10) was not decreased by the addition of 10% serum in A549 cells. In addition, cytotoxicity of the polyplex with Man-α-CDE (G3, DSM10) was not observed in A549 and NIH3T3 cells even in a charge ratio of 200(carrier/pDNA). However, the gene transfer activity of Man-α-CDE (G3, DSM10) was independent of the expression of mannose receptors. Interestingly, Alexa-pDNA complex with TRITC-Man-α-CDE (G3, DSM10), but not the complex with TRITC-α-CDE (G3), was showed nuclear translocation at 24 h after incubation in A549 cells. Interestingly, HVJ-E vector including mannan, but neither the vector alone nor the vector including dextran, decreased the nuclear localization of TRITC-Man-α-CDE (G3, DSM10) to a striking degree after 24 h incubation in A549 cells. These results strongly suggest that Man-α-CDE (G3, DSM10) has less cytotoxicity and prominent gene transfer activity through not only its serum resistant and endosomal escaping abilities but also nuclear localization ability, although Man-α-CDE (G3, DSM10) did not elicit antigen presenting cell-specific gene delivery as Gal-α-CDE (G2, DSG4).

3.4.2 LACTOSYLATED DENDRIMER/α-CYCLODEXTRIN CONJUGATE (LAC-α-CDE)

As described above, Gal-α-CDE (G2, DSG4) did not possess hepatocyte-specific gene transfer activity. Here, we envisaged that phenyl isothiocyanate as a spacer between galactose moiety and dendrimer in Gal-α-CDE (G2, DSG4) may be involved in the lack of cell-specific gene delivery owing to the short length and nonflexible of the spacer. Thereby, Arima et al. (2010) prepared lactose-appended α-CDEs (Lac-α-CDE, G2) with the various degree of substitution of the α-lactose moiety (DSL) (Figure 3.2). In the case of Lac-α-CDEs (G2), lactose, a disaccharide consisting of α-glucose and α-galactose, was directly attached to primary amino residues of α-CDE (G2). In the Lac-α-CDEs (G2), the α-glucose moiety plays a role as a spacer. Lac-α-CDE (G2, DSL3) was found to have much higher gene transfer activity than dendrimer (G2), α-CDE (G2) and Lac-α-CDEs (G2, DSL1, 5, 6 and 10) in HepG2 cells (ASGPR1 (+)), but not A549 cells (ASGPR1 (–)). Unlike Gal-α-CDE (G2, DSG4), luciferase gene transfer activity of Lac-α-CDE (G2, DSL3) was markedly suppressed by adding AF, but not bovine serum albumin (BSA) in HepG2 cells. A similar competitive effect of AF was observed in the polyplex with jetPEI™-Hepatocyte, a commercially available hepatocyte-specific transfection reagent. On the other hand, neither AF nor BSA inhibited gene transfer activity of α-CDE (G2). Additionally, the flow cytometric analysis showed that cellular association of pDNA complex with Lac-α-CDE (G2, DSL3) was also decreased by addition of AF, not BSA, in HepG2 cells. To make sure whether Lac-α-CDE (G2, DSL3) was actually recognized by ASGPR1, the molecular interactions of peanut lectin (PNA), a galactose lectin, with α-CDE (G2) and Lac-α-CDE (G2, DSL3) were determined using the SPR method. The association constant of Lac-α-CDE (G2, DSL3) increased ca. 100-fold that of α-CDE (G2). Therefore, these results suggest that Lac-α-CDE (G2, DSL3) provided hepatocyte-specific gene transfer activity through the binding of the carrier to ASGPR1 on HepG2 cells. Moreover, TRITC-Lac-α-CDE (G2, DSL3)/Alexa-pDNA complex, but not TRITC-α-CDE (G2) complex, showed nuclear translocation in HepG2 cells, implying the lactose-mediated nuclear localization. Furthermore, no cytotoxicity of pDNA complexes with α-CDE (G2) and Lac-α-CDE (G2, DSL3) was observed in HepG2 cells and A549 cells up to the charge ratio of at least 150 (carrier/pDNA). Importantly, severe cytotoxicity of pDNA complex with jetPEI™-Hepatocyte was observed even at a charge ratio of 10. These results suggest that Lac-α-CDE (G2, DSL3) has great advantages as a nonviral vector, i.e., superior transfection efficiency and less cytotoxicity.

    To evaluate the potential use of Lac-α-CDE (G2, DSL3) as a hepatocyte-specific carrier, Arima et al. (2010) evaluated in vivo gene transfer activity of pDNA complex with Lac-α-CDE (G2, DSL3) in mice after intravenous administration. Luciferase activity was much higher in the kidney than in other tissues, followed by the liver 12 h after administration. It should be noted that in the liver Lac-α-CDE (G2, DSL3) provided strikingly higher gene transfer activity than α-CDE (G2) and jetPEI™-Hepatocyte, while in the spleen Lac-α-CDE (G2, DSL3) provided lower gene transfer activity than α-CDE (G2). Moreover, luciferase activity in hepatic parenchymal cells was found to be much higher than that in hepatic nonparenchymal cells in the Lac-α-CDE (G2, DSL3) complex system. Remarkably, luciferase activity in hepatic parenchymal cells in the Lac-α-CDE (G2, DSL3) complex system was significantly higher than that in the jetPEI™-Hepatocyte complex system. From the viewpoint of safety, Arima et al. (2010) confirmed that blood chemistry value such as AST, ALT, CRE, BUN and LDH after intravenous administration of the pDNA complexes with α-CDE (G2), Lac-α-CDE (G2, DSL3) and jet-PEI™-Hepatocyte in mice. The ALT value in the Lac-α-CDE (G2, DSL3) system slightly increased, but it was not statistically different, compared to control (5% mannitol solution). The similar results of blood chemistry values were observed in the pDNA complex with jetPEI™-Hepatocyte, possibly due to a low charge ratio (carrier/pDNA). The other parameters in the Lac-α-CDE (G2, DSL3) system were almost equivalent to those in the control system. These results strongly suggest that the pDNA complex with Lac-α-CDE (G2, DSL3) has a good safe profile even in vivo. Hence, these results hold promise for the potential use of Lac-α-CDE (G2, DSL3) as a novel hepatocyte-specific nonviral vector with negligible cytotoxicity.

Recently, in an attempt to improve hepatocyte-specific gene transfer activity of Lac-α-CDEs (G2), Motoyama et al. (2011) prepared higher generation carrier system, Lac-α-CDE (G3) (Figure 3.2). Lac-α-CDE (G3, DSL1) was found to have much higher gene transfer activity than α-CDE (G3), Lac-α-CDE (G2, DSL3) and Lac-α-CDEs (G3, DSL3, 4 and 6) in HepG2 cells. Higher gene transfection efficiency of Lac-α-CDEs (G3) may be dependent on ASGPR1 expression of cell-surface. Notably, Lac-α-CDE (G3, DSL1) provided negligible cytotoxicity up to a charge ratio of 100(carrier/pDNA) in HepG2 cells, suggesting the potential use of Lac-α-CDE (G3, DSL1) as a nonviral vector for gene delivery into hepatocytes (Motoyama et al., 2011).

To clarify the potential use of Lac-α-CDEs (G3) as novel hepatocyte-specific siRNA carriers, Hayashi et al. evaluated the RNAi effect of siRNA complexes with Lac-α-CDEs (G3) both in vitro and in vivo (Hayashi et al., 2012). Transthyretin (TTR)-related familial amyloid polyneuropathy (FAP), which is induced by amyloidogenic TTR (ATTR), is an autosomal dominant form of fatal hereditary amyloidosis characterized by systemic accumulation of amyloid fibrils in peripheral nerves and other organs (Ando et al., 2005). So, we targeted TTR gene expression, which mainly produced in hepatocytes. Lac-α-CDE (G3, DSL1) complex with TTR siRNA (siTTR) had the potent RNAi effect against TTR gene and protein expression through adequate physicochemical properties, ASGPR1-mediated cellular uptake, efficient endosomal escape and the delivery of the siRNA complex to cytoplasm, but not nucleus, with no cytotoxicity. The Lac-α-CDE (G3, DSL1) complex with siTTR had the potential to exert the in vivo RNAi effect after intravenous administration in the liver of mice. The blood chemistry values of Lac-α-CDE (G3, DSL1) system were almost equivalent to those of control system. Thus, these results suggest that Lac-α-CDE (G3, DSL1) has the potential for novel hepatocyte-specific siRNA carriers in vitro and in vivo, and has a possibility as a therapeutic tool for FAP. As described above, Lac-α-CDEs have the potential for a novel hepatocyte-specific DNA and siRNA carrier in vitro and in vivo (Hayashi et al., 2012). However, DNA and siRNA transfer activity of Lac-α-CDEs decreased in the presence of high concentration of serum in vitro. In addition, hepatocytes-specific transfer activity of Lac-α-CDE was not enough satisfactory in vivo, although it was superior to that of jetPEI™-Hepatocyte (Arima et al., 2010).

3.4.3   PEG-APPENDED LACTOSYLATED DENDRIMER/ α-CYCLODEXTRIN CONJUGATE (PEG-LAC-α-CDE)

Most recently, Arima and colleagues developed multifunctional Lac-α-CDEs (G3), PEG-appended Lac-α-CDE (G3) (PEG-Lac-α-CDE (G3)) (Figure 3.2) (Arima et al., 2013; Hayashi et al., 2013). The PEG-Lac-α-CDE (G3) (G3, degrees of substitution of the PEG moiety (DSP) 2) showed higher gene transfer activity than other PEG-Lac-α-CDEs (G3, DSP4, 6) in HepG2 cells. Additionally, high gene transfer activity of PEG-Lac-α-CDE (G3, DSP2) was maintained even in the presence of 50% serum, although gene transfer activity of Lac-α-CDE (G3, DSL1) severely decreased in the presence of 20% serum. In vivo gene transfer activity of Lac-α-CDE (G3, DSL1) was approximately 3-fold higher than that of α-CDE (G3) in the liver, although there was not statistically significant difference among these carriers. However, luciferase activity was also observed in the spleen, lung and heart after intravenous administration of Lac-α-CDE (G3, DSL1) complex with pDNA complex. Fortunately, luciferase activity in the liver in the PEG-Lac-α-CDE (G3, DSP2) system was approximately 1.5-fold higher than that in the α-CDE (G3) system. In addition, luciferase expression in the spleen was decreased to less than 1% after intravenous administration of PEG-Lac-α-CDE (G3, DSP2) complex with pDNA, compared to that of the other carrier ones. Moreover, PEG-Lac-α-CDE (G3, DSP2) did not show gene transfer activity in the kidney, lung or heart under the present experimental conditions. Importantly, the ratio of luciferase activity of PEG-Lac-α-CDE (G3, DSP2) complex with pDNA in the liver to the total activity in various organs was significantly higher than that of α-CDE (G3) and Lac-α-CDE (G3, DSL1), but lower that in the spleen. Furthermore, PEG-Lac-α-CDE (G3, DSP2) showed selective gene transfer activity to hepatic parenchymal cells rather than hepatic nonparenchymal cells. Taken together, these results suggest the potential use of PEG-Lac-α-CDE (G3, DSP2) as a hepatocytes-selective DNA carrier in vitro and in vivo (Hayashi et al., 2013). Furthermore, PEG-Lac-α-CDE (G3, DSP2) complex with siTTR elicited the prominent in vivo RNAi effect after intravenous administration at a lower dose of siRNA than Lac-α-CDE (G3, DSL1)/siTTR complex. Notedly, these siTTR complexes showed negligible cytotoxicity up to a charge ratio of 20 even in vivo (manuscript in preparation).

3.5 CONCLUSION

In this review, we have demonstrated the potential of α-CDE-based hepatocyte-targeted DNA and siRNA delivery systems. Hence, the principle properties of α-CDE-based hepatocyte-targeted α-CDEs are summarized in Table 3.1. Firstly, we prepared Gal-α-CDE (G2) as hepatocyte-targeted α-CDEs. Gal-α-CDE (G2) having a phenylisothiocyanate as a spacer has various advantages such as superior gene transfer activity to dendrimer (G2) and α-CDE (G2), serum resistance and no cytotoxicity, but does not show ASGPR1-mediated gene transfer activity. Secondly, to achieve α-CDE-based hepatocyte-specific gene transfer carrier, we developed Lac-α-CDEs

(G2, G3) having a α-glucose as a spacer. Actually, Lac-α-CDEs (G2, G3) showed ASGPR1-mediated gene and ODN transfer activity by changing the spacer type. After intravenous administration of pDNA and siRNA complexes in mice, hepatic parenchymal cell-selective gene expression and the potent RNAi effect were observed. However, pDNA and siRNA transfer activity of Lac-α-CDEs (G2, G3) lowered in the presence of high concentration of serum in vitro. Unexpectedly, hepatocyte-specificity of

TABLE 3.1 Principal Properties of Gal-α-CDE (G2), Lac-α-CDE (G2, G3), PEG-Lac-α-CDE (G3) As a Gene and Oligonucleotide Carrier

System	Applicable gene and oligonucleotide	Property
Gal-α-CDE (G2)	pDNA	• Higher gene transfer activity than α-CDE (G2) • Lack of ASGPR1 selectivity • Serum resistance (in the presence of 10% serum) • No cytotoxicity
Lac-α-CDE (G2, G3)	pDNA, siRNA	• ASGPR1-mediated pDNA and siRNA transfer activity • High nuclear localization ability of pDNA complex • Negligible cytotoxicity • Hepatic parenchymal cell-selective gene transfer activity in vivo • Prominent in vivo RNAi effect after intravenous administration
PEG-Lac-α-CDE (G3)	pDNA, siRNA	• ASGPR1-mediated pDNA and siRNA transfer activity • Serum resistance (in the presence of 50% serum) • Negligible cytotoxicity • Hepatic parenchymal cell-selective gene transfer activity in vivo • Prominent in vivo RNAi effect after intravenous administration

Gal-α-CDE: Galactosylated dendrimer/α-cyclodextrin conjugate. Lac-α-CDE: Lactosylated dendrimer/α-cyclodextrin conjugate. PEG-Lac-α-CDE: PEG-appended lactosylated dendrimer/α-cyclodextrin conjugate. pDNA: Plasmid DNA. siRNA: Small interfering RNA. ASGPR1: Asialoglycoprotein receptor 1. RNAi: RNA interference.

Lac-α-CDEs (G2, G3) was not enough in vivo. Thirdly, in order to overcome a drawback of Lac-α-CDEs (G2, G3), Arima and colleagues developed PEG-Lac-α-CDEs (G3). Fortunately, PEG-Lac-α-CDEs (G3) have great advantages such as ASGPR1-mediated pDNA and siRNA transfer activity, serum resistance in the presence of 50% serum, negligible cytotoxicity, hepatocyte-specific pDNA and siRNA transfer activity after intravenous administration. Nevertheless, gene expression and the RNAi effect of the pDNA and siRNA complexes with PEG-Lac-α-CDEs (G3), respectively, in hepatocytes are not enough to therapeutic use. Thereby, further investigations regarding more efficient hepatocyte-specific gene and ODN delivery system should be necessary.

Many attempts have been made to design and evaluate CyD conjugates with polymers for gene and ODN carriers. In fact, we developed various α-CDE-based systems such as mannose receptor-targeted, fucose receptor-targeted, folate receptor-targeted and sustained release system, etc. However, their clinical use may be still very limited, so we have sought to extend the function of α-CDEs. Further investigations are required to develop novel carriers for various gene and ODN drugs such as antisense DNA, decoy DNA, shRNA, siRNA, ribozyme and aptamers. The future should see certain clinical applications using CyD-containing carriers for gene and ODN.

ACKNOWLEDGEMENTS

We would like to express sincere thanks K. Uekama and F. Hirayama, Faculty of Pharmaceutical Sciences, Sojo University, for their valuable advice, warm support and kind help. We thank H. Kihara, K. Wada, T. Tsutsumi, Y. Chihara, S. Yamashita and Y. Mori, Graduate School of Pharmaceutical Sciences, Kumamoto University, K. Hattori, NanoDex Inc., T. Takeuchi, Tokyo Polytechnic University, Y. Katayama, Faculty of Engineering, Kyushu University, and T. Niidome, Graduate School of Science and Technology, Kumamoto University, for their excellent contribution to this study. This work was partially supported by Grant-in-Aid for Scientific Research (C) from Japan Society for the Promotion of Science (16590114, 18590144, 20590037).

KEYWORDS

- applications in gene delivery
- asialoglycoprotein receptor
- cyclodextrin
- gene and RNAi therapy
- polyamidoamine dendrimer
- RNA interference

REFERENCES

Abdou, S.; Collomb, J.; Sallas, F.; Marsura, A.; Finance, C. β-Cyclodextrin derivatives as carriers to enhance the antiviral activity of an antisense oligonucleotide directed toward a coronavirus intergenic consensus sequence. *Arch. Virol.*, **1997**, *142*, 1585–1602.

Ando, Y.; Nakamura, M.; Araki, S. Transthyretin-related familial amyloidotic polyneuropathy. *Arch. Neurol.*, **2005**, *62*, 1057–1062.

Arima, H.; Chihara, Y.; Arizono, M.; Yamashita, S.; Wada, K.; Hirayama, F.; Uekama, K. Enhancement of gene transfer activity mediated by mannosylated dendrimer/α-cyclodextrin conjugate (generation 3, G3). *J. Control. Release*, **2006**, *116*, 64–74.

Arima, H.; Kihara, F.; Hirayama, F.; Uekama, K. Enhancement of gene expression by polyamidoamine dendrimer conjugates with α-, β-, and γ-cyclodextrins. *Bioconjug. Chem.*, **2001**, *12*, 476–484.

Arima, H.; Motoyama, K. Recent findings concerning PAMAM dendrimer conjugates with cyclodextrins as carriers of DNA and RNA. *Sensors*, **2009**, *9*, 6346–6361.

Arima, H.; Motoyama, K.; Higashi, T. Polyamidoamine dendrimer conjugates with cyclodextrins as novel carriers for DNA, shRNA and siRNA. *Pharmaceutics*, **2012**, *4*, 130–148.

Arima, H.; Motoyama, K.; Higashi, T. Potential use of polyamidoamine dendrimer conjugates with cyclodextrins as novel carriers for siRNA. *Pharmaceuticals*, **2012**, *5*, 61–78.

Arima, H.; Motoyama, K.; Higashi, T. Sugar-appended polyamidoamine dendrimer conjugates with cyclodextrins as cell-specific nonviral vectors. *Adv. Drug Deliv. Rev.*, **2013**, *65*, 1204–1214.

Arima, H.; Tsutsumi, T.; Yoshimatsu, A.; Ikeda, H.; Motoyama, K.; Higashi, T.; Hirayama, F.; Uekama, K. Inhibitory effect of siRNA complexes with polyamidoamine dendrimer/α-cyclodextrin conjugate (generation 3, G3) on endogenous gene expression. *Eur. J. Pharm. Sci.*, **2011**, *44*, 375–384.

Arima, H.; Yamashita, S.; Mori, Y.; Hayashi, Y.; Motoyama, K.; Hattori, K.; Takeuchi, T.; Jono, H.; Ando, Y.; Hirayama, F.; Uekama, K. In vitro and in vivo gene delivery mediated by Lactosylated dendrimer/α-cyclodextrin conjugates (G2) into hepatocytes. *J. Control. Release*, **2010**, *146*, 106–117.

Bermejo, J. F.; Ortega, P.; Chonco, L.; Eritja, R.; Samaniego, R.; Mullner, M.; de Jesus, E.; de la Mata, F. J.; Flores, J. C.; Gomez, R.; Munoz-Fernandez, A. Water-soluble carbosilane dendrimers: synthesis biocompatibility and complexation with oligonucleotides; evaluation for medical applications. *Chemistry*, **2007**, *13*, 483–495.

Braun, C. S.; Fisher, M. T.; Tomalia, D. A.; Koe, G. S.; Koe, J. G.; Middaugh, C. R. A stopped-flow kinetic study of the assembly of nonviral gene delivery complexes. *Biophys. J.;* **2005**, *88*, 4146–4158.

Challa, T.; Stefanovic, B. Dendrimers: A novel polymer for drug delivery. *Int. J. Pharm. Sci. Rev. Res.*, **2011**, 9, 88–99.

Cowles, E. A.; Moutsatsos, I. K.; Wang, J. L.; Anderson, R. L. Expression of carbohydrate binding protein 35 in human fibroblasts: comparisons between cells with different proliferative capacities. *Exp. Gerontol.*, **1989**, *24*, 577–585.

Croyle, M. A.; Roessler, B. J.; Hsu, C. P.; Sun, R.; Amidon, G. L. β-Cyclodextrins enhance adenoviral-mediated gene delivery to the intestine. *Pharm. Res.*, **1998**, *15*, 1348–1355.

Davis, M. E. The first targeted delivery of siRNA in humans via a self-assembling, cyclodextrin polymer-based nanoparticle: from concept to clinic. *Mol. Pharm.*, **2009**, *6*, 659–668.

Davis, M. E.; Pun, S. H.; Bellocq, N. C.; Reineke, T. M.; Popielarski, S. R.; Mishra, S.; Heidel, J. D. Self-assembling nucleic acid delivery vehicles via linear, water-soluble, cyclodextrin-containing polymers. *Curr. Med. Chem.*, **2004**, *11*, 179–197.

Davis, M. E.; Zuckerman, J. E.; Choi, C. H.; Seligson, D.; Tolcher, A.; Alabi, C. A.; Yen, Y.; Heidel, J. D.; Ribas, A. Evidence of RNAi in humans from systemically administered siRNA via targeted nanoparticles. *Nature*, **2010**, *464*, 1067–1070.

Dufes, C.; Uchegbu, I. F.; Schatzlein, A. G. Dendrimers in gene delivery. *Adv. Drug Deliv. Rev.*, **2005**, *57*, 2177–2202.

Dutta, T.; Jain, N. K.; McMillan, N. A.; Parekh, H. S. Dendrimer nanocarriers as versatile vectors in gene delivery. *Nanomedicine*, **2010**, *6*, 25–34.

Fauvelle, F.; Debouzy, J. C.; Crouzy, S.; Goschl, M.; Chapron, Y. Mechanism of α-cyclodextrin-induced hemolysis. 1. The two-step extraction of phosphatidylinositol from the membrane. *J. Pharm. Sci.*, **1997**, *86*, 935–943.

Gonzalez, H.; Hwang, S. J.; Davis, M. E. New class of polymers for the delivery of macromolecular therapeutics. *Bioconjug. Chem.*, **1999**, *10*, 1068–1074.

Gotto, A. M.; Jr.; Moon, J. E. Pharmacotherapies for lipid modification: beyond the statins. *Nat. Rev. Cardiol.*, **2013**, *10*, 560–570.

Hara, T.; Aramaki, Y.; Takada, S.; Koike, K.; Tsuchiya, S. Receptor-mediated transfer of pSV2CAT DNA to mouse liver cells using asialofetuin-labeled liposomes. *Gene Ther.*, **1995**, *2*, 784–788.

Hayashi, Y.; Higashi, T.; Motoyama, K.; Mori, Y.; Jono, H.; Ando, Y.; Arima, H. Design and evaluation of polyamidoamine dendrimer conjugate with PEG, α-cyclodextrin and lactose as a novel hepatocyte-selective gene carrier in vitro and in vivo. *J. Drug Target.*, **2013**, *21*, 487–496.

Hayashi, Y.; Mori, Y.; Higashi, T.; Motoyama, K.; Jono, H.; Sah, D. W.; Ando, Y.; Arima, H. Systemic delivery of transthyretin siRNA mediated by lactosylated dendrimer/α-cyclodextrin conjugates into hepatocyte for familial amyloidotic polyneuropathy therapy. *Amyloid*, **2012**, *19*, 47–49.

Hayashi, Y.; Mori, Y.; Yamashita, S.; Motoyama, K.; Higashi, T.; Jono, H.; Ando, Y.; Arima, H. Potential use of lactosylated dendrimer (G3)/α-cyclodextrin conjugates as hepatocyte-specific siRNA carriers for the treatment of familial amyloidotic polyneuropathy. *Mol. Pharm.*, **2012**, *9*, 1645–1653.

Heidel, J. D.; Yu, Z.; Liu, J. Y.; Rele, S. M.; Liang, Y.; Zeidan, R. K.; Kornbrust, D. J.; Davis, M. E. Administration in nonhuman primates of escalating intravenous doses of targeted nanoparticles containing ribonucleotide reductase subunit M2 siRNA. *Proc. Natl. Acad. Sci. U.S.A.* **2007**, *104*, 5715–5721.

Hu, H. M.; Zhang, X.; Zhong, N. Q.; Pan, S. R. Study on galactose-poly(ethylene glycol)-poly(L-lysine) as novel gene vector for targeting hepatocytes in vitro. *J. Biomater. Sci. Polym. Ed.*, **2012**, *23*, 677–695.

Irie, T.; Otagiri, M.; Sunada, M.; Uekama, K.; Ohtani, Y.; Yamada, Y.; Sugiyama, Y. Cyclodextrin-induced hemolysis and shape changes of human erythrocytes in vitro. *J. Pharmacobiodyn.*, **1982**, *5*, 741–744.

Kawakami, S.; Hirayama, R.; Shoji, K.; Kawanami, R.; Nishida, K.; Nakashima, M.; Sasaki, H.; Sakaeda, T.; Nakamura, J. Liver- and lobe-selective gene transfection following the instillation of plasmid DNA to the liver surface in mice. *Biochem. Biophys. Res. Commun.*, **2002**, *294*, 46–50.

Kihara, F.; Arima, H.; Tsutsumi, T.; Hirayama, F.; Uekama, K. Effects of structure of polyamidoamine dendrimer on gene transfer efficiency of the dendrimer conjugate with α-cyclodextrin. *Bioconjug. Chem.*, **2002**, *13*, 1211–1219.

Kihara, F.; Arima, H.; Tsutsumi, T.; Hirayama, F.; Uekama, K. In vitro and in vivo gene transfer by an optimized α-cyclodextrin conjugate with polyamidoamine dendrimer. *Bioconjug. Chem.*, **2003**, *14*, 342–350.

Kim, J. B.; Choi, J. S.; Nam, K.; Lee, M.; Park, J. S.; Lee, J. K. Enhanced transfection of primary cortical cultures using arginine-grafted PAMAM dendrimer, PAMAM-Arg. *J. Control. Release*, **2006**, *114*, 110–117.

Klajnert, B.; Bryszewska, M. Dendrimers: properties and applications. *Acta Biochim. Pol.*, **2001**, *48*, 199–208.

Kukowska-Latallo, J. F.; Bielinska, A. U.; Johnson, J.; Spindler, R.; Tomalia, D. A.; Baker, J. R.; Jr. Efficient transfer of genetic material into mammalian cells using Starburst polyamidoamine dendrimers. *Proc. Natl. Acad. Sci. U.S.A.;* **1996**, *93*, 4897–4902.

Liu, F. T.; Patterson, R. J.; Wang, J. L. Intracellular functions of galectins. *Biochim. Biophys. Acta*, **2002**, *1572*, 263–273.

Liu, F.; Huang, L. Noninvasive gene delivery to the liver by mechanical massage. *Hepatology*, **2002**, *35*, 1314–1319.

Liu, J.; Zhou, J.; Luo, Y. SiRNA delivery systems-based on neutral cross-linked dendrimers. *Bioconjug. Chem.*, **2012**, *23*, 174–183.

Liu, X.; Huang, H.; Wang, J.; Wang, C.; Wang, M.; Zhang, B.; Pan, C. Dendrimers-delivered short hairpin RNA targeting hTERT inhibits oral cancer cell growth in vitro and in vivo. *Biochem. Pharmacol.*, **2011**, *82*, 17–23.

McMahon, A.; O'Neill, M. J.; Gomez, E.; Donohue, R.; Forde, D.; Darcy, R.; O'Driscoll, C. M. Targeted gene delivery to hepatocytes with galactosylated amphiphilic cyclodextrins. *J. Pharm. Pharmacol.*, **2012**, *64*, 1063–1073.

Merkel, O. M.; Mintzer, M. A.; Librizzi, D.; Samsonova, O.; Dicke, T.; Sproat, B.; Garn, H.; Barth, P. J.; Simanek, E. E.; Kissel, T. Triazine dendrimers as nonviral vectors for in vitro and in vivo RNAi: the effects of peripheral groups and core structure on biological activity. *Mol. Pharm.*, **2010**, *7*, 969–983.

Merkel, O. M.; Mintzer, M. A.; Simanek, E. E.; Kissel, T. Perfectly shaped siRNA delivery. *Ther. Deliv.*, **2010**, *1*, 737–742.

Morgan, D. M.; Larvin, V. L.; Pearson, J. D. Biochemical characterization of polycation-induced cytotoxicity to human vascular endothelial cells. *J. Cell Sci.*, **1989**, *94*, 553–559.

Motoyama, K.; Mori, Y.; Yamashita, S.; Hayashi, Y.; Jono, H.; Ando, Y.; Hirayama, F.; Uekama, K.; Arima, H. In vitro gene delivery mediated by lactosylated dendrimer/α-cyclodextrin conjugate (generation 3, G3) into hepatocytes. *J. Incl. Phenom. Macro. Chem.*, **2011**, 333–338.

Nguyen, T. H.; Ferry, N. Liver gene therapy: advances and hurdles. *Gene Ther.*, **2004**, *11*, 76–84.

Nie, C.; Liu, C.; Chen, G.; Dai, J.; Li, H.; Shuai, X. Hepatocyte-targeted psiRNA delivery mediated by galactosylated poly(ethylene glycol)-graft-polyethylenimine in vitro. *J. Biomater. Appl.*, **2011**, *26*, 255–275.

O'Neill, M. J.; Guo, J.; Byrne, C.; Darcy, R.; O'Driscoll, C. M. Mechanistic studies on the uptake and intracellular trafficking of novel cyclodextrin transfection complexes by intestinal epithelial cells. *Int. J. Pharm.*, **2011**, *413*, 174–183.

Ohsaki, M.; Okuda, T.; Wada, A.; Hirayama, T.; Niidome, T.; Aoyagi, H. In vitro gene transfection using dendritic poly(L-lysine). *Bioconjug. Chem.*, **2002**, *13*, 510–517.

Ohtani, Y.; Irie, T.; Uekama, K.; Fukunaga, K.; Pitha, J. Differential effects of α-, β- and γ-cyclodextrins on human erythrocytes. *Eur. J. Biochem.*, **1989**, *186*, 17–22.

Ortiz Mellet, C.; Garcia Fernandez, J. M.; Benito, J. M. Cyclodextrin-based gene delivery systems. *Chem. Soc. Rev.*, **2011**, *40*, 1586–1608.

Perez, A. P.; Romero, E. L.; Morilla, M. J. Ethylendiamine core PAMAM dendrimers/siRNA complexes as in vitro silencing agents. *Int. J. Pharm.*, **2009**, *380*, 189–200.

Pun, S. H.; Davis, M. E. Development of a nonviral gene delivery vehicle for systemic application. *Bioconjug. Chem.*, **2002**, *13*, 630–639.

Qin, Z.; Liu, W.; Li, L.; Guo, L.; Yao, C.; Li, X. Galactosylated N-2-hydroxypropyl methacrylamide-b-N-3-guanidinopropyl methacrylamide block copolymers as hepatocyte-targeting gene carriers. *Bioconjug. Chem.*, **2011**, *22*, 1503–1512.

Raper, S. E.; Wilson, J. M. Gene therapy for human liver disease. *Prog. Liver Dis.*, **1995**, *13*, 201–230.

Roessler, B. J.; Bielinska, A. U.; Janczak, K.; Lee, I.; Baker, J. R.; Jr. Substituted β-cyclodextrins interact with PAMAM dendrimer-DNA complexes and modify transfection efficiency. *Biochem. Biophys. Res. Commun.*, **2001**, *283*, 124–129.

Ruponen, M.; Honkakoski, P.; Ronkko, S.; Pelkonen, J.; Tammi, M.; Urtti, A. Extracellular and intracellular barriers in nonviral gene delivery. *J. Control. Release*, **2003**, *93*, 213–217.

Sahebkar, A.; Watts, G. F. New LDL-cholesterol lowering therapies: pharmacology, clinical trials, and relevance to acute coronary syndromes. *Clin. Ther.*, **2013**, *35*, 1082–1098.

Sato, Y.; Murase, K.; Kato, J.; Kobune, M.; Sato, T.; Kawano, Y.; Takimoto, R.; Takada, K.; Miyanishi, K.; Matsunaga, T.; Takayama, T.; Niitsu, Y. Resolution of liver cirrhosis using vitamin A-coupled liposomes to deliver siRNA against a collagen-specific chaperone. *Nat. Biotechnol.*, **2008**, *26*, 431–442.

Shah, N.; Steptoe, R. J.; Parekh, H. S. Low-generation asymmetric dendrimers exhibit minimal toxicity and effectively complex DNA. *J. Pept. Sci.*, **2011**, *17*, 470–478.

Shcharbin, D.; Shakhbazau, A.; Bryszewska, M. Poly(amidoamine) dendrimer complexes as a platform for gene delivery. *Expert Opin. Drug Deliv.*, **2013**, *10*, 1687–1698.

Shimizu, K.; Kawakami, S.; Hayashi, K.; Kinoshita, H.; Kuwahara, K.; Nakao, K.; Hashida, M.; Konishi, S. In vivo site-specific transfection of naked plasmid DNA and siRNAs in mice by using a tissue suction device. *PLoS One*, **2012**, *7*, e41319.

Singha, K.; Namgung, R.; Kim, W. J. Polymers in small-interfering RNA delivery. *Nucleic Acid Ther.*, **2011**, *21*, 133–147.

Symens, N.; Mendez-Ardoy, A.; Diaz-Moscoso, A.; Sanchez-Fernandez, E.; Remaut, K.; Demeester, J.; Fernandez, J. M.; De Smedt, S. C.; Rejman, J. Efficient transfection of hepatocytes mediated by mRNA complexed to galactosylated cyclodextrins. *Bioconjug. Chem.*, **2012**, *23*, 1276–1289.

Szente, L.; Szejtli, J. Highly soluble cyclodextrin derivatives: chemistry, properties, and trends in development. *Adv. Drug. Deliv. Rev.*, **1999**, *36*, 17–28.

Tomalia, D. A.; Baker, H.; Dewald, J.; Hall, M.; Kallos, G.; Martin, S.; Roeck, J.; Ryder, J.; Smith, P. A new class of polymers: Starburst-dendritic macromolecules. *Polym. J.*, **1985**, *17*, 117–132.

Troiber, C.; Wagner, E. Nucleic acid carriers-based on precise polymer conjugates. *Bioconjug. Chem.*, **2011**, *22*, 1737–1752.

Tsutsumi, T.; Hirayama, F.; Uekama, K.; Arima, H. Evaluation of polyamidoamine dendrimer/α-cyclodextrin conjugate (generation 3, G3) as a novel carrier for small interfering RNA (siRNA). *J. Control. Release*, **2007**, *119*, 349–359.

Tsutsumi, T.; Hirayama, F.; Uekama, K.; Arima, H. Potential use of polyamidoamine dendrimer/α-cyclodextrin conjugate (generation 3, G3) as a novel carrier for short hairpin RNA-expressing plasmid DNA. *J. Pharm. Sci.*, **2008**, *97*, 3022–3034.

Uekama, K. Design and evaluation of cyclodextrin-based drug formulation. *Chem. Pharm. Bull.*, **2004**, *52*, 900–915.

Uekama, K.; Hirayama, F.; Irie, T. Cyclodextrin drug carrier systems. *Chem. Rev.*, **1998**, *98*, 2045–2076.

Wada, K.; Arima, H.; Tsutsumi, T.; Chihara, Y.; Hattori, K.; Hirayama, F.; Uekama, K. Improvement of gene delivery mediated by mannosylated dendrimer/α-cyclodextrin conjugates. *J. Control. Release*, **2005**, *104*, 397–413.

Wada, K.; Arima, H.; Tsutsumi, T.; Hirayama, F.; Uekama, K. Enhancing effects of galactosylated dendrimer/α-cyclodextrin conjugates on gene transfer efficiency. *Biol. Pharm. Bull.*, **2005**, *28*, 500–505.

Wang, J. L.; Gray, R. M.; Haudek, K. C.; Patterson, R. J. Nucleocytoplasmic lectins. *Biochim. Biophys. Acta*, **2004**, *1673*, 75–93.

Wang, J. L.; Laing, J. G.; Anderson, R. L. Lectins in the cell nucleus. *Glycobiology*, **1991**, *1*, 243–252.

Wang, J.; Lu, Z.; Wientjes, M. G.; Au, J. L. Delivery of siRNA therapeutics: barriers and carriers. *AAPS J.;* **2010**, *12*, 492–503.

Westerlind, U.; Westman, J.; Tornquist, E.; Smith, C. I.; Oscarson, S.; Lahmann, M.; Norberg, T. Ligands of the asialoglycoprotein receptor for targeted gene delivery, part 1: Synthesis of and binding studies with biotinylated cluster glycosides containing N-acetylgalactosamine. *Glycoconj. J.;* **2004**, *21*, 227–241.

Wu, C. H.; Wu, G. Y. Receptor-mediated delivery of foreign genes to hepatocytes. *Adv. Drug Deliv. Rev.*, **1998**, *29*, 243–248.

Wu, G. Y.; Wu, C. H. Receptor-mediated gene delivery and expression in vivo. *J. Biol. Chem.*, **1988**, *263*, 14621–14624.

Yuan, Q.; Lee, E.; Yeudall, W. A.; Yang, H. Dendrimer-triglycine-EGF nanoparticles for tumor imaging and targeted nucleic acid and drug delivery. *Oral Oncol.*, **2010**, *46*, 698–704.

Yuan, X.; Naguib, S.; Wu, Z. Recent advances of siRNA delivery by nanoparticles. *Expert. Opin. Drug Deliv.*, **2011**, *8*, 521–536.

Zhang, G.; Budker, V.; Wolff, J. A. High levels of foreign gene expression in hepatocytes after tail vein injections of naked plasmid DNA. *Hum. Gene Ther.*, **1999**, *10*, 1735–1737.

CHAPTER 4

PRECISE PREPARATION OF FUNCTIONAL POLYSACCHARIDES NANOPARTICLES BY AN ENZYMATIC APPROACH

JUN-ICHI KADOKAWA

Department of Chemistry, Biotechnology, and Chemical Engineering, Graduate School of Science and Engineering, Kagoshima University, 1-21-40 Korimoto, Kagoshima 890-0065, Japan

CONTENTS

ABSTRACT

In this chapter, precision synthesis of functional polysaccharide nanoparticles and nanomaterials by enzymatic approach is presented. Polysaccharides are one of the major classes of biological macromolecules, which often have very complicated structures composed of a wide variety of monosaccharide units and glycosidic linkages. Such structural variety strongly contributes to serving a whole range of their functions in the host organism, and a subtle change in the structure of monosaccharide residue and the type of glycosidic linkage have profound effects on their functions. An enzymatic approach has been understood as a very powerful tool to synthesize polysaccharides with well-defined structure. Two types of enzymes have been efficiently employed for the precision synthesis of nanostructured polysaccharides, that is, phosphorylase and sucrase. After the brief description on the characteristics of phosphorylase and sucrase catalyses, this chapter mainly deals with phosphorylase- and sucrase-catalyzed synthesis of functional polysaccharides with controlled nanostructures such as nanoparticles.

4.1 INTRODUCTION

Polysaccharides are one of three major classes of biological macromolecules present in the plant, animal, and microbial kingdom, which are vital materials for important in vivo functions, for example, acting as a structural material, providing an energy source, and conferring specific biological property (Schuerch, 1986). They generally have the structures composed of monosaccharide residues bound through glycosidic linkages. A glycosidic linkage is a type of covalent bond that joins a monosaccharide residue to another group, which is typically another saccharide residue (Paulsen, 1982; Schmidt, 1986; Toshima and Tatsuta, 1993). Polysaccharides thus consisting of the glycosidic linkages have very complicated structures owing to not only a structurally variety of the monosaccharide residues, but also the differences in stereo- and regiofashions of the glycosidic linkages. In contrast, the other two major biological macromolecules, i.e., nucleic acids and proteins, have relatively simple structures compared with polysaccharides because these substrates are constructed by a type of specific linkage between several kinds of nucleotides and 20 kinds of amino acids, respectively. A great variety of the polysaccharide structures contributes to serving a whole range of their functions in the host organism, and a subtle change in the structure of monosaccharide residue and the type of glycosidic linkage have profound effects on their properties and functions (Yalpani, 1988). Furthermore, because polysaccharides are nontoxic and ecofriendly substances, they have practically been used as functional materials in biomedical and tissue engineering applications. Accordingly, the synthesis of polysaccharides with well-defined structure has increasingly attracted much attention because of their potential to apply as practical materials in the field of medicine, pharmaceutics, cosmetics, and food industries (Schatz and Lecommandoux, 2010; Stern and Jedrzejas, 2008). In particular, the fabrication of nanostructured polysaccharides such as nanoparticles leads to practical applications in environmental nanotechnology, as example, which have a potential to be employed as matrix substances for drug delivery (Huang et al., 2015; Liu et al., 2008; Posocco et al., 2015; Saravanakumar et al., 2012; Shelke et al., 2014).

The reaction for the formation of glycosidic linkage is called 'glycosylation' (Paulsen, 1982; Schmidt, 1986; Toshima and Tatsuta, 1993). Polysaccharides are theoretically synthesized by polymerization based on the repeated glycosylations (Kadokawa and Kobayashi, 2010; Kobayashi and Makino, 2009; Kobayashi et al., 2001). For the occurrence of such 'polyglycosylation' efficiently, the starting substrates should be designed appropriately according to the reaction manner. A typical reaction scheme of the glycosylation between two glucose substrates, a glycosyl donor and a glycosyl acceptor is shown in Figure 4.1 for the possible formation of $\alpha(1\rightarrow4)$- and $\beta(1\rightarrow4)$-linked glucose dimers, maltose and cellobiose. For design of the glycosyl donor and acceptor, an anomeric carbon (C-1) of the former substrate is activated by introducing a leaving group, and a hydroxy group in the latter substrate, which takes part in the glycosylation, is employed as a free form (e.g., C-4 in Figure 4.1), whereas the rest of hydroxy groups in both the substrates are protected. For the stereo- and regioselective formation of a glycosidic linkage, not only the leaving group (X) and protective groups (R, R'), but also a catalyst and a solvent should appropriately be selected. The perfect control in stereo- and

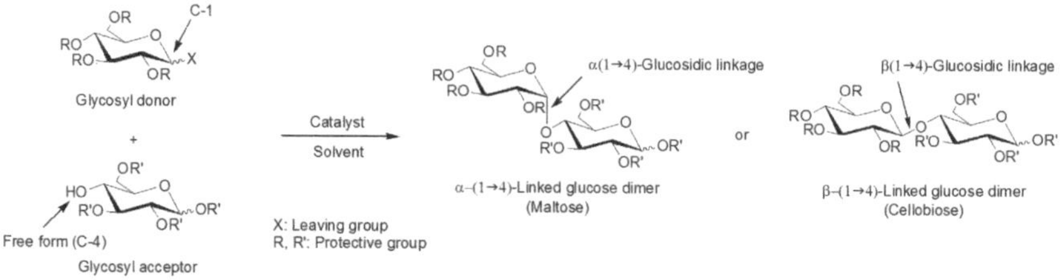

FIGURE 4.1   Typical reaction manner of glycosylation between two glucose substrates.

regiofashions of glycosidic linkages still remains as difficult barrier in the general chemical glycosylations.

Alternative to such chemical glycosylations, in vitro reaction approach by enzymatic catalysis has been significantly developed in recent years, so-called 'enzymatic glycosylation' because enzymatic reaction is the superior method in terms of stereo- and regioselectivities compared with that by chemical catalysis (Shoda, 2001; Shoda et al., 2003). Similar to the general chemical glycosylation, the enzymatic glycosylation for the formation of a glycosidic linkage between an anomeric carbon of a monosaccharide and one of hydroxy groups of the other monosaccharide can be realized by the reaction of a activated glycosyl donor at the anomeric position with a glycosyl acceptor, where these substrates can be employed in their unprotected forms beside the anomeric position of the glycosyl donor (Figure 4.2). In the enzymatic glycosylation, first, the glycosyl donor is recognized by a catalytic center of enzyme to form a glycosyl-enzyme complex. Then, the complex is attacked by the hydroxy group of the glycosyl acceptor in the stereo- and regioselective manners according to specificity of each enzyme, leading to the direct formation of the unprotected glycoside. Thus, repetition of the enzymatic glycosylations, i.e., enzymatic polymerization, forms polysaccharides with well-defined structure. Amongst the six main classes of enzymes, i.e., oxidoreductase (EC1), transferase (EC2), hydrolase (EC3), lyase (EC4), isomerase (EC5), and ligase (EC6), transferase (glycosyl transferase) and hydrolase (glycosyl hydrolase) have been practically applied as catalysts for the in vitro enzymatic synthesis of polysaccharides via the glycosylations (Blixt and Razi, 2008; Thiem and Thiem, 2008). The former glycosyl transferase is mainly subclassified into synthetic enzymes (Leloir glycosyl transferases), phosphorolytic enzymes (phosphorylases), and sucrases (Seibel et al., 2006a, 2006b). Leloir glycosyl transferases

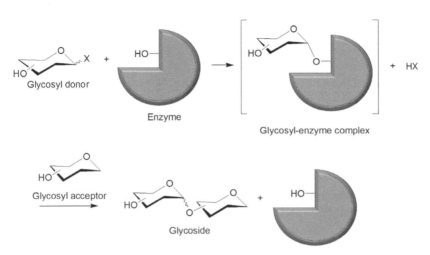

FIGURE 4.2 General scheme for enzymatic glycosylation.

are the biologically important catalysts because they conduct the role to synthesize saccharide chains in vivo. However, Leloir glycosyl transferases are generally transmembrane-type proteins, present in fewer amounts in nature, and unstable for isolation and purification. Therefore, in vitro approach by Leloir glycosyl transferase catalysis has hardly been employed for practical synthesis of polysaccharides.

Phosphorylases and sucrases have been used as the catalysts for the synthesis of well-defined polysaccharides via enzymatic polymerization, which have been applied further to fabrication of environmentally benign nanostructured materials (Kadokawa, 2011a). On the basis of the above viewpoints and backgrounds, in this chapter, the precision synthesis of functional polysaccharide nanoparticles and nanomaterials by phosphorylase- and sucrase-catalyzed enzymatic polymerizations is presented.

4.2 PRECISION SYNTHESIS OF POLYSACCHARIDES BY PHOSPHORYLASE- AND SUCRASE-CATALYZED ENZYMATIC POLYMERIZATION

Phosphorylases are the enzymes that catalyze phosphorolytic cleavage of a respective glycosidic linkage in the saccharide chain in the presence of inorganic phosphate (Pi) to produce the corresponding monosaccharide

1-phosphate and the saccharide chain with one smaller degree of polymerization (DP) (Figure 4.3) (Kitaoka and Hayashi, 2002; Nakai et al., 2013; Puchart, 2015). Because the bond energy of a phosphate ester in the monosaccharide 1-phosphate product is comparable with that of the glycosidic linkage, the reversibility of the phosphorylase-catalyzed reactions is conceived. Accordingly, phosphorylases have been employed as catalysts in the practical synthesis of saccharide chains via enzymatic glycosylations. In the reactions, monosaccharide 1-phosphates are used as glycosyl donors and the monosaccharide residue is transferred from the donor to the nonreducing end of a specific glycosyl acceptor to form a stereo- and regiocontrolled glycosidic linkage accompanied with liberating Pi. Of the phosphorylases, which have been known so far, α-glucan phosphorylase (glycogen phosphorylase, starch phosphorylase, hereafter, this enzyme is simply called 'phosphorylase' in this chapter) is the most extensively studied and used as a catalyst for the enzymatic preparation of polysaccharide materials based on the α(1→4)-glucan structure (Kadokawa and Kaneko, 2013; Seibel et al., 2006b).

Phosphorylase is the enzyme that catalyzes the reversible phosphorolysis of α(1→4)-glucans at the nonreducing end, such as starch and glycogen, in the presence of Pi to produce α-D-glucose 1-phosphate (Glc-1-P) (Figure 4.4a). Depending on the reaction conditions, phosphorylase catalyzes the enzymatic glucosylation using Glc-1-P as a glycosyl donor to form α(1→4)-glucosidic linkage (Figure 4.4a). As glycosyl acceptors, oligo-α(1→4)-glucans, that is, maltooligosaccharides, are used. In the glycosylation, accordingly, a glucose residue is transferred from Glc-1-P to the nonreducing end of a maltooligosaccharide to produce the α(1→4)-glucan chain with one larger DP. Phosphorylases specifically have the smallest DP value in α(1→4)-glucan chains to recognize depending on their sources, and accordingly, maltooligosaccharides with DPs higher

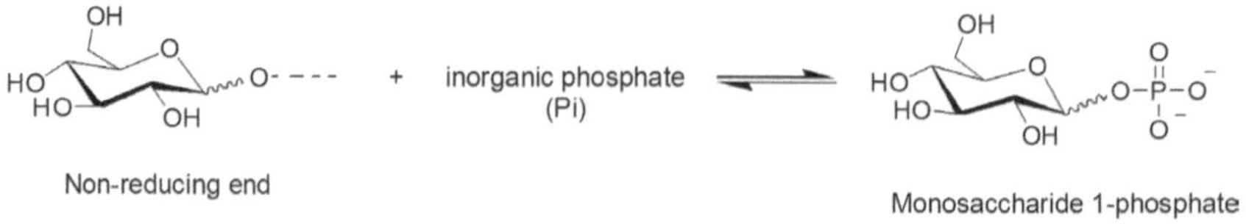

Non-reducing end

Monosaccharide 1-phosphate

FIGURE 4.3 Phosphorolysis reaction by phosphorylase catalysis in the presence of inorganic phosphate.

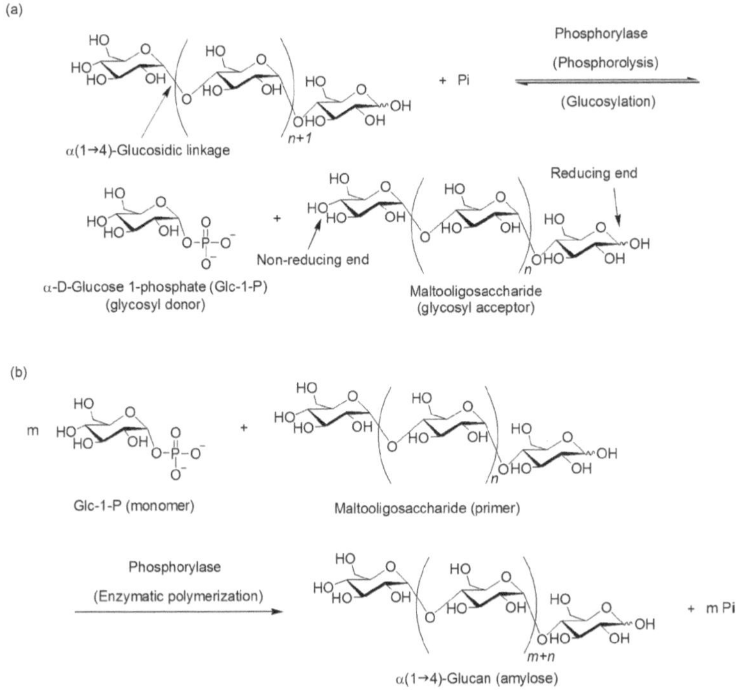

FIGURE 4.4 Phosphorylase-catalyzed (a) phosphorolysis and glycosylation and (b) enzymatic polymerization.

than the smallest one should be used as glycosyl acceptors. The smallest substrates for phosphorolysis and glycosylation recognized by phosphorylase isolated from potato (potato phosphorylase) are maltopentaose (Glc_5) and maltotetraose (Glc_4), respectively. On the other hand, those for the former and latter reactions recognized by phosphorylase isolated from thermophilic bacteria sources (thermostable phosphorylase) are Glc_4 and maltotriose (Glc_3), respectively (Boeck and Schinzel, 1996; Takaha et al., 2001; Yanase et al., 2005, 2006). These observations suggest the different recognition specificities of phosphorylases for the substrates depending on their sources.

When the excess molar ratio of Glc-1-P to the maltooligosaccharide acceptor is present in the reaction system, phosphorylase catalyzes an enzymatic polymerization of Glc-1-P as a monomer according to the reaction manner of successive glycosylations to produce a polysaccharide composed of α(1→4)-glucan chain, that is, amylose (Figure 4.4b) (Fujii et al., 2003; Ohdan et al., 2006; Yanase et al., 2006; Ziegast and Pfannemüller, 1987). The phosphorylase-catalyzed polymerization is exactly initiated at the nonreducing end of the maltooligosaccharide acceptor and the propagation is progressed from the elongating non-reducing end of the produced α(1→4)-glucan chain. Accordingly, the polymerization manner belongs to chain-growth polymerization and the maltooligosaccharide acceptor is often called a 'primer' of the polymer-ization. The phosphorylase-catalyzed enzymatic polymerization proceeds analogously in the living polymerization manner because of no occurrence of termination and chain-transfer reaction. Therefore, DPs of the produced amyloses can be controlled by the monomer/primer feed ratios and their polydispercities are typically narrow (Kitamura, 1996). On the basis of the reaction manner, furthermore, maltooligosaccharides immobilized on other substances at the reducing ends can be used as primers for the phos-phorylase-catalyzed enzymatic polymerization because the reducing end does not participate in the reaction (Figure 4.5) (Izawa and Kadokawa, 2009; Kadoakwa, 2012, 2013, 2014, 2015a; Kaneko and Kadokawa, 2009; Kitamura et al., 1982; Omagari and Kadokawa, 2011). This type of the material typically serves multifunction due to the presence of the plural maltooligosaccharide chains on the substance.

Because phosphorylase often shows loose specificity for recognition of the substrate structure, the allowance of analog glycosyl donor structures of Glc-1-P in the phosphorylase-catalyzed enzymatic glycosylation has

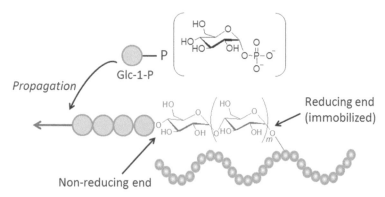

FIGURE 4.5 Phosphorylase-catalyzed enzymatic polymerization using modified primer.

been investigated to obtain nonnatural oligosaccharides having the different monosaccharide residues at the nonreducing end (Kadokawa, 2011b, 2013, 2015b). It has been reported that potato phosphorylase recognizes α-D-xylose, 2-deoxy-α-D-glucose, α-D-mannose, α-D-glucosamine, and N-formyl-α-D-glucosamine 1-phosphates (Xyl-1-P, dGlc-1-P, Man-1-P, GlcN-1-P, and GlcNF-1-P, respectively) as glycosyl donors in the glycosylations using Glc$_4$ as a glycosyl acceptor to produce nonnatural pentasaccharides having Xyl, dGlc, Man, GlcN, and GlcNF units, respectively (Figure 4.6) (Evers et al., 1994; Evers and Thiem, 1997; Kawazoe et al., 2010; Nawaji et al., 2008a, 2008b; Percival and Withers, 1988). Particularly, it is noted that the enzymatic glycosylation using GlcN-1-P (glucosaminylation) produces a basic oligosaccharide having an amino group at C-2 position of the GlcN unit.

Although potato phosphorylase-catalyzed glycosylation using α-D-glucuronic acid 1-phosphate (GlcA-1-P) as a glycosyl donor (glucuronylation) with Glc$_4$ was attempted to obtain an acidic oligosaccharide having carboxylic acid group at C-5 position of the GlcA unit, the enzyme did not recognize this analog substrate. Interestingly, it was found that thermostable phosphorylase from *Aquifex aeolicus* VF5 (Bhuiyan et al., 2003) recognized GlcA-1-P and catalyzed the glucuronylation of Glc$_3$, which was a smallest glycosyl acceptor for this enzyme, to produce an acidic tetrasaccharide having a GlcA unit at the nonreducing end (Figure 4.7) (Umegatani et al., 2012).

Sucrases are the enzymes that catalyze the transfer reactions either of glucose or of fructose moiety of sucrose, resulting in glucose-based polysaccharides (glucans) or fructose-based polysaccharides (fructans) of

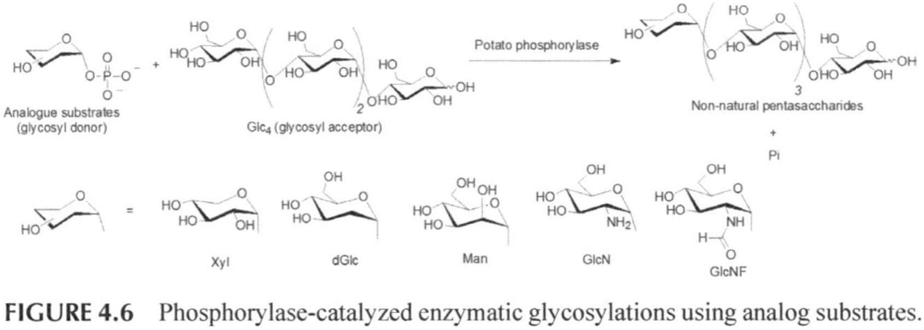

FIGURE 4.6 Phosphorylase-catalyzed enzymatic glycosylations using analog substrates.

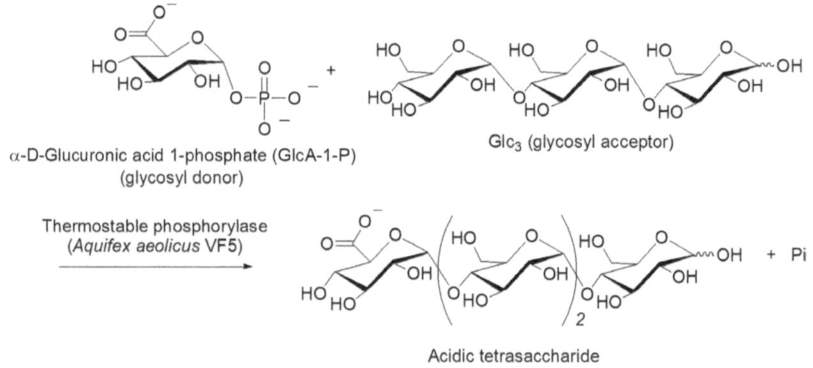

FIGURE 4.7 Phosphorylase-catalyzed enzymatic glucuronylation to give acidic tetrasaccharide.

different types with respect to glycosidic linkages and side chains (Seibel et al., 2006a, 2006b). The simplified reaction schemes are represented for glucansucrases; n sucrose → glucan + n fructose and for fructansucrases; n sucrose → fructan + n glucose.

Glucansucrases transfer glucose residues of sucrose onto polysaccharides, or appropriate acceptors with release of fructose. Accordingly, dextran (α(1→6)-glucosidic linkage), mutan (α(1g3)-), alternan (alternating α(1g3)- and α(1→6)-), reuteran (α(1→4)- and α(1→6)-), and amylose (α(1→4)-) are typically obtained by the following glucansucrases, dextran-, mutan-, alternan-, reuteran-, and amylosucrases-catalyzed glucozyl transfer reactions from sucrose, respectively (Korakli and Vogel, 2006; Remaud-Simeon et al., 2000, 2003; van Hijum et al., 2006). In addition, the modifications with respect to the glucosidic linkage pattern in glucan synthesis have been investigated.

Amylosucrase is the most extensively studied glucansucrase (Albenne et al., 2004; van der Veen et al., 2004; van Leeuwen et al., 2009) The catalytic properties of the highly purified amylosucrase from *Neisseria polysaccharea* were investigated, which revealed that in the presence of sucrose alone, several reactions took place by this enzyme catalysis in addition to the amylose synthesis, which were sucrose hydrolysis, maltose and maltotriose synthesis by successive transfers of the glucose moiety of sucrose onto the released glucose, and finally turanose and trehalulose synthesis obtained by glucose transfer onto fructose (de Montalk et al.,

2000a). In the presence of glycogen as an acceptor, on the other hand, the sucrose hydrolysis decreased strongly with increasing the concentration of glycogen, as resulting in the occurrence of oligosaccharide synthesis, by glucose transfer onto glucose and fructose by the amylosucrase catalysis (de Montalk et al., 2000b). The glucose units consumed were then preferentially used for the elongation of glycogen chains. Polysaccharides with α(1→4)- or α(1→4)- and α(1→6)-glucosidic linkages have been found to act as acceptors for the amylosucrase-catalyzed chain-elongation (Rolland-Sabaté et al., 2004). It was reported that recombinant amylosucrase produced amylose with DP of 35–58 from sucrose without use of an acceptor (Potocki-Véronèse et al., 2005).

Fructansucrases transfer fructose moiety of sucrose onto polysaccharides, or appropriate acceptors with release of glucose (Seibel et al., 2006a, 2006b). Enzymatically produced fructans, therefore, are either levan composed of β(2→6)-linked fructose residues by levansucrase catalysis and inulin composed of β(2g1)-linked fructose residues by inulosucrase catalysis (Korakli and Vogel, 2006; van Hijum et al., 2006).

4.3 PHOSPHORYLASE-CATALYZED SYNTHESIS OF POLYSACCHARIDE NANOPARTICLES

The preparation of polysaccharide nanoparticles (or microparticles) and their hierarchically conversion further into functional polysaccharide materials have been investigated by the phosphorylase-catalyzed enzymatic reactions (Kadoakwa, 2011a, 2012, 2013; Kadokawa and Kaneko, 2013). For example, by means of the thermostable phosphorylase-catalyzed glycosylations using analog substrates, amphoteric polysaccharide nanoparticles have been synthesized. For this purpose, a highly branched cyclic dextrin (glucan dendrimer, GD) and glycogen were employed as multifunctional glycosyl acceptors having nanoparticle fashions. GD is a water-soluble dextrin nanoparticle, that is produced by cyclization of amylopectin catalyzed by branching enzyme (BE, *Bacillus stearothermophilus*) (Figure 4.8) (Takata et al., 1996a, 1996b, 1997). In this molecule, α(1→4)-glucan chains form highly branched structures by interlinked α(1→6)-glucosidic linkages, which are mostly originated from the heavily branched region of amylopectin. Furthermore, one α(1→6)-glucosidic linkage takes part in the cyclic

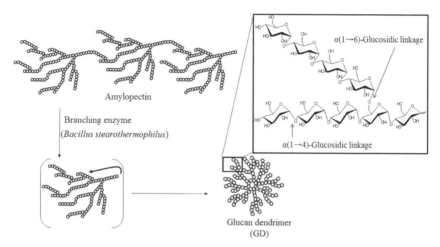

FIGURE 4.8 Action of branching enzyme on amylopectin to produce glucan dendrimer (GD).

structure, which is newly formed by the BE-catalyzed reaction. Glycogen is a water soluble polysaccharide nanoparticle composed of α(1→4)-linked linear glucan chains interlinked by α(1→6)-glucosidic linkages, which has the same chemical structure as that of GD, but the molecular weight is much higher than GD (Calder, 1991; Manners, 1991). GD and glycogen act as multifunctional polymeric primers for the phosphorylase-catalyzed enzymatic polymerization because of the presence of a number of the nonreducing α(1→4)-glucan chain ends owing to the highly branched structure. For example, when the potato phosphorylase-catalyzed enzymatic polymerization of Glc-1-P from glycogen was carried out in aqueous acetate buffer solution, followed by standing further at room temperature for 24 h, the reaction mixture turned into a hydrogel form (Figure 4.9) (Izawa et al., 2009). The enzymatically elongated amylose chains among glycogen molecules formed double helicies (Eisenhaber and Schulz, 1992; Hinrichs et al., 1987), which acted cross-linking points for hierarchically production of the hydrogel from glycogen nanoparticles. The hydrogels were facilely converted into porous cryogels by lyophilization. The porous morphology was based on the cross-linked network structures in the hydrogels.

The thermostable phosphorylase-catalyzed glucuronylation of GD using GlcA-1-P was carried out to produce acidic GD nanoparticles having GlcA residues at the nonreducing ends (Figure 4.10) (Takemoto et al., 2013).

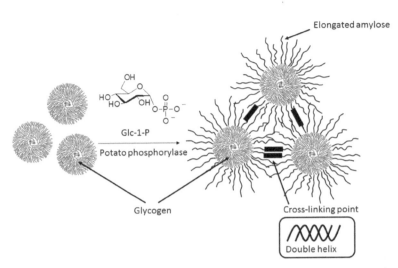

FIGURE 4.9 Preparation of glycogen hydrogel by phosphorylase-catalyzed enzymatic polymerization.

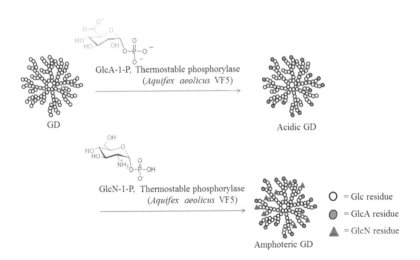

FIGURE 4.10 Preparation of amphoteric GD nanoparticle by phosphorylase-catalyzed successive glucuronylation and glucosaminylation.

Amphoteric GD nanoparticles having both GlcA and GlcN residues at the nonreducing ends were then prepared by the subsequent thermostable phosphorylase-catalyzed glucosaminylation of the acidic products using

GlcN-1-P (Figure 4.10) (Takata et al., 2014). The same procedures including the glucuronylation and glucosaminylation were performed on glycogen as a glycosyl acceptor to obtain amphoteric glycogen nanoparticles having GlcA/GlcN residues at the nonreducing ends (Takata et al., 2015). The GlcA/GlcN unit ratios in the amphoteric products from both GD and glycogen were depended on the GlcA-1-P/GlcN-1-P feed ratios in the two reactions (Table 4.1). The ζ-potential values of the amphoteric products were changed from positive to negative in terms of the pH change from acidic and basic. The values of inherent isoelectric points of the amphoteric products were calculated by the pH values, where ζ-potential became zero charge, which were reasonably dependent upon the GlcA/GlcN ratios in the products (Table 4.1).

The surface charges on the amphoteric GD nanoparticles affected their assembling behavior in water (Table 4.1). The neutralized amphoteric products in water at pH = pI assembled to form the large aggregates with average diameters of ca. 660–670 nm in all cases, which were estimated by dynamic light scattering (DLS) measurement. At pH values shifted from pI, the products disassembled to exhibit average diameters of ca. 10–14 nm, which were almost identical with those of the original GD sample at all the pH values. At pH = pI, the amphoteric GD nanoparticles are more

TABLE 4.1 Synthetic Conditions and Characterizations of Amphoteric GD and Glycogen

Substrate	Feed equivalent of glycosyl acceptors for the number of nonreducing ends[a]		Functionalities[b] (%)		pI[c]	Average diameter at pH = pI[d] (nm)	Average diameter at pH shifted from pI[d] (nm)
	GlcA-1-P	GlcN-1-P	GlcA	GlcN			
GD[e]	6	1	84	16	1.0	668.7	12.4–13.8
GD[e]	1	1	26	65	5.0	668.8	10.3–13.6
GD[e]	0.5	1	14	86	6.2	659.4	10.1–14.5
Glycogen	3.2	0.2	85	13	3.0	—[f]	590.1–615.8
Glycogen	1.8	0.6	57	41	5.7	—[f]	610.8–615.1
Glycogen	0.9	0.8	26	65	7.3	—[f]	617.1–628.6

[a]Enzymatic reactions were performed in 200 mM sodium acetate buffer (pH 6.2) at 55°C for 8 h. [b]Determined by ^{1}H NMR spectra. [c]Determined by ζ-potential measurement. [d]Determined by DLS measurement. [e]Data from reference (Takata et al., 2014). [f]Precipitated.

hydrophobic and less stable due to absence of intermolecular repulsive forces, leading to assembly at their pI values. On the other hand, the electrostatic repulsion between the cationic/cationic or anionic/anionic groups on the amphoteric GD nanoparticles at the different pH values from pI appeared, resulting in prevention of assembly. The similar pH-responsive assembling/disassembling properties were observed from the amphoteric glycogen nanoparticles, in which they disassembled to show average diameters of ca. 600 nm, which were almost identical with those of the original glycogen. At pH = pI, the materials assembled to form large precipitates. Therefore, average diameters of the aggregates could not be estimated by the DLS measurement due to their agglomeration in the solutions.

The phosphorylase-catalyzed enzymatic polymerization of Glc-1-P from nonfunctionalized nonreducing ends of the amphoteric glycogen nanoparticles was carried out to hierarchically fabricate the amphoteric glycogen hydrogels owing to cross-linking by the double helix formation of the elongated amylose chains (Takata et al., 2015). Because of the amphoteric structure, the produced hydrogels exhibited the pH-responsive property. Under alkaline conditions (pH = 12) by addition of aqueous NaOH solution, the hydrogels were solubilized because of the dissociation of the amylose double helical conformation under alkaline conditions (Figure 4.11a). When the resulting solutions were gradually changed to acidic by addition of aqueous acetic acid solution, the solutions returned to hydrogel form at pH = 9. During acidification to pH = 9 by addition of aqueous acetic acid, the amylose chains reformed double helices, which act as cross-linking points for rehydrogelation (Figure 4.11b). By further addition of aqueous acetic acid solution to be the lower pH, the resulting hydrogels were shrunk at pH = pI, owing to neutral charges on the surfaces and the absence of intermolecular repulsive forces (Figure 4.11c). Then, the hydrogels were swollen again at pH values below the pI by further addition of aqueous acetic acid solution (Figure 4.11d). Under weakly basic and acidic conditions shifted from pI, the surfaces of glycogens in the hydrogels exhibit negative and positive charges, respectively, resulting in swelling of the hydrogels owing to anionic/anionic and cationic/cationic electrostatic repulsions on the surfaces of glycogens.

A poly(l-lysine) bearing pendant maltooligosaccharide and cholesterol groups formed relatively monodisperse and positively charged nanogels

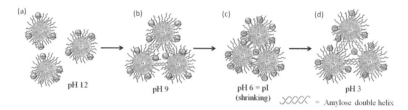

FIGURE 4.11   Plausible mechanism for pH-responsive behavior of amphoteric glycogen hydrogel.

(~50 nm) via self-assembly in water (Morimoto et al., 2013). The nanogels contained the outer maltooligosaccharide hair and the cationic poly(l-lysine) with cholesterol core. The nanogels underwent a secondary structural transition from a random coil to an α-helix in response dependent upon pH. The chain elongation from the maltooligosaccharide primer through the phosphorylase-catalyzed enzymatic polymerization according to Figure 4.12 shielded the positive charge of the nanogels. The enzymatically elongated amylose chains recognized hydrophobic molecules to form inclusion complexes.

A series of amylose-based star polymers (1, 2, 4, and 8 arms) was synthesized by a click reaction and the phosphorylase-catalyzed enzymatic polymerization of specific primers (Nishimura et al., 2015). N-hydroxysuccinimide ester modified poly(ethylene glycol) (PEG) derivatives were initially functionalized with azide propyl amine. In the second step, Glc_5-functionalized PEGs were prepared by treating alkyne-functionalized Glc_5 with azide-functionalized PEGs. A hydrogel was obtained from the 8-arm primer due to effective cross-linking between the multiarmed structures through the phosphorylase-catalyzed enzymatic polymerization. The star polymers with DP = ca. 60 per arm acted as an allosteric multivalent host for hydrophobic molecules by helical formation. A cationic 8-arm star polymer catalyzed DNA strand exchange as nucleic acid chaperone.

4.4 SUCRASE-CATALYZED SYNTHESIS OF POLYSACCHARIDE NANOPARTICLES

The recombinant amylosucrase from *Neisseria polysaccharea* was used to glucosylated glycogen particles in the presence of sucrose as a glycosyl

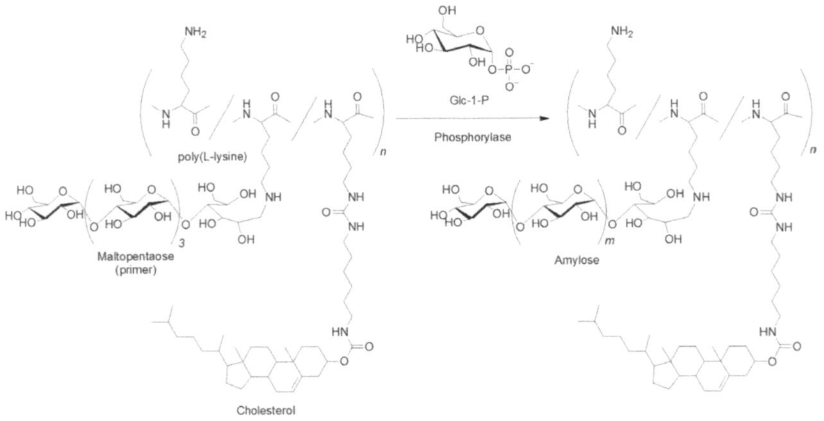

FIGURE 4.12 Phosphorylase-catalyzed enzymatic polymerization using primer-modified cholesteryl poly(l-lysine).

donor (Figure 4.13) (Putaux et al., 2006). The morphology and structure of the resulting insoluble products were depended on the sucrose/glycogen feed ratios. For the lower ratio, all glucose molecules enzymatically produced from sucrose were transferred onto glycogen, giving rise to a slight elongation of the external chains and their organization into small crystallites at the surface of the glycogen particles. With a high initial feed ratio, on the other hand, the external glycogen chains were elongated by amylosucrase catalysis, leading to dendritic nanoparticles with a diameter 4–5 times that of the initial particle.

The enzymatic preparation of amylose microbeads was investigated by the catalysis of the recombinant amylosucrase from *Deinococcus gelthermalis* (Lim et al., 2014). The produced microbeads in the enzymatic reaction system were precipitated when a critical concentration and molecular weight were reached. The microbeads were formed through self-association of the synthesized amylose molecules. Single walled carbon nanotubes (SWCNTs) were incorporated spontaneously into the amylose microstructure during the enzymatic reaction to form well-defined amylose-SWCNT composite microbeads through a self-assembly process via hydrophobic interaction.

Amylose magnetic microbeads were also prepared by the amylosucrase catalysis in the presence of sucrose and iron oxide magnetic nanoparticles (Lim et al., 2015). The products had a well-defined spherical shape with

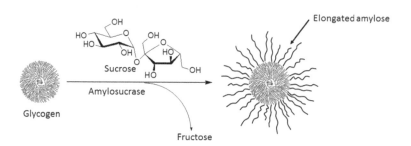

FIGURE 4.13 Amylosucrase-catalyzed enzymatic chain-elongation from glycogen primer.

a diameter of ca. 2 μm, which exhibited excellent magnetic responses and dispersibility in aqueous solutions. The products possessed a high specificity and purification capacity for the maltose binding protein.

Another glucansucrase, dextransucrase, was used as a catalyst to prepare dextran nanoparticles (Semyonov et al., 2014). Under the optimal conditions, the enzyme generated spherical dextran nanoparticles with 100–400 nm sizes. The entrapping of a hydrophobic nutraceutical, the isoflavone genistein, with the product was investigated by the two inclusion complexation methods, dimethyl sulfoxide (DMSO) dilution in water and acidification. The DMSO method was found to be more suitable for inclusion of genistein in dextran, giving rise to a higher genistein load, and a higher % of nanosized particles (85%, 105–400 nm). For both protocols, addition of a freezing-drying step exerted a positive effect presumably due to the formation of new hydrogen bonds and van der Waals interactions.

4.5 CONCLUSIONS

This chapter overviewed that the phosphorylase- and sucrose-catalyzed enzymatic reactions efficiently provide various functional polysaccharide nanoparticles and nanomaterials with precisely controlled structure. Such nanostructured polysaccharide materials exhibit specific and unique functions which are much different from those of synthetic polymers. Furthermore, the enzymatic reactions have been identified as a greener and sustainable process because it is generally conducted in aqueous media under mild conditions. Accordingly, the enzymatic methods described herein have a potential to be practically applied to the production of additional useful functional polysaccharide nanomaterials in various fields such as environmentally benign, biomedical, and tissue engineering.

KEYWORDS

- **enzymatic approach**
- **glycosylation**
- **polymerization**
- **polysaccharide**

REFERENCES

Albenne, C.; Skov, L. K.; Mirza, O.; Gajhede, M.; Feller, G.; D'Amico, S.; Andre, G.; Potocki-Véronèse, G.; van der Veen, B. A.; Monsan, P.; Remaud-Simeon, M. Molecular basis of the amylose-like polymer formation catalyzed by *Neisseria polysaccharea* amylosucrase. *J. Biol. Chem.* **2004**, *279*, 726–734.

Bhuiyan, S. H.; Rus'd, A. A.; Kitaoka, M.; Hayashi, K. Characterization of a hyperthermo-stable glycogen phosphorylase from *Aquifex aeolicus* expressed in *Escherichia coli. J. Mol. Cat. B: Enzym.* **2003**, *22*, 173–180.

Blixt, O.; Razi, N. Enzymatic glycosylation by transferases. In: *Glycoscience Chemistry and Chemical Biology*; 2nd ed.; Fraser-Reid, B. O.; Tatsuta, K.; Thiem, J.; Eds.; Springer: Cham; New York, 2008, pp. 1361–1385.

Boeck, B.; Schinzel, R. Purification and characterization of an α-glucan phosphorylase from the thermophilic bacterium *Thermus thermophilus. Eur. J. Biochem.* **1996**, *239*, 150–155.

Calder, P. C. Glycogen structure and biogenesis. *Int. J. Biochem.* **1991**, *23*, 1335–1352.

de Montalk, G. P.; Remaud-Simeon, M.; Willemot, R. M.; Sarçabal, P.; Planchot, V.; Monsan, P. Amylosucrase from *Neisseria polysaccharea*: novel catalytic properties. *FEBS Lett.* **2000a**, *471*, 219–223.

de Montalk, G. P.; Remaud-Simeon, M.; Willemot, R. M.; Monsan, P. Characterization of the activator effect of glycogen on amylosucrase from *Neisseria polysaccharea. Fems Microbiol. Lett.* **2000b**, *186*, 103–108.

Eisenhaber, F.; Schulz, W. Monte-carlo simulation of the hydration shell of double-helical amylose – a left-handed antiparallel double helix fits best into liquid water-structure. *Biopolymers* **1992**, *32*, 1643–1664.

Evers, B.; Mischnick, P.; Thiem, J. Synthesis of 2-deoxy-α-D-arabino-hexopyranosyl phosphate and 2-deoxy-maltooligosaccharides with phosphorylase. *Carbohydr. Res.* **1994**, *262*, 335–341.

Evers, B.; Thiem, J. Further syntheses employing phosphorylase. *Bioorg. Med. Chem.* **1997**, *5*, 857–863.

Fujii, K.; Takata, H.; Yanase, M.; Terada, Y.; Ohdan, K.; Takaha, T.; Okada, S.; Kuriki, T. Bioengineering and application of novel glucose polymers. *Biocatal. Biotransform.* **2003**, *21*, 167–172.

Hinrichs, W.; Buttner, G.; Steifa, M.; Betzel, C.; Zabel, V.; Pfannemuller, B.; Saenger, W. An amylose antiparallel double helix at atomic resolution. *Science* **1987**, *238*, 205–208.

Huang, G. L.; Chen, Y. L.; Li, Y.; Huang, D.; Han, J.; Yang, M. Two important polysaccharides as carriers for drug delivery. *Mini Rev. Med. Chem.* **2015**, *15*, 1103–1109.

Izawa, H.; Kadokawa, J. Preparation of functional amylosic materials by phosphorylase-catalyzed polymerization. In: *Interfacial Researches in Fundamental and Material Sciences of Oligo- and Polysaccharides*; Kadoakwa, J.; Ed.; Transworld Research Network: Trivandrum, 2009, pp. 69–86.

Izawa, H.; Nawaji, M.; Kaneko, Y.; Kadokawa, J. Preparation of glycogen-based polysaccharide materials by phosphorylase-catalyzed chain elongation of glycogen. *Macromol Biosci* **2009**, *9*, 1098–1104.

Kadoakwa, J. Synthesis of amylose-grafted polysaccharide materials by phosphorylase-catalyzed enzymatic polymerization. In: *Biobased Monomers, Polymers, and Materials*; Smith, P. B.; Gross, R. A.; Eds.; ACS Symposium Series 1105; American Chemical Society: Washington, DC, 2012, pp. 237–255.

Kadoakwa, J. Synthesis of new polysaccharide materials by phosphorylase-catalyzed enzymatic α-glycosylations using polymeric glycosyl acceptors. In: *Green Polymer Chemistry: Biocatalysis and Materials II*; Cheng, H. N.; Gross, R. A.; Smith, P. B.; Eds.; ACS Symposium Series 1144; American Chemical Society: Washington, DC, 2013, pp. 141–161.

Kadokawa, J. Chemoenzymatic synthesis of amylose-grafted cellulose derivatives. In: *Cellulose and Cellulose Derivatives*; Mondal, M. I. H.; Ed.; Nova Science Publishers, Inc.: Hauppauge, NY, 2015a, pp. 299–311.

Kadokawa, J. Chemoenzymatic synthesis of functional amylosic materials. *Pure Appl. Chem.* **2014**, *86*, 701–709.

Kadokawa, J. Enzymatic synthesis of nonnatural oligo- and polysaccharides by phosphorylase-catalyzed glycosylations using analog substrates. In: *Green Polymer Chemistry: Biobased Materials and Biocatalysis*; Cheng, H. N.; Gross, R. A.; Smith, P. B.; Eds.; ACS Symposium Series 1192; American Chemical Society: Washington, DC, 2015b, pp. 87–99.

Kadokawa, J. Facile synthesis of unnatural oligosaccharides by phosphorylase-catalyzed enzymatic glycosylations using new glycosyl donors. In: *Oligosaccharides: Sources, Properties and Applications*; Gordon, N. S.; Ed.; Nova Science Publishers, Inc.: Hauppauge, NY, 2011b, pp. 269–281.

Kadokawa, J. Precision polysaccharide synthesis catalyzed by enzymes. *Chem. Rev.* **2011a**, *111*, 4308–4345.

Kadokawa, J. Synthesis of nonnatural oligosaccharides by α-glucan phosphorylase-catalyzed enzymatic glycosylations using analog substrates of α-D-glucose 1-phosphate. *Trends Glycosci. Glycotechnol.* **2013**, *25*, 57–69.

Kadokawa, J.; Kaneko, Y. *Engineering of Polysaccharide Materials – by Phosphorylase-Catalyzed Enzymatic Chain-Elongation*; Pan Stanford Publishing Pvt Ltd.: Singapore, 2013.

Kadokawa, J.; Kobayashi, S. Polymer synthesis by enzymatic catalysis. *Curr. Opin. Chem. Biol.* **2010**, *14*, 145–153.

Kaneko, Y.; Kadokawa, J. Chemoenzymatic synthesis of amylose-grafted polymers. In: *Handbook of Carbohydrate Polymers: Development, Properties and Applications*; Ito, R.; Matsuo, Y.; Eds.; Nova Science Publishers, Inc.: Hauppauge, NY, 2009, pp. 671–691.

Kawazoe, S.; Izawa, H.; Nawaji, M.; Kaneko, Y.; Kadokawa, J. Phosphorylase-catalyzed *N*-formyl-α-glucosaminylation of maltooligosaccharides. *Carbohydr. Res.* **2010**, *345*, 631–636.

Kitamura, S. Starch polymers, natural and synthetic. In: *The Polymeric Materials Encyclopedia, Synthesis, Properties And Applications*; Salamone, C.; Ed.; CRC Press: New York, 1996; Vol. 10, pp. 7915–7922.

Kitamura, S.; Yunokawa, H.; Mitsuie, S.; Kuge, T. Study on polysaccharide by the fluorescence method. 2. Micro-brownian motion and conformational change of amylose in aqueous-solution. *Polym. J.* **1982**, *14*, 93–99.

Kitaoka, M.; Hayashi, K. Carbohydrate-processing phosphorolytic enzymes. *Trends Glycosci. Glycotechnol.* **2002**, *14*, 35–50.

Kobayashi, S.; Makino, A. Enzymatic polymer synthesis: An opportunity for green polymer chemistry. *Chem. Rev.* **2009**, *109*, 5288–5353.

Kobayashi, S.; Uyama, H.; Kimura, S. Enzymatic polymerization. *Chem. Rev.* **2001**, *101*, 3793–3818.

Korakli, M.; Vogel, R. F. Structure/function relationship of homopolysaccharide producing glycansucrases and therapeutic potential of their synthesized glycans. *Appl. Microbiol. Biotechnol.* **2006**, *71*, 790–803.

Lim, M. C.; Lee, G. H.; Huynh, D. T. N.; Letona, C. A. M.; Seo, D. H.; Park, C. S.; Kim, Y. R. Amylosucrase-mediated synthesis and self-assembly of amylose magnetic microparticles. *Rsc Adv.* **2015**, *5*, 36088–36091.

Lim, M. C.; Seo, D. H.; Jung, J. H.; Park, C. S.; Kim, Y. R. Enzymatic synthesis of amylose nanocomposite microbeads using amylosucrase from *Deinococcus geothermalis*. *Rsc Adv.* **2014**, *4*, 26421–26424.

Liu, Z. H.; Jiao, Y. P.; Wang, Y. F.; Zhou, C. R.; Zhang, Z. Y. Polysaccharides-based nanoparticles as drug delivery systems. *Adv. Drug Deliv. Rev.* **2008**, *60*, 1650–1662.

Manners, D. J. Recent developments in our understanding of glycogen structure. *Carbohydr. Polym.* **1991**, *16*, 37–82.

Morimoto, N.; Yamazaki, M.; Tamada, J.; Akiyoshi, K. Polysaccharide-hair cationic polypeptide nanogels: Self-assembly and enzymatic polymerization of amylose primer-modified cholesteryl poly(L-lysine). *Langmuir* **2013**, *29*, 7509–7514.

Nakai, H.; Kitaoka, M.; Svensson, B.; Ohtsubo, K. Recent development of phosphorylases possessing large potential for oligosaccharide synthesis. *Curr. Opin. Chem. Biol.* **2013**, *17*, 301–309.

Nawaji, M.; Izawa, H.; Kaneko, Y.; Kadokawa, J. Enzymatic synthesis of α-D-xylosylated maltooligosaccharides by phosphorylase-catalyzed xylosylation. *J. Carbohydr. Chem.* **2008b**, *27*, 214–222.

Nawaji, M.; Izawa, H.; Kaneko, Y.; Kadokawa, J. Enzymatic α-glucosaminylation of maltooligosaccharides catalyzed by phosphorylase. *Carbohydr. Res.* **2008a**, *343*, 2692–2696.

Nishimura, T.; Mukai, S.; Sawada, S.; Akiyoshi, K. Glyco star polymers as helical multivalent host and biofunctional nano-platform. *Acs Macro Lett.* **2015**, *4*, 367–371.

Ohdan, K.; Fujii, K.; Yanase, M.; Takaha, T.; Kuriki, T. Enzymatic synthesis of amylose. *Biocatal. Biotransform.* **2006**, *24*, 77–81.

Omagari, Y.; Kadokawa, J. Synthesis of heteropolysaccharides having amylose chains using phosphorylase-catalyzed enzymatic polymerization. *Kobunshi Ronbunshu* **2011**, *68*, 242–249.

Paulsen, H. Advances in selective chemical syntheses of complex oligosaccharides. *Angew. Chem., Int. Ed. Engl.* **1982**, *21*, 155–173.

Percival, M. D.; Withers, S. G. Applications of enzymes in the synthesis and hydrolytic study of 2-deoxy-α-D-glucopyranosyl phosphate. *Can. J. Chem.* **1988**, *66*, 1970–1972.

Posocco, B.; Dreussi, E.; de Santa, J.; Toffoli, G.; Abrami, M.; Musiani, F.; Grassi, M.; Farra, R.; Tonon, F.; Grassi, G.; Dapas, B. Polysaccharides for the Delivery of Antitumor Drugs. *Materials* **2015**, *8*, 2569–2615.

Potocki-Véronèse, G.; Putaux, J. L.; Dupeyre, D.; Albenne, C.; Remaud-Simeon, M.; Monsan, P.; Buleon, A. Amylose synthesized in vitro by amylosucrase: Morphology, structure, and properties. *Biomacromolecules* **2005**, *6*, 1000–1011.

Puchart, V. Glycoside phosphorylases: Structure, catalytic properties and biotechnological potential. *Biotechnol. Adv.* **2015**, *33*, 261–276.

Putaux, J. L.; Potocki-Véronèse, G.; Remaud-Simeon, M.; Buleon, A. α-D-Glucan-based dendritic nanoparticles prepared by in vitro enzymatic chain extension of glycogen. *Biomacromolecules* **2006**, *7*, 1720–1728.

Remaud-Simeon, M.; Albenne, C.; Joucla, G.; Fabre, E.; Bozonnet, S.; Pizzut, S.; Escalier, P.; Potocki-Vèronèse, G.; Monsan, P. Glucansucrases: Structural basis, mechanistic aspects, and new perspectives for engineering. In: *Oligosaccharides in Food and Agriculture*; Eggleston, G.; Côté, G. L.; Eds.; ACS Symposium Series 849; American Chemical Society: Washington, DC, 2003, pp. 90–103.

Remaud-Simeon, M.; Willemot, R. M.; Sarcabal, P.; de Montalk, G. P.; Monsan, P. Glucansucrases: Molecular engineering and oligosaccharide synthesis. *J. Mol. Cat. B: Enzym.* **2000**, *10*, 117–128.

Rolland-Sabaté, A.; Colonna, P.; Potocki-Véronèse, G.; Monsan, P.; Planchot, V. Elongation and insolubilization of α-glucans by the action of *Neisseria polysaccharea* amylosucrase. *J. Cer. Sci.* **2004**, *40*, 17–30.

Saravanakumar, G.; Jo, D. G.; Park, J. H. Polysaccharide-based nanoparticles: A versatile platform for drug delivery and biomedical imaging. *Curr. Med. Chem.* **2012**, *19*, 3212–3229.

Schatz, C.; Lecommandoux, S. Polysaccharide-containing block copolymers: synthesis, properties and applications of an emerging family of glycoconjugates. *Macromol. Rapid Commun.* **2010**, *31*, 1664–1684.

Schmidt, R. R. New methods for the synthesis of glycosides and oligosaccharides—are there alternatives to the Koenigs-Knorr method? *Angew. Chem., Int. Ed. Engl.* **1986**, *25*, 212–235.

Schuerch, C. Polysaccharides. In: *Encyclopedia of Polymer Science and Engineering*; 2nd ed.; Mark, H. F.; Bilkales, N.; Overberger, C. G.; Eds.; John Wiley & Sons: New York, 1986; Vol. 13, pp. 87–162.

Seibel, J.; Beine, R.; Moraru, R.; Behringer, C.; Buchholz, K. A new pathway for the synthesis of oligosaccharides by the use of non-Leloir glycosyltransferases. *Biocatal. Biotransform.* **2006a**, *24*, 157–165.

Seibel, J.; Jordening, H. J.; Buchholz, K. Glycosylation with activated sugars using glycosyltransferases and transglycosidases. *Biocatal. Biotransform.* **2006b**, *24*, 311–342.

Semyonov, D.; Ramon, O.; Shoham, Y.; Shimoni, E. Enzymatically synthesized dextran nanoparticles and their use as carriers for nutraceuticals. *Food Funct.* **2014**, *5*, 2463–2474.

Shelke, N. B.; James, R.; Laurencin, C. T.; Kumbar, S. G. Polysaccharide biomaterials for drug delivery and regenerative engineering. *Polym. Adv. Technol.* **2014**, *25*, 448–460.

Shoda, S. Enzymatic glycosylation. In: *Glycoscience Chemistry and Chemical Biology*; Fraser-Reid, B. O.; Tatsuta, K.; Thiem, J.; Eds.; Springer-Verlag: Berlin, Heidelberg, New York, 2001; Vol. II, pp. 1465–1496.

Shoda, S.; Izumi, R.; Fujita, M. Green process in glycotechnology. *Bull. Chem. Soc. Jpn.* **2003**, *76*, 1–13.

Stern, R.; Jedrzejas, M. J. Carbohydrate polymers at the center of life's origins: The importance of molecular processivity. *Chem. Rev.* **2008**, *108*, 5061–5085.

Takaha, T.; Yanase, M.; Takata, H.; Okada, S. Structure and properties of *Thermus aquaticus* α-glucan phosphorylase expressed in *Escherichichia coli*. *J. Appl. Glycosci.* **2001**, *48*, 71–78.

Takata, H.; Takaha, T.; Nakamura, H.; Fujii, K.; Okada, S.; Takagi, M.; Imanaka, T. Production and some properties of a dextrin with a narrow size distribution by the cyclization reaction of branching enzyme. *J. Ferment. Bioeng.* **1997**, *84*, 119–123.

Takata, H.; Takaha, T.; Okada, S.; Hizukuri, S.; Takagi, M.; Imanaka, T. Structure of the cyclic glucan produced from amylopectin by *Bacillus stearothermophilus* branching enzyme. *Carbohydr. Res.* **1996a**, *295*, 91–101.

Takata, H.; Takaha, T.; Okada, S.; Takagi, M.; Imanaka, T. Cyclization reaction catalyzed by branching enzyme. *J. Bacteriol.* **1996b**, *178*, 1600–1606.

Takata, Y.; Shimohigoshi, R.; Yamamoto, K.; Kadokawa, J. Enzymatic synthesis of dendritic amphoteric α-glucans by thermostable phosphorylase catalysis. *Macromol Biosci* **2014**, *14*, 1437–1443.

Takata, Y.; Yamamoto, K.; Kadokawa, J. Preparation of pH-responsive amphoteric glycogen hydrogels by α-glucan phosphorylase-catalyzed successive enzymatic reactions. *Macromol. Chem. Phys.* **2015**, *216*, 1415–1420.

Takemoto, Y.; Izawa, H.; Umegatani, Y.; Yamamoto, K.; Kubo, A.; Yanase, M.; Takaha, T.; Kadokawa, J. Synthesis of highly branched anionic α-glucans by thermostable phosphorylase-catalyzed α-glucuronylation. *Carbohydr. Res.* **2013**, *366*, 38–44.

Thiem, J.; Thiem, J. Enzymatic glycosylation by glycohydrolases and glycozynthases. In: *Glycoscience Chemistry and Chemical Biology*; 2nd ed.; Fraser-Reid, B. O.; Tatsuta, K.; Thiem, J.; Eds.; Springer: Cham; New York, 2008, pp. 1387–1409

Toshima, K.; Tatsuta, K. Recent progress in *O*-glycosylation methods and its application to natural products synthesis. *Chem. Rev.* **1993**, *93*, 1503–1531.

Umegatani, Y.; Izawa, H.; Nawaji, M.; Yamamoto, K.; Kubo, A.; Yanase, M.; Takaha, T.; Kadokawa, J. Enzymatic α-glucuronylation of maltooligosaccharides using α-glucuronic acid 1-phosphate as glycosyl donor catalyzed by a thermostable phosphorylase from *Aquifex aeolicus* vf5. *Carbohydr. Res.* **2012**, *350*, 81–85.

van der Veen, B. A.; Potocki-Véronèse, G.; Albenne, C.; Joucla, G.; Monsan, P.; Remaud-Simeon, M. Combinatorial engineering to enhance amylosucrase performance: Construction, selection, and screening of variant libraries for increased activity. *FEBS Lett.* **2004**, *560*, 91–97.

van Hijum, S. A. F. T.; Kralj, S.; Ozimek, L. K.; Dijkhuizen, L.; van Geel-Schutten, I. G. H. Structure-function relationships of glucansucrase and fructansucrase enzymes from lactic acid bacteria. *Microbiol. Mol. Biol. Rev.* **2006**, *70*, 157–176.

van Leeuwen, S. S.; Kralj, S.; Eeuwema, W.; Gerwig, G. J.; Dijkhuizen, L.; Kamerling, J. P. Structural characterization of bioengineered α-D-glucans produced by mutant glucansucrase GTF180 enzymes of *Lactobacillus reuteri* strain 180. *Biomacromolecules* **2009**, *10*, 580–588.

Yalpani, M. *Polysaccharides: Syntheses, Modifications, and Structure/Property Relations*; Elsevier: Amsterdam; New York, 1988.

Yanase, M.; Takaha, T.; Kuriki, T. α-Glucan phosphorylase and its use in carbohydrate engineering. *J. Sci. Food Agric.* **2006**, *86*, 1631–1635.

Yanase, M.; Takata, H.; Fujii, K.; Takaha, T.; Kuriki, T. Cumulative effect of amino acid replacements results in enhanced thermostability of potato type L α-glucan phosphorylase. *Appl. Environ. Microbiol.* **2005**, *71*, 5433–5439.

Ziegast, G.; Pfannemüller, B. Linear and star-shaped hybrid polymers. 4. Phosphorolytic syntheses with di-functional, oligo-functional and multifunctional primers. *Carbohydr. Res.* **1987**, *160*, 185–204.

CHAPTER 5

POLYSACCHARIDE-DRUG NANOCONJUGATES AS TARGETED DELIVERY VEHICLES

SABYASACHI MAITI

Department of Pharmaceutics, Gupta College of Technological Sciences, Asansol, West Bengal, India, E-mail: sabya245@rediffmail.com

CONTENTS

ABSTRACT

In recent years, the delivery of drugs using natural polysaccharides has attracted considerable attention of the pharmaceutical scientists especially due to their inert, biocompatible and biodegradable nature. Moreover, the presence of a variety of functional groups in biopolymer structure allows a number of chemical conjugation reactions. Chemical conjugation of drugs to native or modified biopolymers may lead to the formation of stable bonds such as ester, amide, and disulfide that can retard drug release during its transport before the cellular localization of the drug. In addition, polymer-drug conjugates offer a number of advantages over their small therapeutic precursors. Polymer-drug conjugates can enhance water solubility and bioavailability of drug, provide protection to the drug from deactivation and preserve drug activity during circulation, transport to targeted organ or tissue and intracellular trafficking. The conjugates can also be used to alter pharmacokinetics and biodistribution of drugs, which exhibit short blood plasma half-lives. Furthermore, the conjugation approach provides an opportunity for passive or active targeting of drugs specifically to the site of action. Therefore, the polymeric prodrug conjugates are being widely investigated to solve different bio-performance related issues for a variety of drugs. This overview focuses on some water-soluble natural polymers such as alginate, pullulan, dextran and pectin; and their drug conjugates. The achievements in the field of biopolymer-drug conjugates are discussed herein.

5.1 INTRODUCTION

Polymer–drug covalent conjugation is one of the major approaches of drug modifications to enhance solubility, permeability and stability, biodistribution, and thus bioactivity of small drug moieties (Nichifor and Mocanu, 2006). The conjugates are hydrolyzed under the influence of chemicals or enzymes; and the active drug is released by disruption of the chemical bond between polymer and drug (Grodzinski, 1999).

Ideally, the conjugates should be water-soluble, nontoxic, and nonimmunogenic (Godwin et al., 2001). The polymer should possess functional groups amenable to chemical modifications for direct covalent conjugation with drug or via a degradable/nondegradable linker/spacer. Besides, they can also be conveniently used to anchor physicochemical or biological modifiers such as solubilizing molecules, and environment-sensitive moieties and targeting agents. The rate of breakage of covalent linkages controls the drug release rate at the target site (Duncan et al., 2006).

The conjugate systems are designed for passive or active targeting of drugs. Passive targeting is based on natural distribution pattern of a drug-carrier in vivo, and enhanced permeability and retention (EPR) effect

(Matsumura and Maeda, 1986). The EPR effect is usually attributed to the high tumor vascularization, increased permeability of the tumor vessels, defective tumor vasculature, and inappropriate lymphatic drainage in the tumor interstitium (Minko et al., 2000). On contrary, the active targeting relies upon the selective localization and interaction of a ligand at cell-specific receptors. Thus, the polymer-anticancer drug conjugates, if properly designed, may improve the half-life, therapeutic index, aqueous solubility of the drugs and also their exposure time to the tumor vasculature, thereby reducing their systemic toxicity.

The polymers featured with multiple functionalities can be conveniently used to anchor high drug amounts via releasable bonds and, eventually, to conjugate physicochemical or biological modifiers such as solubilizing molecules, environmentally sensitive moieties and targeting agents.

Small molecule chemotherapeutics usually have undesired physiochemical and pharmacological properties, such as low solubility, severe side effects, and narrow therapeutic index (Langer, 1998) and thus limit their clinical applications. In the past 2–3 decades, two strategies have been adapted to address these drawbacks (Li and Wallace, 2008). One approach is to design new derivatives of existing chemotherapeutics with improved physiochemical and pharmacological properties to modulate the molecular processes and pathways associated with tumor progression (Collins and Workman, 2006). The next approach is to use drug delivery technology to modify biopharmaceutical behaviors of the existing agents (Duncan, 2006; Peer et al., 2007). Different drug delivery systems such as polymeric micelles, liposomes, nanoparticles that can control and selectively release drugs in the disease site have been developed, thus opening new perspectives for drug therapy (Davis et al., 2008; Lammers et al., 2008; Peer et al., 2007).

Currently, the polymer-drug conjugate is being increasingly popular for the delivery of hydrophobic drugs due to some obvious reasons such as improved aqueous solubility, therapeutic efficacy by controlling drug release at the target sites in optimum dose, and reduced side effects (Alexis et al., 2008; Van et al., 2010; Zhou et al., 2010). The concept of polymer-drug covalent conjugation was proposed by Ringsdorf (1975). According to Ringsdorf model, an ideal polymer–drug conjugate should have a biocompatible polymeric vehicle and bioactive substances for attachment to the polymer scaffolds *via* a bio-responsive linker. If required, a targeting ligands or a solubilizer can be introduced into the conjugate to further improve the therapeutic efficacy. More often, the use of hydrophilic polymers changes the physicochemical and biological behavior of the parent drugs. A dramatic improvement in aqueous solubility is of significant relevance because it has been estimated that 40% of drugs that are currently being developed or under development exhibit poor bioavailability due to inadequate aqueous solubility (Lipinski, 2002). The conjugates can protect the entrapped or covalently linked drugs from degradation leading to increased bioavailability (Xiao et al., 2011). Compared to nontargeted ones, ligand-mediated polymer-drug conjugates accumulate at the target tissues/cells and show higher toxicity to the specific cells while being less toxic to others (Maeda et al., 2000; Scomparin et al., 2011). The conjugates that are internalized *via* endocytosis pathway are transferred into lysosomes, digested by lysosomal enzymes at low pH, and released the attached load of small drug molecules (Chan et al., 2006).

Polymer-drug conjugates prepared from biocompatible, biodegradable, and water-soluble polymers have received tremendous interest for targeted release of drugs. In case the polymer is nonbiodegradable, its molecular size must be lower than the renal threshold to allow excretion and prevent accumulation in the body. However, the maximum molecular weight (MW) that will allow renal excretion depends on polymer characteristics such as conformations in solution and levels of hydration. In general, for nonbiodegradable polymers, the size of conjugates should be within 40kDa (Greco and Vicent, 2008). The linker needs to be stable in the blood circulation but ensure drug-release at the target sites after a precise biological trigger such as an enzyme or pH variations. The drug has to bear a chemical group that allows conjugation to the polymer.

Some synthetic polymers such as N-(2-hydroxypropyl) methacrylamide (HPMA) copolymers (Chytil et al., 2006; Kopecek et al., 2000), poly (ethylene glycol)(PEG)(Greenwald et al., 2003), have been studied extensively as carriers for drugs, especially for anticancer agents. However, the synthetic conjugates designed for lysosomotropic delivery, have failed in early clinical development due to some degree of unacceptable toxicity.

To resolve this problem, polysaccharides are currently being investigated as the most promising carriers for targeted delivery of drugs (Azzam et al., 2002; Kaneo et al., 2001). Due to their excellent physicochemical properties, polysaccharides can favor the design of conjugates with desirable biopharmaceutical properties and therapeutic performance compared to parent drugs (Caliceti et al., 2010; Hashida et al., 1984; Kojima et al., 1980; Pawar et al., 2008). In fact, these biomaterials display high degree of biocompatibility, biodegradability, and multiple functional groups to endow derivatives with tunable, bio-responsive targeting properties, or to combine molecules with synergistic therapeutic effect.

The biodegradable nature of natural polysaccharides facilitates metabolic removal from the body after administration, and prevents accumulation in the body (Duncan, 1992). However, the premature degradation of the conjugate may limit its biopharmaceutical performance due to presence of glycosidases in the circulation. Rather, covalent conjugation of pendent groups has been shown to reduce the susceptibility of normally biodegradable polymers to enzymatic attack, e.g., modification of dextran reduces its rate of enzymatic hydrolysis by dextranases (Crepon et al., 1991; Vercauteren et al., 1990). Further, their pharmacokinetics is influenced by molecular weight, electric charge and degree of substitution, polydispersity and branching. Other challenges of using polysaccharides include the difficult conjugation chemistry and purification. Nevertheless, most of the reported studies using polysaccharides are limited to preclinical investigations (Danhauser-Riedl et al., 1993; Wente et al., 2005). Tumor targeting and selective anticancer drug delivery can in fact, be accomplished by receptor-mediated uptake or passive fluid-phase endocytosis of polysaccharide conjugates.

Choosing the exact polymer in one hand can enhance the safer systemic circulation time of a conjugate and on the other hand, controlling parameters like size, functionality, charge, etc. can amend the effect of biological half-life of a drug in a conjugate (Li and Huang, 2008). Various types of natural and synthetic polymers have been conjugated with hydrophobic drugs for the development of long circulating polymer-drug conjugates.

Indeed, the task of acquiring a successful polymer–drug conjugate seems intricate. The design of an appropriate carrier is strongly influenced by its proposed route, frequency of administration and dose. Other factors like desired target, type of conjugation (direct or indirect), linker chemistry and the MW should also be considered during design of the conjugates. The conjugates can self-assemble into nano-micellar morphological structures in aqueous phase. The polymer–drug conjugates thus start a new era of drug delivery systems.

This chapter reviews the achievements and state-of-the-art in polysaccharide-based nanoconjugate systems emphasizing sodium alginate, pullulan, dextran and pectin for targeted release of small drug molecules, followed by a brief discussion on future perspectives to speed up their translation into the clinic. The different synthetic strategies have also been incorporated for easy understanding of the novel polysaccharide-drug conjugate systems.

5.2 POLYSACCHARIDE-DRUG NANOCONJUGATES

5.2.1 ALGINATE-DRUG NANOCONJUGATES

Sodium alginate, an anionic algal polysaccharide consists of $(1{\rightarrow}4)$ linked β-D-mannuronic acid and α-L-guluronic acid residues (Sarmento et al., 2007). This block copolymer is biodegradable, biocompatible, nontoxic, and nonimmunogenic in nature. Further, two secondary hydroxyl groups at C-2 and C-3 and one carboxyl group at C-6 are the attractive sites for the desired chemical modifications to develop alginate-drug conjugates.

Morgan et al.(1995) prepared alginate-5-aminosalicylic acid (5-ASA) conjugates via 6-aminohexanamide-L-phenylalanine linker for α-chymotrypsin enzyme-sensitive delivery of 5-ASA. The solution of 6-aminohexanamide-L-phenylalanyl-5ASA formate in DMSO/NaOH was reacted with propylene glycol alginate (pH10) at 4°C for 72h. The alginate-drug conjugate was precipitated with acetone, redissolved in NaOH/ distilled water (pH 10), and dialyzed for removal of unbound drug prior to lyophilization. The 5ASA was bound to alginate via the peptidyl spacer 6-aminohexanamide-L-phenylalanine. The reaction scheme for the synthesis of alginate conjugates is shown in Figure 5.1.

The alginate-5ASA conjugate contained 1.7% (w/w) 5ASA. This conjugate released only 48% drug in the presence of α-chymotrypsin in 24h. The initial drug release rate was about 6-fold slower than that from the low molecular weight analog, 6-aminohexanamide-L-phenylalanyl-5-ASA in Tris buffer (pH 7.8). The conjugate did not release any amount of drug in buffer system in absence of this enzyme. It was speculated that the

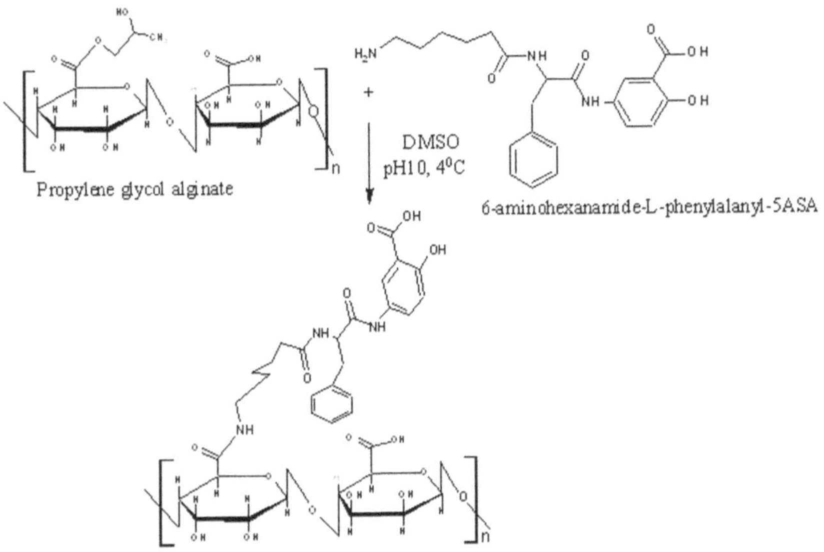

FIGURE 5.1 Reaction scheme for the synthesis of alginate-5-ASA conjugates via 6-aminohexanamide-L-phenylalanine spacers.

polymer imparted steric hindrance in the rapid formation of enzyme-substrate complex and slowed drug release rate from the alginate conjugate. The chymotrypsin, or chymotrypsin-like activity is found in the colon and small intestine (Woodley, 1991). Henceforth, this type of drug delivery could be suitable for the treatment of inflammatory lesions occurring in these segments.

Al-Shamkhani et al. (1995) used an acid-sensitive linker (*cis*-aconityl) for covalent binding of daunomycin (DNM) to high (250 kDa) and low (61 kDa) MW alginates. The primary amine groups were introduced into alginate by excess ethylenediamine treatment. The reaction of DNM with *cis-aconitic* anhydride produced *N-cis*-aconityl-DNM, which was then covalently liked with the amino-modified alginate using carbodiimide reaction. The reaction scheme is illustrated in Figure 5.2.

The drug loading of high MW alginate-DNM (HMW-alginate-DNM) conjugate was 0.8% w/w compared with 1.3% w/w for the low MW (LMW-alginate-DNM) conjugate. The conjugates contained <1% free DNM in relation to the total DNM content. The drug release from the conjugate was tested in citrate/phosphate buffers of different pH values

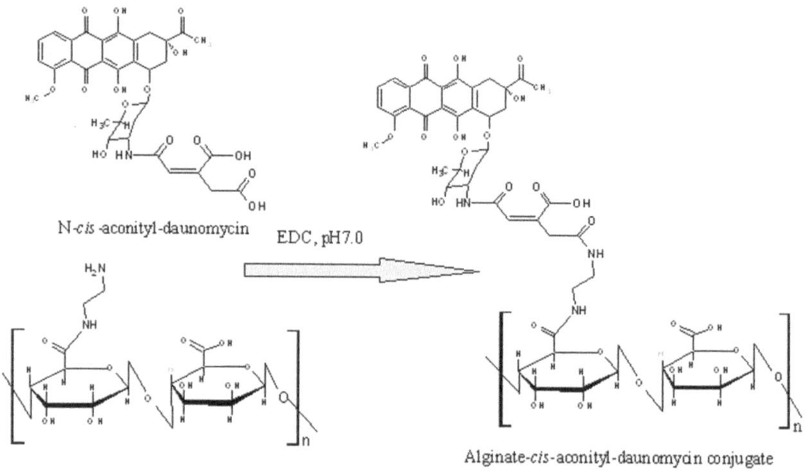

FIGURE 5.2 The possible pathway of alginate-*cis*-aconityl-daunomycin conjugate synthesis.

(5.0, 6.0 and 7.0). At pH 7.0, only 3.1% DNM was released over 48 h. The drug release rate was about 2.5-fold faster at pH 5.0 than that at pH6.0. HMW-alginate-DNM conjugate discharged a maximum of 22.0% and 8.7% at pH 5.0 and6.0, respectively. Unlike HMW-alginate-DNM, LMW-alginate-DNM was found to release a drug derivative in addition to DNM. At pH 5.0, DNM/DNM-derivative was released from LMW-alginate-DNM whereas; the conjugate released only 1.6% loads over 48 h, demonstrating its stability at pH 7.0. A total of 61.9% DNM/DNM-derivative released at pH5.0 after 48 h. The use of rat liver lysosomal enzymes in pH 5.0 buffer did not speed up the release of DNM/DNM derivative.

HMW-alginate-DNM conjugate displayed dose-dependent cytotoxicity against B16F10 cells with an IC50 value of approximately 0.5 mg/ml. On comparison, HMW-alginate conjugate exhibited nearly 700-fold less cytotoxicity than the free drug. The antitumor activity of LMW-alginate-DNM was evaluated in mice bearing B16 subcutaneous tumors after single intraperitoneal injection. The less viscous solution of LMW conjugate in PBS was more suitable for in vivo administration (~30 mg/ml) than the HMW-alginate-DNM. The animals treated with LMW-alginate-DNM conjugate (~5 mg/kg) exhibited a slower rate of tumor growth. Treatment with free DNM or DNM/LMW alginate did not cause significant delay in tumor growth at the same dose. The conjugate increased the lifespan of animals by 14.1%, compared to 6.3% and 4.2% for DNM and DNM/alginate mixture, respectively. It was noteworthy that the mean survival time for the treated groups did not differ significantly with regards to control group. However, the drug-related toxicity was observed in 40% animals treated with free drug. Conversely, there were no signs of toxicity to the animals treated with LMW-alginate-DNM or DNM/alginate mixture.

The clinical use of cisplatin (CP) in ovarian cancer treatment is limited primarily due to nonspecific or systemic side effects, including acute nephrotoxicity, adverse gastrointestinal (GI) reactions and chronic neurotoxicity (Uchino et al., 2005). Liposomes are among the most widely studied novel drug delivery systems. However, the poor aqueous solubility of CP leads to its low entrapment in conventional CP-liposome formulations, resulting in insufficient delivery of CP to the tumors (Hirai et al., 2010). Wang and coworkers (2014) paid their attention to enhance therapeutic specificity and efficacy of CP by designing a targeted liposomal formulation.

CP-sodium alginate (SA) conjugates (CS) were formed by gentle stirring of SA into aqueous solution of CP and incubating for 24 h in the dark. Un-reacted CP was removed from the conjugates by dialysis. A coordination bond between the platinum (II) atom in CP and the carboxyl group in the side chain of alginate resulted in the spontaneous formation of water soluble CS conjugates. They prepared basic liposomes and targeting liposomes by thin film hydration method. Briefly, egg yolk phosphatidylcholine (EPC), cholesterol, DSPE (1,2-distearoyl-*sn*-glycero-3-phosphoethanolamine)-mPEG2000 and DSPE-PEG2000-amine (molar ratios, 4.6:1.3:0.1:0.01) were dissolved in chloroform. The organic solvent was removed by rotary evaporation and the resulting film was dried under vacuum. After film hydration, the solution was sonicated at 50W using a probe sonicator. EDC/sulfo-NHS (N-hydroxysulfosuccinimide) was added under nitrogen flow for 30 min and epidermal growth factor (EGF)was then added to form the EGF-modified liposome (EGF-Lip). DSPE-PEG2000-Amine was not used in the formulation of conventional PEGylated liposomes (PEG-Lip). The CP-loaded liposomes (CS-PEG-Lip and CS-EGF-Lip) were prepared by adding CS to water prior to the hydration step. The EGF conjugation efficiency was 95.56±1.02%. The size of CS-PEG-Lip was 79.25± 0.44 nm and 112.37± 0.52 nm for CS-EGF-Lip. The drug entrapment efficiency was 86.81±5.46% and 89.92±4.28%for CS-PEG-Lip and CS-EGF-Lip, respectively. CS-PEG-Lip and CS-EGF-Lip exhibited almost similar drug release profiles at pH 7.4 and pH 5.5 over 168 h without any burst effect.

EGF was used as ligands due to overexpression of its receptors in human ovarian cancer cells. Moreover, EGF is a stable protein with well-defined reaction sites for conjugation, undergoes receptor-mediated endocytosis, and transports the ligands-receptor complex into the cells (Tseng et al., 2009; Wang et al., 2013). The cellular uptake of EGF-modified liposomes was greater in EGFR-positive SKOV3 human ovarian cancer cell lines than that of PEG liposomes and firmly supported the hypothesis regarding receptor-mediated cellular internalization.

The CP drug formulations (CP, CS-PEG-Lip and CS-EGF-Lip) inhibited SKOV3 growth in concentration-and time-dependent manner. CS-EGF liposomes demonstrated the strongest cytotoxicity. It was correlated with greater cellular uptake of CS-EGF-Lip, which delivered more CP into the cells and consequently, inhibited the cell growth. Furthermore, a significant antitumor activity was noted with CS-EGF vesicles in SKOV3 tumor bearing mice, with reduced nephrotoxicity, compared to CP, CS-PEG-Lip at an equivalent dose of 5 mg/kg.

Curcumin (CUR) is a natural polyphenol with anticancer activities probably due to presence of methoxy groups in the structure (Aggarwal et al., 2003; Ravindran et al., 2009; Shi et al., 2006). However, low solubility and bioavailability hinders its clinical application in cancer treatments (Anand et al., 2007). Besides hydrophobic nature, rapid metabolism and hydrolytic degradation may contribute to poor bioavailability of this compound.

Dey and Sreenivasan (2014) synthesized covalently bound polymer-drug conjugates to redress these drawbacks. They focused on the conjugation of phenolic OH group of curcumin with the C-6 carboxyl functionality of hydrophilic SA via an ester linkage to produce SA-CUR conjugate. The conjugate self-assembled in aqueous solution and formed spherical micelles with an average hydrodynamic diameter of 459±0.32 nm. The same was 62.5±11 nm on TEM images. Thus, it can be said that the size of SA-CUR conjugate micelles depends on the methods of measurement. In dynamic light scattering, the hydrodynamic diameter is measured in solution phase; whereas in TEM the nanoparticles are visualized in completely dry state (Manju and Sreenivasan, 2011). Thus, the SA-CUR micelles remained in swollen state due to hydrophilic alginate polymer and appeared much larger than in dry condition. The possible structural configuration is presented in Figure 5.3.

On an average, one unit of curcumin was conjugated to every 200 hexuronic acid residues. The aqueous solubility of SA-CUR conjugate was >10 mg/ml, which corresponded to 109 µg/ml curcumin and was higher than un-conjugated CUR. This core-shell structure shielded curcumin from hydrolytic degradation to large extent at PBS (pH 7.4). In this medium, the micelles were stable due to higher negative zeta potential (–45.43±0.2 mV). In SA-CUR conjugate micelles, the outer shell is composed of strongly hydrophilic polymer alginate and hence, negative surface charge was encountered. Hence, SA-CUR micelles may evade protein adsorption to considerable extent, giving rise to their prolonged circulation time and elevated EPR effect to promote passive targeting to tumor.

The cytotoxic potential of CUR improved following SA conjugation in a dose dependent manner. A small amount of SA-CUR conjugate was sufficient to produce considerable cytotoxicity in L-929 mouse fibroblast cells. A low level of cytotoxicity of free curcumin could be attributed to ineffective exposure time. In case of SA-CUR conjugates, the exposure time was higher due to enhanced solubility, better cellular internalization,

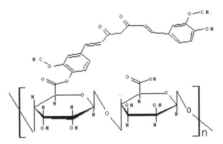

Alginate-Curcumin conjugate

FIGURE 5.3   The possible chemical structures of alginate-curcumin conjugate.

and stability at physiological pH, thus showing relatively more cytotoxicity to cells. Other hydrophobic drug molecules can also be loaded into hydrophobic micellar core of the conjugates. Hence, this system could also serve as a nanosized delivery vehicle for other hydrophobic drugs. Sarika and groups (2015) also studied the effect of conjugation on the solubility and anticancer activity of CUR using gum Arabic as hydrophilic polysaccharide. The self-assembled nanomicelles (270±5 nm) caused ~900-fold greater solubility than free CUR. A profound activity against human hepatocellular carcinoma (HepG2) cells was observed than that against human breast carcinoma (MCF-7) cells. Moreover, the galactose moieties in the gum facilitated toxicity in HepG2 cells, thus exhibiting their potential as hepatic drug targeting vehicles.

5.2.2 PULLULAN-DRUG CONJUGATES

Pullulan (PU) is a linear polysaccharide produced by *aureobasidium pullulans* with maltotriosyl repeating units joined by α-(1,6)-linkages (Shingel, 2004). It has been examined for preparing conjugates with different drugs, especially for liver targeting of drugs. It has been stated that pullulan has the affinity towards hepatic ASGP-R, that expressed on the hepatocytes and its conjugates can be cell-internalized via receptor-mediated endocytosis (Kaneo et al., 2001; Tanaka et al., 2005). Therefore, pullulan could be an ideal material for designing active liver-targeting carriers. Some of them are discussed herein to gain appreciation of this biopolymer towards covalent conjugation. Sarika and colleagues (2015) adapted a complicated strategy for the conjugation of CUR to pullulan. Through a series of reaction steps, they developed galactosylated pullulan (GP)–succinylcurcumin (SC) conjugate (GPSC) for specific delivery of curcumin to hepatocarcinoma cells. The reaction schemes are illustrated in Figure 5.4. Pullulan was oxidized to pullulan aldehyde (PU-ALD) based on a reported procedure using sodium periodate (Bruneel and Schacht, 1993). The amount of aldehyde produced after oxidation was found to be 7.4×10^{-3} mol/g. The modification of lactobionic acid to lactobionic amine (LANH$_2$) followed the similar method as described by Haensler and Schuber (1988).

     The aldehyde group of PU-ALD was conjugated with amino group of ethylenediamine- modified lactobionic acid (LANH$_2$) by Schiff's base

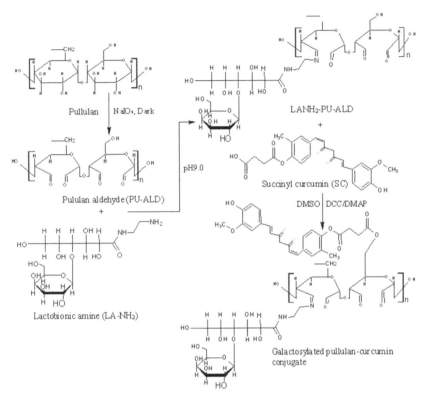

FIGURE 5.4 Schematic presentation of the reactions involved in the synthesis of galactosylated pullulan-curcumin conjugate.

reaction to obtain LANH$_2$-PU-ALD. Succinylcurcumin (SC) was also prepared by a method reported earlier with slight modification (Yang et al., 2012). Briefly, the phenolic hydroxyl group of the curcumin was modified with succinic anhydride to introduce acid functional groups. Lastly, SC was fixed to LANH$_2$-PU-ALD or PU and the un-reacted curcumin was removed after dialysis and freeze-dried to obtain GPSC. The synthesis of PU-SC did not involve oxidation and galactosylation reactions.

Galactosylated conjugates have already been shown to be effective for the delivery of drugs to hepatocarcinoma cells via asialoglycoprotein-mediated endocytosis (Tao et al., 2014; Zheng et al., 2012; Zhou et al., 2013). Galactose-terminal molecules are selectively recognized by asialoglycoprotein receptors on sinusoidal surface of the hepatocytes and transported to lysosomes inside the liver cells (Li et al., 2008). Thus, galactose-containing lactobionic acid was selected as the targeting ligands.

GPSC and PU-SC conjugates self-assembled in water with hydrodynamic diameters of 355±9 nm and 363±10 nm, respectively. Both conjugates showed spherical micellar morphology with enhanced CUR stability in physiological pH. Compared to PU-SC, the toxicity and cellular internalization of GPSC was predominant in HepG2 cells. Thus, the superior uptake of GPSC conjugate was said to occur via asialoglycoprotein receptor (ASGPR)-mediated endocytosis mechanism.

CUR displays an aqueous solubility of 2.8 µg/ml (Manju et al., 2011; Thomsen et al., 2011), insufficient to produce significant tumor-inhibition activity. The curcumin conjugation with $LANH_2$-PU-ALD and PU caused ~200-fold and 314-fold increase in solubility, respectively. Higher CUR release was estimated at acidic pH than that observed at pH 7.4. The faster release was attributed to the easy disruption of ester linkages, in addition to higher sensitivity of the Schiff's base formed between $LANH_2$ and PU-ALD (Xin et al., 2012). At both pH values, GPSC ensured higher CUR release than PU-SC. Indeed, this may be the consequence of greater rupture of relatively more hydrolysable groups present in GPSC conjugate.

HepG2 cells were exposed to both the conjugates and curcumin at different curcumin concentration over a range of 25–1.56 µg/ml for 24 h, and dose-dependent toxicity was found. At all the tested concentrations, GPSC elevated the toxicity to HepG2 cells compared to PU-SC. The antitumor activity of PU-SC conjugate was greatly enhanced after galactosylation, possibly due to increased uptake of GPSC via ASGPR mediated endocytosis (Craparo et al., 2013; Yang et al., 2011). IC50 value decreased from 11.7 µg/ml to 4.8 µg/ml after galactosylation and was lower than that of free CUR. $LANH_2$-PU-ALDwas nontoxic to HepG2 cells at all equivalent concentrations. Thus, at the same equivalent dose of CUR, GPSC conjugate was more effective than that of nongalactosylated PU-SC and supported the influence of galactose moiety in inducing elevated toxicity to HepG2cells.

Cell uptake of GPSC and PU-SC conjugates was performed in HepG2 cells for a period of 24 h to evaluate the impact of ASGPR in cellular internalization. The green fluorescence properties of curcumin allowed qualitative analysis of bio-distribution of the conjugates in HepG2 by confocal laser scanning microscopy (CLSM). The fluorescent intensity of CUR was very less in HepG2, compared to PU-SC and GPSC. Low fluorescence intensity of CUR could be due to its rapid metabolism in HepG2 cells. GPSC conjugate exhibited intense fluorescence in HepG2 cells than PU-SC, thus highlighting the importance of galactosylationin internalization and accumulation of CUR in HepG2 via ASGPR mediated pathway. Both galactosylated and nongalactosylated conjugate micelles were suitable for the selective delivery of curcumin towards hepatocarcinoma cells.

According to the report of Wang et al. (2013), pullulan can be modified to N-urocanylpullulan (URCP) and subsequently, URCP can be chemically bonded to methotrexate (MT) via DCC/DMAP schemes for the generation of MT-URCP nanoconjugates. MT was successfully conjugated to hydroxyl group of URCP via ester bond formation. The degree of urocanyl group substitution in URCP was 5.2% and MT content ranged from 7.5% to 25.0%. The MT-URCPs demonstrated amphiphilic characters due to hydrophilic pullulan backbone, and the hydrophobic urocanyl and MT moieties. They used this self-assembled micellar system for physical loading of another antiangiogenic drug, combretastatin A4 (CA4). MT-URCP nanoparticles had 187.1±15.2 nm size. However, the size of CA4/MT-URCP nanoparticles ranged between 200 and 250 nm.

According to the earlier reports (Jin et al., 2008; Kim et al., 2003; Wang et al., 2008), the urocanyl grafted group can exhibit pH-responsiveness owing to its imidazole ring system (the pK_a about 6.5). In weakly acidic tumors, urocanyl group are protonated and results in swelling of MT-URCP nanoparticles due to electrostatic charge repulsion. This contributed to the rapid release of CA4 from the nanoparticles at pH 6.0 (early endosome) and 5.0 (late endosome) than that observed at pH 7.4 phosphate buffered saline (PBS) solution. Only 25.6% CA4 was released at pH 7.4 over a 24 h period, but that reached upto 68.5% and 85.3% at pH 6.0 and pH 5.0, respectively. However, the release of MT became albeit relatively slower, depending on the disruption of the chemical bonds. Hence, two antitumor drugs exhibited different release mechanisms at the target site.

MT-URCP nanoparticles were readily internalized into human liver adenocarcinoma HepG2 cells as was evident from confocal microscopy. Therefore, MT-URCP nanoparticles was said to have a specific affinity for liver tumor cells. Moreover, a delayed inhibitory effect of MTX-URPA nanoparticles on cell proliferation was observed with regards to free MT, which may be due to sustained release of the drug. The bio-distribution of CA4 and MT in PLC/PRF/5 tumor-bearing nude mice was further investigated after IV administration of CA4/MT-URCP nanoparticles. CA4 concentrations in heart, liver, spleen, kidney, and lung, decreased during 0.5–4 h period. However, CA4 concentration in tumor tissues gradually increased within same duration, probably due to EPR effect of nanoparticles.

The tissue concentration of MT for CA4/MT-URCP nanoparticles was entirely different from free MT, reached highest value at 12h, and attained relatively high levels in the livers, spleens and tumors even at 24h post administration. In addition, the distributions of CA4 and MT were significantly higher in liver than in other tissues, suggesting strong liver targeting property of this polymeric drug carrier. Hence, CA4/MT-URCP nanoconjugates system can be used for combined delivery of drugs for hepatocellular carinoma therapy, which is not achievable with free CA4 and MT.

Doxorubicin (DOX) is a DNA interacting drug used for the treatment of ovarian, breast, prostate, cervix and lung cancers. However, its short half-life, GI toxicity and heart failure limit clinical application in addition to drug resistance (Al-Shabanah et al., 2000; Petit, 2004). The folic acid receptor constitutes a useful target for tumor-specific drug delivery because of its up-regulation in many human cancers, and higher density as the stage/grade of the cancer worsens (Lu and Low, 2002). Thus, the cancers, difficult to treat by classical methods may be targeted with folate-linked therapeutics. Some workers relied upon enzymatically degradable spacers and pH-sensitive hydrazone bonds for synthesizing pullulan-DOX conjugates (Lu et al., 2009; Nogusa et al., 2000). The multistep reactions are often involved in the design of suitable polymer-drug conjugates, which lead to low drug-loading capacity. A DOX content of only 3.18% has been reported by Lu et al. (2009). Thus, the improvement of drug loading capacity still constitutes a bottleneck problem. In this scenario, Zhang et al. (2011) developed folic acid (FA)–maleilated pullulan (MP)–DOX conjugate to enhance the therapeutic potential of DOX by reducing systemic side effects. The native polymer was converted into maleilated pullulan (MP), which contained vinyl carboxylic acid groups. DOX was then chemically conjugated to MP via primary amide bonds using EDC/NHS and TEA as coupling agents. Folic acid was finally bonded to pendent hydroxyl groups of MP to produce FA–MP–DOX. Excess reagents and un-conjugated DOX were removed by dialysis method. The MP precursor had a degree of substitution (DS) of 51, which was much larger than that reported by Lu et al. (2009). This laid down a solid foundation for conjugating DOX up to 10%. The DOX and FA content in the FA–MP–DOX conjugate were 8.9% and 1.7%, respectively. Both MP–DOX and FA–MP–DOX conjugates produced effective DOX concentrations of 10.5 mg/ml and 8.2 mg/ml, respectively due to their good water solubility. These data were similar to the water solubility of doxorubicin hydrochloride (10 mg/ml).

The DOX release from the MP–DOX conjugate and FA–MP–DOX conjugate reached 65% and 68.71% at pH 2.5; whereas the conjugates released only 24.45% and 26% drug at pH 7.4 in 30 h, respectively. The folate cojugation accelerated the DOX release rate. The ester bonds were hydrolyzed as time elapsed and lowered pH of the medium and then, speeded up the hydrolysis of amide bond. The conjugates selectively degraded and released the drug by sensing low intracellular pH in tumors, cell cytoplasm, endosomes (pH 5–6) or lysosomes (pH 4–5) (Fan et al., 2010; Tong and Cheng, 2007; Torchilin, 2007).

The cytotoxicity of the DOX-conjugates was evaluated on ovarian carcinoma A2780 cells. The cell viability decreased with increasing drug concentration upto 0.5 mg/l. Regardless of free DOX or MP–DOX conjugate, the FA–MP–DOX conjugates were more toxic against A2780 cells with an IC50 value of 0.036 mg/l. As FA–MP conjugate did not show any significant cytotoxic effects, it was suggested that the folate conjugation was responsible for enhanced cytotoxic potential of the FA–MP–DOX conjugates, rather than FA–MP backbone. Ovarian cancer cell line A2780 was an FA-receptor expressing cancer cell line. DOX-conjugates demonstrated much higher cellular uptake efficiency than the free drug. The superior cytotoxicity was apparently due to the increased cell-internalization efficiency of FA, and the increased retention time in cells by the EPR effect of MP.

The fluorescence microscopy revealed that the FA–MP–DOX conjugate was endocytosed via folate receptor-mediated mechanism in ovarian carcinoma A2780 cells. The free DOX were found mainly present within the nucleus, due to their low molecular weight. For nonfolated MP, the red fluorescence was observed probably in endosomes, which was observed much less in the nucleus. The conjugate showed an intense red fluorescence around the nucleus and the fluorescence was widely distributed in the cytoplasm. With the increase of incubation time from 0.5 h to 2 h, the cytoplasm fluorescence increased significantly and nuclei turned purple. This observation also supported that the FA–MP–DOX conjugate was endocytosed via a folate receptor-mediated mechanism.

Nogusa and team (1995) used a tetrapeptide spacer for DOX amino group attachment to the carboxyl group of carboxymethylpullulan (CMP), having DS value of 0.6. They used Gly-Gly-Phe-Gly, Gly-Phe-Gly-Gly, and Gly-Gly-Gly-Glytetrapeptide spacers. The DOX contents of the conjugates lied in the range of 6.1–7.1%. Self-assembled CMP-DOX nanoconjugates linked through Gly-Gly-Phe-Gly spacer released 35% DOX over 24h in PBS (pH7.4), enriched with rat liver lysosomal enzymes. CMP-DOX conjugate without spacer did not liberate any free DOX. Compared to DOX, the use of Gly-Gly-Phe-Gly and Gly-Phe-Gly-Gly spacers significantly suppressed the tumor growth in rats bearing Walker 256 carcinoma cells. The Gly-Gly-Gly-Gly spacer lowered the antitumor effect than DOX. No in vivo antitumor effect was seen for CMP-DOX conjugate devoid of spacer, even at a dose of 20 mg/kg.

5.2.3   DEXTRAN-DRUG CONJUGATES

Dextran is composed of α-(1,6) linked D-glucose units with varying branches depending on the dextran-producing bacterial strain. The biocompatibility, biodegradability, nonimmunogenic and nonantigenic properties qualify dextran as a physiologically harmless biopolysaccharide (Cadée et al., 2000). It undergoes depolymerization by dextranases occurring in liver, spleen, kidney, and lower GI tract. The prevalence of hydroxyl groups facilitates the introduction of drugs into the polymer backbone. This polysaccharide has been functionalized with various drugs, like naproxen (Harboe et al., 1989), daunorubicin (Levi-Schaffer et al., 1982), mitomycin C (Kojima et al., 1980), and cisplatin (Ohya et al., 2001) yielding efficient conjugates. Hornig et al.(2009) prepared hydrophobic dextran derivatives by esterification of the hydroxyl groups

with carboxylic acids of ibuprofen/naproxen via *in situ* activation with
N, N'-carbonyldiimidazole (CDI) in DMSO. Then, dextran was allowed
to react with activated drug solution for 24 h at 80°C. The maximum DS
values of 2.08 and 1.62, was attained for the ibuprofen and naproxen dextran esters, respectively. The nanoparticles were prepared by dialysis and
nano-precipitation techniques. The DS and the manufacturing techniques
strongly influenced the size, size distribution, and stability of the resulting nanoparticles. The high DS particle suspensions remained stable at
low electrolyte concentration in pH 4–11 over months. Ibuprofen esters
were more stable against hydrolysis and ionic strength than low substituted naproxen esters. The drug content of the dextran-ibuprofen esters
increased from 37% to 71% as DS was increased from 0.5 to 2.08. This
was 35.1–68.3% for naproxen conjugates as the DS was increased from
0.41 to 1.62. Smaller particle size (102–287 nm) was obtained with nanoprecipitation than dialysis technique (177–387 nm). Highly substituted
conjugates produced less uniform, larger particles. In an experiment, the
number average molar mass of ibuprofen-dextran esters with DS of 0.50
reduced from 38.3 to 1.3 kDa following dextranase treatment for 4.5h.
Thus, the conjugates were thought susceptible to enzymatic degradation
via dextranase, an essential prerequisite for their therapeutic application.
The low molar mass fractions became consequently more susceptible to
other enzymes like esterases and hydrolases (Vercauteren et al., 1990).
The natural molecular recognition of the biopolymer due to specific
receptors in certain cells, or the attachment of bio-specific ligands could
be advantageous to direct the particles for targeting the tissues or organs.

Ulcerative colitis invariably affects the rectum and may extend proxi-
mally in a confluent pattern to involve a part of or the entire colon (Friend,
2005; Knigge, 2002). Severe adverse effect profiles associated with the
administration of systemic glucocorticoids limit their application in ulcerative colitis (Rodriguez et al., 2001). A microbially triggered dextran-
drug system, stable in upper GI tract could constitute an effective colon
therapy because the system can release the drug only after hydrolysis by
colonic dextranases with reduced side effects (Sinha and Kumria, 2001).
This property coupled with its limited inter- and intra-species variations
in dextranase activity makes it a suitable carrier for colon specific drug
delivery (Lee et al., 2001; Mehvar, 2000). Keeping this in mind, Varshosaz

et al. (2009) designed dextran–budesonide conjugates using succinate as spacer for colon specific delivery.

Dextran conjugates were synthesized according to the method reported for dexamethasone (Pang et al., 2002). Briefly, a solution of triethylamine and dextran was added to the solution of budesonide (BSD)-21-hemisuccinate and 1,1′-carbonyldiimidazole in anhydrous dimethylsulfoxide (DMSO) and the reaction mixture was continued for 21 h at 25°C in tight-closed containers. Budesonide–dextran conjugates of varying molecular weights (10, 70 and 500 kDa) were prepared. DS values of 19.33±0.84, 14.29±0.41 and 11.60±0.37 mg/100 mg of were estimated for BSD-10, BSD-70 and BSD-500, respectively. The solubility was dependent on the MW of the conjugates. Highest solubility (53.552 mg/ml) was achieved with the BSD-70 conjugates, which corresponded to ~10.389 mg budesonide/ml and was higher than budesonide (0.04 mg/ml).

The drug release study was conducted in presence of rat gastric (pH4.4), small intestinal (pH7.4) and colonic contents (pH6.8), respectively (Lee et al., 2001; McLeod et al., 1993). Less than 10% of the drug was released in presence of gastric and intestinal contents. The drug release suddenly boosted up following transfer of the conjugates in the buffer containing rat caecal and colonic contents from simulated intestinal fluid. The drug release eventually decreased with increasing MW of the polymer. Only for BSD-70, the trend of drug release in presence of caecal and colonic contents was different to that released in the small intestinal contents. The hydrolysis of drug-spacer and polymer-spacer ester bonds was more pro-found in acidic media and as the pH increased, the drug release rate slowed down. Further, the drug release can be modulated by increasing MW of the polymer. The bulky structure of the polymer may hinder the hydrolysis of high MW conjugates. Of all the conjugates, BSD-70 showed great prom-ise for colon specific delivery of budesonide for the treatment of ulcerative colitis.

In a study by Ahmad et al. (2006), dextran dialdehydes were coupled with the α-amino ($-NH_2$) group of 5-ASA. A high degree of oxidation (93%) provided maximum conjugation with 5-ASA (49.1 mg/100 mg product) but stability to dextranase hydrolysis. Less oxidized dextrans (12%) conjugated proportionally less 5-ASA (15.1 mg/100 mg product) but were susceptible to dextranase enzyme hydrolysis, suggesting their potential for the treatment of Crohn's disease in the distal ileum and proximal colon. The efficiency of the 5-ASA conjugation was maximal at pH7.0. The molecular weight of dextran interfered with the overall drug release process because different molecular-weight dextrans (6–500 kDa) were hydrolyzed to the same extent, but at different rates.

The rate and extent of hydrolysis decreased as the degree of oxidation was increased. The dextran–5-ASA remained stable in 0.5 or 1.0 M HCl but partially hydrolyzed in 2.0 or 4.0 M HCl. The 5-ASA–dextran conjugates with greatest degrees of conjugation were most resistant to enzymatic hydrolysis. Hence, the degree of conjugation was also a controlling factor for dextranase hydrolysis of the polysaccharide. The native, oxidized and conjugated dextrans were stable to pancreatin enzyme hydrolysis. This confirmed that dextran–5-ASA conjugates would potentially be stable in the duodenum and jejunum. The results suggested that dextran-5-ASA conjugates could be a potential therapeutic tool for5-ASA in Crohn's disease.

The use of immunosuppressive agents has been associated with a significant degree of toxicity (Boitard and Bach, 1989). Recently, Mehvar and groups (1997, 1999, 2000) investigated the possible use of MP-dextran conjugates (70 kDa) for the reduction of drug toxicity. Such a conjugate is expected to accumulate in RES such as the liver and spleen, which are major organs responsible for the immune response. Previous studies have shown that the dextran-drug conjugates may be pharmacologically active either in the conjugated from or in nonconjugated form in vivo. For example, the pharmacologic effects of mitomycin C-dextran conjugates have been attributed to the released fraction of the drug (Kojima et al., 1980). Again, the tacrolimus conjugate showed in vitro immunosuppressive activity almost equal to that of free drug (Yura et al., 1999). Therefore, Rensberger et al. (2000) evaluated in vitro immunosuppressive activity of dextran-methylprednisolone succinate (DEX-MPS) conjugates, and compared with the free drug, using rat blood and spleen lymphocytes. DEX-MPS carried 8 mg MP per 100 mg conjugate. Lymphocyte proliferation assay indicated that DEX-MPS was substantially less effective than free MP at equivalent concentration with regards to immunomodulating effects. The low immunomodulation may be due to a direct effect of the conjugate and/partial release of free MP from DEX-MPS.

Previous studies have shown that MP is extracted from DEX-MPS at physiologic pH of 7.4 (Mehvar et al., 2000). At low pH (pH5.0–7.0), the rate of hydrolysis of DEX-MPS decreased by more than 20-fold (McLeod et al., 1993). At the beginning of the lymphocyte proliferation assay, the cells are incubated at pH7.4. However, the media pH becomes progressively acidic as cells proliferate. Therefore, it is likely that a small fraction of DEX-MPS is hydrolyzed during the incubation period, resulting in the low immunosuppressive activity of DEX-MPS. The immunomodulating effects observed at high in vitro concentration (50 nM) are likely attributed to the free MP released during the incubation period, not due to direct effects of the conjugate. IC_{50} (concentration producing half of maximum inhibition) was 3.03nM and 52.1 nM, respectively for MP and DEX-MPS for the inhibition of spleen lymphocytes, respectively. The same was 1.38 nM and 39.8 nM for inhibition of blood lymphocytes, respectively for MP and DEX-MPS.

The significant difference between the in vitro IC50 values of free and dextran-conjugated MP observed in this study was in contrast with the data obtained on dextran-tacrolimus conjugates (Yura et al., 1999). The tacrolimus conjugate by itself was almost as effective as the free drug. The discrepancy in the results may be thought as follows. Rensberger and groups used neutral dextran in their study. In contrast, a negatively charged carboxypentyl dextran was used by Yura et al. (1999) for conjugation. Henceforth, electric charge may affect the release of free drug and interaction of the conjugate with the receptor, if any. In addition, although succinic acid was used in both studies as a linkage between the drugs and dextran, the tacrolimus–dextran linkage also contained a carboxypentyl and an N-hydroxysuccinamide group in the spacer arm (Yura et al., 1999). The longer spacer arm for the tacrolimus–dextran conjugate, compared with DEX-MPS, may also contribute to the difference in the effects of the conjugates. Because the immunosuppressive activity on blood and spleen lymphocyte proliferation was minimal after conjugation, it becomes necessary that DEX-MPS should release MP in the body for *in vivo* efficacy.

The nanometric carriers can facilitate accumulation of the drug–polymer conjugate in the tumor tissues through EPR effect (Maeda et al., 1992). The prolonged blood circulation of hydrophilic nanocarriers would also facilitate their extravasations and passive targeting to the tumor tissue. It is well understood that small sized polymeric hydrogel nanoparticles can show extremely low RES uptake and remain highly stable in biological milieu as drug delivery vehicles (Gaur et al., 2000). Considering the extra benefit of such systems, Mitra et al. (2001) encapsulated the polyaldehyde dextran-doxorubicin conjugate (PAD–DOX) into chitosan nanoparticles. It was found that mole of glucosidic residues can bind 2–3 moles of DOX. Highly mono-dispersed, spherical particles of 100±10 nm diameter were obtained by microemulsion method. This reiterated the possibility of drug retention and delivery to solid tumor at the target site.

The relative efficacy of free drug, PAD–DOX and PAD–DOX-encapsulated chitosan nanoparticles was recorded over 90 days after inoculation of murine macrophage cells into Balb/c mice. The free drug, PAD–DOX, and conjugate-loaded nanoparticles were dosed at 8 mg/kg, 15 mg/kg and 15 mg/kg body weight (BW), respectively.

Treatment of tumor bearing mice with DOX alone resulted in slow, gradual increase of tumor size (466±25 mm^3) on day 45. Comparatively, the group treated with PAD–DOX as well as the conjugate-encapsulated particles showed initial increase in tumor volume up to 45 days of tumor implantation, and thereafter, gradual regression in mean tumor volume was noted. The group treated with conjugate-encapsulated particles showed faster regression in tumor volume (from 514±6 to 170±7.3 mm^3) compared to that of PAD–DOX conjugate (from 453.6±19.99 to 284±11.5 mm^3) on day 90-post implantation. Thus, PAD–DOX conjugates showed more regression effects after encapsulation. In addition, there was significant difference in survival rate following the two treatments. The animals treated with nanoparticles showed 50% survival rate, even up to 90 days while those with PAD–DOX conjugate had a survival rate of only 25% during the same timeframe.

Considering antitumor efficacy and survival time of tumor bearing mice, both the PAD–DOX conjugate and nanoparticles showed effective tumor regression and increased the life span of animals. The relatively better performance of the nanoparticles-treated group could be attributed to higher particle accumulation and gradual release of drug at the target tumor tissues. DOX was found to retain its antitumor activity even in the form of conjugate. The dextran conjugation effectively reduced the toxicity of free DOX, resulting in optimum drug efficacy.

Suprofen and flurbiprofen were individually reacted with carbonyldiimidazole to form acylimidazole, which in turn, formed drug-dextran conjugates after reaction with dextran. The mean degree of substitution of flurbiprofen and suprofen was between 8–9.5% and 7.5–9.0%, respectively. The analgesic activity of flurbiprofen-dextran conjugate (FD) and suprofen-dextran conjugate (SD) (64.23% and 41.50%, respectively) was comparable to that of the parent drugs-flurbiprofen (72.60%) and suprofen (44.30%). The deep ulceration, swelling and intense perforation in the gastric mucosa were seen after 7 days of administration of flurbiprofen and suprofen in albino rats. The ulcer indices were reduced from of 29.69 to 5.88 and 31.0 to 6.06, respectively for FD and SD. Shrivastava et al. (2009) recommended dextran as a promoiety or carrier for the delivery of flurbiprofen and suprofen with lower ulcer indices.

Vyas et al. (2007) tested another NSAID, ketorolac for making dextran ester conjugates using N-acylimidazole spacer. In vivo biological screening in mice and rats indicated that the conjugates retained analgesic and anti-inflammatory activities with significantly reduced GI side effects compared to the parent drug.

A liver-selective antiviral prodrug was developed by Chimalakonda et al. (2007). Lamivudine (LV) was coupled to dextran (25 kDa) using succinate linker. The degree of drug substitution was 6.5% for the conjugate. The conjugate slowly released LV in presence of rat liver lysosomes, but stopped releasing the drug in the corresponding buffer. In rat model, the dextran conjugation resulted in 40- and 7-fold lowering of the clearance and volume of distribution of the drug, respectively. However, the accumulation of the conjugated drug in the liver was 50-fold higher than that of the parent drug. The high liver accumulation of the conjugate was associated with the sustained discharge of LV in organ and thus, showed its potential for liver-specific drug delivery.

Vittorio et al. (2014) reported that catechin-dextran conjugates with commercial magnetic nanoparticles (Endorem) can increase intracellular concentration of the drug and can induce apoptosis in 98% of pancreatic tumor cells, driven by magnetic field.

5.2.4 PECTIN-DRUG CONJUGATES

Pectin, a natural, water-soluble anionic polysaccharide has been studied as a vehicle material for drug delivery due to its good biocompatibility. The basic pectin backbone comprises a linear chain of α-(1,4)-linked D-galacturonic acid, referred to as homogalacturonan (Corredig et al., 2000; Ridley et al., 2001; Willats et al., 2001). There is an abundance of carboxyl and hydroxyl groups on the α-D-galacturonic acid residues. The carboxyl groups of pectin can be condensed with the amino groups of drugs through dehydration to yield pectin–drug conjugates with high drug-loading efficiency. Furthermore, the amide linkage of the conjugates is easily hydrolysable by lysosomal enzymes. Therefore, pectin seems a good polymer for conjugation with anticancer drugs to improve tumor-targeting properties.

Adriamycin (AM) (doxorubicin) as an antitumor agent can induce severe toxic and side effects, such as bone marrow depression, nephrotoxicity, and hepatotoxicity, together with cardiac and GI tract complications. This is further accompanied by multidrug resistance and short half-life *in vivo* (Janes et al., 2001). The polymer conjugation is an approach, which can improve its activity, and decrease the unwanted effects. Tang et al. (2010) modified low-methoxy pectin with AM linking using EDC chemistry to give a pectin–adriamycin (PAM) conjugate. The structure of adriamycin and pectin-adriamycin conjugate are given in Figure 5.5.

At pH <3.5, pectin forms a hydrogel through hydrophobic interaction and hydrogen bonding (Lutz et al., 2009) and interferes with the conjugation process. In order to avoid gel formation during synthesis, the pectin solution was adjusted to pH7.0 before AM and EDC were charged. Because the conjugate was hydrophobic, it was formulated as suspension (~2 mg drug/ml) by a microfluidizer. A *z*-average diameter of 126 nm can avoid trapping of particles via RES and renal clearance and further enables PAM to passively target cancer sites via the EPR effect (Chawla and Amiji, 2003; Hiroshi, 2001; Moghimi et al., 2001; Singh and Lilard, 2009). The drug release from PAM was less than 7.3% in PBS at pH 7.4, pH 5.0, and plasma. However, lysosomal enzymes accelerated the drug release rate up to 33.2%. In the first 15 h, a major fraction of the drug released and after

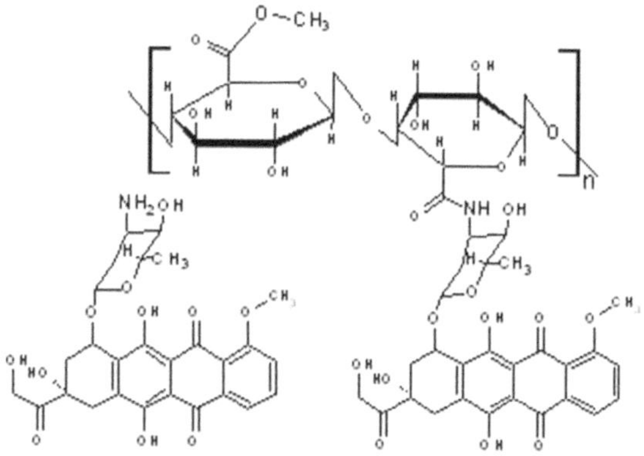

Adriamycin   Pectin-Adriamycin conjugate

FIGURE 5.5   Structures of adriamycin (doxorubicin) and pectin-adriamycin conjugate.

24 h, the release rate declined slowly, probably due to the degradation of AM and inactivation of the lysosomal enzymes. Overall, the amount of AM released from PAM was 6-fold greater than that of the negative control in presence of lysosomes. Therefore, PAM is expected to remain for a long time in systemic circulation, and that, free drug will be released in lysosomes after internalization by tumor cells (Maeda, 2001; Moghimi et al., 2001; Singh and Lillard, 2009).

They also compared intracellular localization of PAM and AM to reveal the mechanism of cytotoxicity of PAM. In majority of A549 cells, incubated with PAM, the red fluorescence was distributed uniformly in the cytoplasm, but in a few cells, the red fluorescence appeared in the nucleus. In contrast, the red fluorescence accumulated in the nucleus with minimal staining of the cytoplasm in A549 cells, treated with free AM. This finding was consistent with the previous report (Dreher et al.,´ 2003; Janes et al., 2001). The results suggested that PAM was successfully internalized by A549 cells.

The accumulation of PAM in tissues at 2 h post injection into mouse assumed the following order: lung > liver > spleen > kidney > tumor > heart. However, the same was found for AM as follows: liver > spleen > lung > tumor > heart > kidney. Thus, the extent of PAM accumulation was more than AM in the tumor cells.

PAM inhibited the growth of B16 cells slightly less than AM or the mixture of AM and pectin over the concentration range equivalent to 0.125–1.000 µg/ml. Moreover, at a concentration equivalent to 2 µg/ml, PAM inhibited the growth of B16 cells by 72.5%, which was less than AM (85.9%) or the mixture of AM and pectin (84.2%). The anomalies can be explained by the fact that endocytosis is a much slower internalization process than simple diffusion. Only 33.2% of AM was released from PAM in fluids containing lysosomal enzymes in 30 h. Therefore, relatively higher concentration of PAM must be established outside the cell to produce the same intracellular effect as the free drug (Khandare and Minko, 2006). PAM showed satisfactory dose-dependent antitumor activities against A549, B16, and 2780cp cells in vitro. The conjugate showed good therapeutic effect on pulmonary metastasis of melanoma (B16 cell line) in C57BL/6 mice and prolonged survival time of the mice remarkably than the free drug. In conclusion, PAM is a potential candidate for development as polymer-drug nano-conjugates for tumor targeting in cancer therapy.

The 5-fluorouracil-pectin (FU-P) conjugate has also been tested for colon-targeted delivery (Wang et al., 2007). 5-FU-P was administered to rats by oral administration at a dose of 22.5 mg/ kg. The different parts of GI tract and plasma were taken at different hours, and the concentration of 5-FU-1-acetic acid in them was estimated. 5-FU-1-acetic acid released from 5-FU-P was mainly localized in the cecum and colon; therefore, the conjugate exhibited a good colon-targeting property.

Verma and Sachin (2008) constructed easily redispersible, self-organized pectin-cisplatin nano-structures (~100 nm) for long blood circulation. Zeta potential data showed the shielding effect on the negative potential of pectin that was about 7 times more than the pectin chains when conjugated with cisplatin. The pectin-cisplatin conjugate system was stable in plasma, and sustained the release of drug without any burst effect. The plasma proteins facilitated drug release (85–89% in 17 days), and only 57% of drug released in 30 days without plasma. The reduced negative charge on the conjugate helped in surface adhesion and subsequent uptake by cells on mouse macrophage cell line (J-774). Nano-conjugates showed long blood retention profile in mice and the cisplatin was traced in circulation even after 24 h. Pharmacokinetic study indicated good efficacy and safety profile of the conjugate than cisplatin.

They further demonstrated that pectin-cisplatin nano-conjugates enhanced the plasma half-life of cisplatin to 11.26 h (Verma et al., 2012). A negligible accumulation of cisplatin was found in kidney that can reduce undesirable nephrotoxic effects of the drug. The conjugates significantly delayed tumor growth and improved survival in tumor bearing C57Bl6 mice. A 3-fold reduction in tumor volume was envisaged, indicating the augmentation of cisplatin activity after pectin conjugation.

The lymphatic targeting potential of pectin-adriamycin conjugates was explored by Cheng et al.(2009). The pectin-adriamycin ester conjugates (PA) was stable in normal saline, but gradually enzymolyzed to release adriamycin in blood plasma and lymph nodes. The results of lymphatic targeting study of PA, demonstrated that the degree of esterification or drug coupling capacity of pectin significantly influenced the lymphatic targeting characteristics of PA. The adriamycin concentration of lymph nodes was 208 times higher than that of plasma after local injection of the PA, of which the adriamycin content was 27.9%.

In addition to these biopolymers, some other biopolymer-based nano-conjugates have also been investigated for their potential use in drug delivery. The drug delivery applications of different nanoconjugates are summarized in Table 5.1.

5.3 CONCLUSION

Although several polymer–drug conjugates have shown considerable promise in clinical trials over the last two decades, the translation into the clinic is the major challenge of the field. The problems such as polymer-related toxicity; limited drug loading, variable drug releasing trends; process validation, purification of conjugates in multistep process, regulatory and pharmacoeconomic considerations are the major hurdles for further development. In this chapter, bio-conjugates of water-soluble hydrophilic polysaccharides with low MW drugs have been discussed. The remarkable progress in imaging techniques that permits noninvasive monitoring of the

TABLE 5.1 Drug Delivery Applications of Different Biopolymer-Based Nanoconjugates

Polysaccharide	Drug	Drug Delivery Attributes	References
Chitosan (CS)	Atorvastatin (AT)	• CS–AT nano-conjugate enhanced about 100-fold solubility compared to pure AT • Sustained release of AT from the conjugate • Significant reduction of acidic degradation of AT • Oral administration to rat exhibited nearly 5-fold increase in bioavailability compared with AT suspension	Anwar et al. (2011)
	Norcantharidin (NCTD)	• NCTD-CS conjugates released <6% drug through the hydrolysis of ester bonds in phosphate-buffered solution (pH 5.0 and 7.4) within 16 days • Low cytotoxic to MGC80–3 cells than that of NCTD • Conjugates arrested MGC80–3 cell cycle at G2/M phase and induced cell death via cell apoptosis similarly to NCTD	Xu et al. (2013)
	Isoniazid (INH)	• Comparable or slightly higher minimum inhibitory concentration against *Mycobacterium tuberculosis* for N-(2-carboxyethyl) chitosan (CEC)-INH, and N-(3-chloro-2-hydroxypropyl) chitosan (CHPC)-INH conjugates than INH • Lower toxicity for the conjugates than for INH by a factor of 3–4 • Studies of acute toxicity in mice revealed a 3–4-fold increase in LD50 for the conjugates compared with INH	Berezina and Skorik (2015)
Carboxymethyl Chitosan (CMCS)	Methotrexate (MTX)	• CMCS-MTX nanoparticles exhibited significant sustained-release behaviors in PBS buffer solution (pH 4.0, 7.2 and 9.0) • Nanoparticles had good in vitro stability and exhibited potential to be used as a novel drug carrier system.	Wang et al. (2011)

TABLE 5.1 (Continued)

Polysaccharide	Drug	Drug Delivery Attributes	References
	Docetaxel (DTX)	• CMCS–DTX conjugates showed good stability in plasma (only 12.46% of DTX was released after incubation in plasma for 48 h) • Significantly cytotoxic against B16 and HepG2 cells • Displayed better antitumor effect than Duopafei® by inhibiting tumor growth and prolonging the survival time of B16 melanoma bearing mice more effectively • Excellent safety profile with a maximum tolerated dose of >250 mg/kg in mice, which was more than 4 fold higher than that of Duopafei®	Liu et al. (2013)
	Melphalan (MEL)	• 40–75% MEL was released in papain and lysosomal enzymes while only about 4–5% was released in plasma, which suggested good plasma stability • Conjugates had 52–70% cytotoxicity against RPMI8226 cells as compared with free MEL, indicating the conjugates did not lose anticancer activity of MEL.	Lu et al. (2013)
Heparin	Doxorubicin (DOX)	• Deoxycholic acid-heparin conjugate had markedly low anticoagulant activity • Conjugate prevented squamous cell carcinoma • DOX-loaded heparin conjugates displayed sustained drug release patterns	Park et al. (2006)
	All-transretinoid acid (ATRA), Paclitaxel (PTX)	• Amphiphilic conjugate had markedly lower anticoagulant activity • Nanoparticles had 93.1% PTX entrapment efficiency	Hou et al. (2011)

TABLE 5.1 (Continued)

Polysaccharide	Drug	Drug Delivery Attributes	References
		• Conjugate was safe material for intravenous administration. • PTX-loaded nanoparticles contributed to an extended circulation of PTX and ATRA and provided opportunity for simultaneous delivery of multiple anticancer drugs	
	Doxorubicin	• Faster drug release rate at pH 5.0 and slow release rate at pH 7.4 indicated pH-sensitive property • Nanoparticles resulted in strong antitumor activity, high antiangiogenesis effects and induced apoptosis on the 4T1 breast tumor model	She et al. (2013)
Hyaluronic acid (HA)	Exendin 4	• HA-exendin conjugates resulted in about 20 times improved in vitro serum stability • Maintained hypoglycemic and glucoregulatory bioactivities of exendin • Excellent glucose-lowering capabilities in type 2 diabetic mice demonstrating protracted hypoglycemic effect upto 3 days after a single subcutaneous injection	Kong et al. (2010)
	Quercetin (QT)	• HA-QT bioconjugates exhibited significant sustained and pH-dependent drug release behaviors without dramatic initial burst • HA-QT micelles were 4 times cytotoxic on MCF-7 cells than free QT • About 20-fold increase in half-life and 5-fold increase in the area-under-the-curve (AUC) of quercetin were achieved for the HA-QT micelles compared with QT • Excellent inhibitory effect on tumor growth in H22 tumor-bearing mice • Hemolytic toxicity and vein irritation assay suggested safe and potent drug delivery system for targeted antitumor therapy	Pang et al. (2014)

TABLE 5.1 (Continued)

Polysaccharide	Drug	Drug Delivery Attributes	References
	Paclitaxel	• HA-PTX was significantly accumulated in tumors and caused 4-fold decrease in tumor volume on day 14 in contrast to PTX alone • The conjugates could enter cells, bypass the lysosomal–endosomal system and improve PTX delivery	Xu et al. (2015)
	Camptothecin (CPT) and DOX	• CPT- and DOX-conjugated HA (CPT–HA–DOX) elicited significant tumor reduction of murine 4T1 breast cancer model after IV administration • Due to potent synergy between CPT and DOX, only low doses of each drug were required to significantly reduce tumor volumes without any visible toxicity in healthy organs	Camacho et al. (2015)
Gum arabic	Curcumin	• Significant enhancement of solubility (900 fold) and stability of curcumin in physiological pH • Conjugate micelles exhibited higher anticancer activity in human hepatocellular carcinoma (HepG2) cells than in human breast carcinoma (MCF-7) cells • Conjugate exhibited enhanced accumulation and toxicity in HepG2 cells due to the targeting efficiency of galactose groups present in gum Arabic	Sarika et al. (2015)

fate of conjugates will undoubtedly contribute to a more rational design of polymer therapeutics. Safety and efficacy of polymer-drug conjugates, as with other traditional therapeutics, is obviously of the utmost concern during the drug development process.

A careful consideration must be given to polymer-drug conjugates because bio-distribution and pharmacokinetic patterns are frequently altered. Further, the release of desired amount of drug at the targets is a prerequisite for eliciting pharmacological response. This is advantageous as a conjugate will be mostly inactive during systemic transport. The overall safety profile of the conjugate will be in trouble, if premature drug release occurs. Therefore, a critical balance between conjugate stability and drug release directly impacts safety and efficacy. Further study is necessary for the structure activity relationship of drugs in conjugated form with the polysaccharides, effect of steric hindrance during chemical conjugation, and enhanced reactivity of spacer-polymers. The literature reports on polysaccharide-based nanoconjugate systems are encouraging due to their exceptional biocompatibility, biodegradability, easy chemical modifications and sometimes, the inherent presence of sugar moieties in the polysaccharide for active targeting. The success of this new bio-polymeric drug delivery system mostly relies on interdisciplinary research efforts.

KEYWORDS

- alginate
- dextran
- drug delivery
- pectin
- polymer-drug conjugates
- polysaccharides
- pullulan

REFERENCES

Aggarwal, B. B.; Kumar, A.; Bharti, A. C. Anticancer potential of curcumin: Preclinical and clinical studies. *Anticancer Res.* **2003**, *23*, 363–398.

Ahmad, S.; Tester, R. F.; Corbett, A.; Karkalas, J. Dextran and 5-aminosalicylic acid (5-ASA) conjugates: synthesis, characterization and enzymic hydrolysis. *Carbohydr. Res.* **2006**, *341*, 2694–2701.

Al-Shamkhani, A.; Duncan, R. Synthesis, controlled release properties and antitumor activity of alginate-*cis*-aconityl-daunomycinconjugates. *Int. J. Pharm.* **1995**, *122*, 107–119.

Al-Shabanah, O. A.; El-Kashef, H. A.; Badary, O. A.; Al-Bekairi, A. M.; Elmazar, M. M. A. Effect of streptozotocin-induced hyperglycaemia on intravenous pharmacokinetics and acute cardiotoxicity of doxorubicin in rats. *Pharmacol. Res.* **2000**, *41*, 31–37.

Alexis, F.; Rhee, J. W.; Richie, J. P.; Moreno, A. F. R.; Langer, R.; Farokhzad, O. C. New frontiers in nanotechnology for cancer treatment. *Urol. Oncol.* **2008**, *26*, 74–85.

Anand, P.; Kunnumakkara, A. B.; Newman, R. A.; Aggarwal, B. B. Bioavailability of curcumin: Problems and promises. *Mol. Pharm.* **2007**, *4*, 807–818.

Anwar, M.; Warsi, M. H.; Mallick, N.; Akhter, S.; Gahoi, S.; Jain, G. K.; Talegaonkar, S.; Ahmad, F. J.; Khar, R. K. Enhanced bioavailability of nano-sized chitosan–atorvastatin conjugate after oral administration to rats. *Eur. J. Pharm. Sci.* **2011**, *44*, 241–249.

Azzam, T.; Eliyahu, H.; Shapira, L.; Linial, M.; Barenholz, Y.; Domb, A. J. Polysaccharide–oligoamine-based conjugates for gene delivery. *J. Med. Chem.* **2002**, *45*, 1817–1824.

Berezina, A. S.; Skorik, Y. A. Chitosan-isoniazid conjugates: Synthesis, evaluation oftuberculostatic activity, biodegradability and toxicity. *Carbohydr. Polym.* **2015**, *127*, 309–315.

Boitard, C.; Bach, J. F. Long-term complications of conventional immunosuppressive treatment. *Adv. Nephrol. Necker. Hosp.* **1989**, *18*, 335–354.

Bruneel, D.; Schacht, E. Chemical modification of pullulan: 1. Periodate oxidation. *Polymer* **1993**, *34*, 2628–2632.

Cadée, J. A. M.; Van Luyn, J. A.; Brouwer, L. A.; Plantinga, J. A.; Van Wachem, P. B.; De Groot, C. J.; Den Otter, W.; Hennink, W. E. In vivo biocompatibility of dextran-based hydrogels. *J. Biomed. Mater. Res.* **2000**, *50*, 397–404.

Caliceti, P.; Salmaso, S.; Bersani, S. Polysaccharide-Based Anticancer Prodrugs. In *Macromolecular Anticancer Therapeutics*; Harivardhan, R. L.; Couvreur, P.; Eds.; Humana Press, New York, 2010; pp. 163–220.

Camacho, K. M.; Kumar, S.; Menegatti, S.; Vogus, D. R.; Anselmo, A. C.; Mitragotri, S. Synergistic antitumor activity of camptothecin–doxorubicin combinations and their conjugates with hyaluronic acid. *J. Control Release.* **2015**, *210*, 198–207.

Chau, Y.; Dang, N. M.; Tan, F. E.; Langer, R. Investigation of targeting mechanism of new dextran-peptide-methotrexate conjugates using biodistribution study in matrix-metalloproteinase-overexpressing tumor xenograft model. *J. Pharm. Sci.* **2006**, *95*, 542–551.

Chawla, J. S.; Amiji, M. M. Cellular uptake and concentrations of tamoxifen upon administration in poly(epsilon-caprolactone) nanoparticles. *AAPS PharmSci.* **2003**, *5*, 28–34.

Cheng, M.; Xie, P.; Tang, X.; Zhang, J.; Xie, Y.; Zheng, K.; He, J. Preparation and lymphatic targeting research of targeting antitumor drug: pectin-adriamycinconjugates. *J. Biomed. Engg.* **2009**, *26*, 569–574.

Chimalakonda, K. C.; Agarwal, H. K.; Kumar, A.; Parang, K.; Mehvar, R. Synthesis, analysis, in vitro characterization, and in vivo disposition of a lamivudine-dextran conjugate for selective antiviral delivery to the liver. *Bioconjug. Chem.* **2007**, *18*, 2097–2108.

Chytil, P.; Etrych, T.; Konak, C.; Sirova, M.; Mrkvan, T.; Rihova, B.; Ulbrich, K. Properties of HPMA copolymer–doxorubicin conjugates with pH-controlled activation: effect of polymer chain modification. *J. Control Release.* **2006**, *115*, 26–36.

Collins, I.; Workman, P. New approaches to molecular cancer therapeutics. *Nat. Chem. Biol.* **2006**, *2*, 689–700.

Corredig, M.; Kerr, W.; Wicker, L. Molecular characterization of commercial pectins by separation with linear mix gel permeation columns in-line with multi-angle light scattering detection. *Food Hydrocolloids*, **2000**, *14*, 41–47.

Craparo, E. F.; Triolo, D.; Pitarresi, G.; Giammona, G.; Cavallaro, G. Galactosylated micelles for a ribavirin prodrug targeting to hepatocytes. *Biomacromolecules*, **2013**, *14*, 1838–1849.

Crepon, B.; Jozefonvicz, J.; Chytry, V.; Rihova, B.; Kopecek, J. Enzymatic degradation and immunogenic properties of derivatizeddextrans. *Biomaterials*, **1991**, *12*, 550–554.

Danhauser-Riedl, S.; Hausmann, E.; Schick, H. D.; Bender, R.; Dietzfelbinger, H.; Rastetter, J.; Hanauske, A. R. Phase-I clinical and pharmacokinetic trial of dextran conjugated doxorubicin (Ad-70 Dox-Oxd). *Invest. New Drugs*, **1993**, *11*, 187–195.

Davis, M. E.; Chen, Z. G.; Shin, D. M. Nanoparticle therapeutics: An emerging treatment modality for cancer. *Nat. Rev. Drug Discov.* **2008**, *7*, 771–782.

Dey, S.; Sreenivasan, K. Conjugation of curcumin onto alginate enhances aqueous solubility and stability of curcumin. *Carbohydr. Polym.* **2014**, *99*, 499–507.

Dreher, M. R.; Raucher, D.; Balu, N.; Colvin, O. M.; Ludeman, S. M.; Chilkoti, A. Evaluation of an elastin-like polypeptide-doxorubicin conjugate for cancer therapy. *J. Control. Release*, **2003**, *91*, 31–43.

Duncan, R. Drug–polymer conjugates: potential for improved chemotherapy. *Anticancer Drugs*, **1992** *3*, 175–210.

Duncan, R. Polymer conjugates as anticancer nanomedicines. *Nat. Rev. Cancer*, **2006**, *6*, 688–701.

Duncan, R.; Ringsdorf, H.; Satchi-Fainaro, R. Polymer therapeutics-polymers as drugs, drug and protein conjugates and gene delivery systems: Past, present and future opportunities. *J. Drug Target.* **2006**, *14*, 337–341.

Fan, L.; Li, F.; Zhang, H. T.; Wang, Y. K.; Cheng, C.; Li, X. Y.; Gu, C. H.; Yang, Q.; Wu, H.; Zhang, S. Y. Co-delivery of PDTC and doxorubicin by multifunctional micellar nanoparticles to achieve active targeted drug delivery and overcome multidrug resistance. *Biomaterials* **2010**, *31*, 5634–5642.

Friend, D. R. New oral delivery systems for treatment of inflammatory bowel disease. *Adv. Drug Deliv. Rev.* **2005**, *57*, 247–265.

Gaur, U.; Sahoo, S. K.; De, T. K.; Ghosh, P. C.; Maitra, A. N.; Ghosh, P. K. Biodistribution of fluoresceinated dextran using novel nanoparticles evading reticuloendothelial system. *Int. J. Pharm.* **2000**, *202*, 1–10.

Godwin, A.; Bolina, K.; Clochard, M.; Dinand, E.; Rankin, S.; Simic, S.; Brocchini, S. New strategies for polymer development in pharmaceutical science-A short review. *J. Pharm. Pharmacol.* **2001**, *53*, 1175–1184.

Greco, F.; Vicent, M. J. Polymer-drug conjugates: Current status and future trends. *Front. Biosci.* **2008**, *13*, 2744–2756.

Greenwald, R. B.; Choe, Y. H.; McGuire, J.; Conover, C. D. Effective drug delivery by PEGylated drug conjugates. *Adv. Drug Deliv. Rev.* **2003**, *55*, 217–250.

Grodzinski, J. J. Biomedical application of functional polymers. *J. React. Funct. Polym.* **1999**, *39*, 99–138.

Haensler, J.; Schuber, F. Preparation of neo-galactosylated liposomes and their interaction with mouse peritoneal macrophages. *Biochim. Biophys. Acta,* **1988,** *946,* 95–105.

Harboe, E.; Larsen, C.; Johansen, M.; Olesen, H. P. Macromolecular Prodrugs. XV. Colon-targeted delivery–bioavailability of naproxen from orally administered dextran-naproxen ester prodrugs varying in molecular size in the pig. *Pharm. Res.* **1989,** *6,* 919–923.

Hashida, M.; Kato, A.; Takakura, Y.; Sezaki, H. Disposition and pharmacokinetics of a polymeric prodrug of mitomycin C, mitomycin C-dextran conjugate, in the rat. *Drug Metab. Dispos.* **1984,** *12,* 492–499.

Hirai, M.; Minematsu, H.; Hiramatsu, Y.; Kitagawa, H.; Otani, T.; Iwashita, S.; Kudoh, T.; Chen, L.; Li, Y.; Okada, M.; Salomon, D. S.; Igarashi, K.;Chikuma, M.; Seno, M. Novel and simple loading procedure of cisplatin into liposomes and targeting tumor endothelial cells. *Int. J. Pharm.* **2010,** *391,* 274–283.

Hornig, S.; Bunjes, H.; Heinze, T. Preparation and characterization of nanoparticles-based on dextran–drug conjugates. *J. Colloid Interface Sci.* **2009,** *338,* 56–62.

Hou, L.; Fan, Y.; Yao, J.; Zhou, J.; Li, C.; Fang, Z.; Zhang, Q. Low molecular weight heparin-all-trans retinoid acid conjugate as a drug carrier for combination cancer chemotherapy of paclitaxel and all-trans retinoid acid. *Carbohydr. Polym.* **2011,** *86,* 1157–1166.

Janes, K. A.; Fresneau, M. P.; Marazuela, A.; Fabra, A.; Alonso, M. J. Chitosan nanoparticles as delivery systems for doxorubicin. *J. Control. Release,* **2001,** *73,* 255–267.

Jin, H.; Xu, C. X.; Kim, H. W.; Chung, Y. S.; Shin, J. Y.; Chang, S. H.; Park, S. J.; Lee, E. S.; Hwang, S. K.; Kwon, J. T.; Minai-Tehrani, A.; Woo, M.; Noh, M. S.; Youn, H. J.; Kim, D. Y.; Yoon, B. I.; Lee, K. H.; Kim, T. H.; Cho, C. S.; Cho, M. H. Urocanic acid modified chitosan-mediated PTEN delivery via aerosol suppressed lung tumorigenesis in K-ras (LA1) mice. *Cancer Gene Ther.* **2008,** *15,* 275–283.

Kaneo, Y.; Tanaka, T.; Nakano, T.; Yamaguchi, Y. Evidence for receptor-mediated hepatic uptake of pullulan in rats. *J. Control. Release,* **2001,** *70,* 365–373.

Khandare, J.; Minko, T. Polymer-drug conjugates: Progress in polymeric prodrugs. *Prog. Polym. Sci.* **2006,** *31,* 359–397.

Kim, T. H.; Ihm, J. E.; Choi, Y. J.; Nah, J. W.; Cho, C. S. Efficient gene delivery by urocanic acid-modified chitosan. *J. Control. Release,* **2003,** *93,* 389–402.

Knigge, K. L. Inflammatory bowel disease. *Clin. Cornerstone,* **2002,** *4,* 49–57.

Kojima, T.; Hashida, M.; Muranishi, S.; Sezaki, H. Mitomycin C-dextran conjugate: A novel high molecular weight prodrug of mitomycin C. *J. Pharm. Pharmacol.* **1980,** *32,* 30–34.

Kong, J.-H.; Oh, E. J.; Chae, S. Y.; Lee, K. C.; Hahn, S. K. Long acting hyaluronate-exendin 4 conjugate for the treatment of type 2 diabetes. *Biomaterials.* **2010,** *31,* 4121–4128.

Kopecek, J.; Kopeckova, P.; Minko, T.; Lu, Z. R. HPMA copolymer–anticancer drug conjugates: Design, activity, and mechanism of action. *Eur. J. Pharm. Biopharm.* **2000,** *50,* 61–81.

Lammers, T.; Hennink, W. E.; Storm, G. Tumor-targeted nano-medicines: Principles and practice. *Br. J. Cancer,* **2008,** *99,* 392–397.

Langer, R. Drug delivery and targeting. *Nature,* **1998,** *392,* 5–10.

Lee, J. S.; Jung, Y. J.; Doh, M. J.; Kim, Y. M. Synthesis and Properties of Dextran–Nalidixic Acid Ester as a Colon-Specific Prodrug of Nalidixic Acid. *Drug Dev. Ind. Pharm.* **2001**, *27*, 331–336.

Levi-Schaffer, F.; Bernstein, A.; Meshorer, A.; Arnon, R. Reduced toxicity of daunorubicin by conjugation to dextran. *Cancer Treat. Rep.* **1982**, *66*, 107–114.

Li, C.; Wallace, S. Polymer–drug conjugates: recent development in clinical oncology. *Adv. Drug. Deliv. Rev.* **2008**, *60*, 886–898.

Li, S. D.; Huang, L. Pharmacokinetics and biodistribution of nanoparticles. *Mol. Pharm.* **2008**, 5, 496–504.

Li, Y.; Huang, G.; Diakur, J.; Wiebe, L. I. targeted delivery of macromolecular drugs: asialoglycoprotein receptor (ASGPR) expression by selected hepatomacell lines used in antiviral drug development. *Curr. Drug Deliv.* **2008**, *5*, 299–302.

Lipinski, C. Poor aqueous solubility – An industry wide problem in drug discovery. *Am. Pharm. Rev.* **2002**, *5*, 82–85.

Liu, F.; Feng, L.; Zhang, L.; Zhang, X.; Zhang, N. Synthesis, characterization and antitumor evaluation of CMCS–DTX conjugates as novel delivery platform for docetaxel. *Int. J. Pharm.* **2013**, *451*, 41–49.

Lu, B.; Huang, D.; Zheng, H.; Huang, Z.; Xu, P.; Xu, H.; Yin, Y.; Liu, X.; Li, D.; Zhang, X. Preparation, characterization, and in vitro efficacy of O-carboxymethyl chitosan conjugate of melphalan. *Carbohydr. Polym.* **2013**, *98*, 36– 42.

Lu, D. X.; Wen, X. T.; Liang, J.; Zhang, X. D.; Gu, Z. W.; Fan, Y. J. A. pH-sensitive nanodrug delivery system derived from pullulan/doxorubicin conjugate. *J. Biomed. Mater. Res.* **2009**, *B89*, 177–183.

Lu, Y.; Low, P. S. Folate-mediated delivery of macromolecular anticancer therapeutic agents. *Adv. Drug Deliv. Rev.* **2002**, *54*, 675–693.

Lutz, R.; Aserin, A.; Wicker, L.; Garti, N. Structure and physical properties of pectins with block-wise distribution of carboxylic acid groups. *Food Hydrocolloids*, **2009**, *23*, 786–794.

Maeda, H. SMANCS and polymer-conjugated macromolecular drugs: Advantages in cancer chemotherapy. *Adv. Drug Deliv. Rev.* **2001**, *46*, 169–185.

Maeda, H.; Seymour, L. W.; Miyamoto, Y. Conjugates of Anticancer Agents and Polymers: Advantages of Macromolecular Therapeutics in Vivo. *Bioconjug. Chem.* **1992**, *3*, 351–362.

Maeda, H.; Wu, J.; Sawa, T.; Matsumura, Y, ; Hori, K. Tumor vascular permeability and the EPR effect in macromolecular therapeutics: A review. *J. Control. Release*, **2000**, *65*, 271–284.

Manju, S.; Sreenivasan, K. Conjugation of curcumin onto hyaluronic acid enhances its aqueous solubility and stability. *J. Colloid Interface Sci.* **2011**, *359*, 318–325.

Matsumura, Y.; Maeda, H. A new concept for macromolecular therapeutics in cancer chemotherapy: mechanism of tumoritropic accumulation of proteins and the antitumor agent SMANCS. *Cancer Res.* **1986**, *46*, 6387–6392.

McLeod, A. D.; Friend, D. R.; Tozer, T. N. Synthesis and chemical stability of glucocorticoid–dextran esters: potential prodrugs for colon-specific delivery. *Int. J. Pharm.* **1993**, *92*, 105–114.

Mehvar, R. Dextrans for Targeted and Sustained Delivery of Therapeutic and Imaging Agents. *J. Control. Release*, **2000**, *69*, 1–25.

Mehvar, R. Simultaneous analysis of dextran-methylprednisolone succinate, methylprednisolone succinate, and methylprednisolone by size-exclusion chromatography. *J. Pharmaceut. Biomed. Anal.* **1999**, *19*, 785–792.

Mehvar, R. Targeted delivery of methylprednisolone using a dextran prodrug. *Pharm. Res.* **1997** *14*, S336.

Mehvar, R.; Dann, R. O.; Hoganson, D. A. Kinetics of hydrolysis of dextran-methylprednisolone succinate, a macromolecular prodrug of methylprednisolone, in rat blood and liver lysosomes. *J. Control. Release*, **2000**, *68*, 53–61.

Minko, T.; Kopecková, P.; Pozharov, V.; Jensen, K. D.; Kopecek, J. The influence of cytotoxicity of macromolecules and of VEGF gene modulated vascular permeability on the enhanced permeability and retention effect in resistant solid tumors. *Pharm. Res.* **2000**, *17*, 505–514.

Mitra, S.; Gaur, U.; Ghosh, P. C.; Maitra, A. N. Tumor targeted delivery of encapsulated dextran–doxorubicin conjugate using chitosan nanoparticles as carrier. *J. Control. Release*, **2001**, *74*, 317–323.

Moghimi, S. M.; Hunter, A. C.; Murray, J. C. Long-circulating and target-specific nanoparticles: Theory to practice. *Pharmacol. Rev.* **2001**, *53*, 283–318.

Morgan, S. M.; Al-Shamkhani, A.; Callant, D.; Schacht, E.; Woodley, J. F.; Duncan, R. Alginates as drug carriers: covalent attachment of alginates to therapeutic agents containing primary amine groups. *Int. J. Pharm.* **1995**, *122*, 121–128.

Nichifor, M.; Mocanu, G. Polysaccharide–Drug Conjugates as Controlled Drug Delivery Systems. In *Polysaccharides for Drug Delivery and Pharmaceutical Applications*; Marchessault, R. H.; Ravenelle, F.; Zhu, X. X.; Ed.; Oxford University Press, New York, 2006; pp. 289–303.

Nogusa, H.; Yamamoto, K.; Yano, T.; Kajiki, M.; Hamana, H.; Okuno, S. Distribution characteristics of carboxymethylpullulan–peptide–doxorubicin conjugates in tumor-bearing rats: different sequence of peptide spacers and doxorubicin contents. *Biol. Pharm. Bull.* **2000**, *23*, 621–626.

Nogusa, H.; Yano, T.; Okuno, S.; Hamana, H.; Inoue, K. Synthesis of carboxymethylpullulan-peptide-doxorubicin conjugates and their properties. *Chem. Pharm. Bull.* **1995**, *43*, 1931–1936.

Ohya, Y.; Oue, H.; Nagatomi, K.; Ouchi, T. Design of macromolecular prodrug of cisplatin using dextran with branched galactose units as targeting moieties to hepatoma cells. *Biomacromolecules*, **2001**, *2*, 927–933.

Pang, X.; Lu, Z.; Du, H.; Yang, X.; Zhai, G. Hyaluronic acid-quercetin conjugate micelles: Synthesis, characterization, in vitro and in vivo evaluation. *Colloids Surf. B Biointerfaces.* **2014**, *123*, 778–786.

Pang, Y. N.; Zhang, Y.; Zhang, Z. R. Synthesis of an enzyme-dependent prodrug and evaluation of its potential for colon targeting. *World J. Gastroenterol.* **2002**, *8*, 913–917.

Park, K.; Lee, G. Y.; Kim, Y.-S.; Yu, M.; Park, R.-W.; Kim, I.-S.; Kim, S. Y.; Byun, Y. Heparin–deoxycholic acid chemical conjugate as an anticancer drug carrier and its antitumor activity. *J Control Release.* **2006**, *114*, 300–306.

Pawar, P.; Jadhav, W.; Bhusare, S.; Borade, R.; Farber, S.; Itzkowitz, D.; Domb, A. Polysaccharides as Carriers of Bioactive Agents for Medical Applications. In *Natural-based Polymers for Biomedical Applications*; Reis, R. L.; Ed.; Woodhead Publishers, New York, 2008; pp. 3–53.

Peer, D.; Karp, J. M.; Hong, S.; Farokhzad, O.; Margalit, R.; Langer, R. Nanocarriers as an emerging platform for cancer therapy. *Nat. Nanotechnol.* **2007**, *2*, 751–760.

Petit, T. Anthracycline-induced cardiotoxicity. *Bull. Cancer*, **2004**, *91*, 159–165.

Ravindran, J.; Prasad, S.; Aggarwal, B. B. Curcumin and cancer cells: How many ways can curry kill tumor cells selectively? *AAPS J.* **2009**, *11*, 495–510.

Rensberger, K. L.; Hoganson, D. A.; Mehvar, R. Dextran-methylprednisolone succinate as a prodrug of methylprednisolone: in vitro immunosuppressive effects on rat blood and spleen lymphocytes. *Int. J. Pharm.* **2000**, *207*, 71–76.

Ridley, B. L.; O'Neill, M. A.; Mohnen, D. Pectins: Structure, biosynthesis, and oligogalacturonide-related signaling. *Phytochemistry*, **2001**, *57*, 929–967.

Ringsdorf, H. Structure and properties of pharmacologically active polymers. *J. Polym. Sci. Polym Symp.* **1975**, *51*, 135–153.

Rodriguez, M.; Antunez, J. A.; Taboada, C.; Seijo, B.; Torres, D. Colon-specific delivery of budesonide from microencapsulated cellulosic cores: evaluation of the efficacy against colonic inflammation in rats. *J. Pharm. Pharmacol.* **2001**, *53*, 1207–1215.

Sarika, P. R.; James, N. R.; Anil Kumar, P. R.; Raj, D. K.; Kumary, T. V. Gum arabic-curcumin conjugate micelles with enhanced loading for curcumin delivery to hepatocarcinoma cells. *Carbohydr. Polym.* **2015**, *134*, 167–174.

Sarika, P. R.; James, N. R.; Nishna, N.; Anil Kumar, P. R.; Raj, D. K. Galactosylatedpullulan–curcumin conjugate micelles for site specific anticancer activity to hepatocarcinoma cells. *Colloids Surf. B. Biointerfaces*, **2015**, *133*, 347–355

Sarmento, B.; Ribeiro, A.; Veiga, F.; Sampaio, P.; Neufeld, R.; Ferreira, D. Alginate/chitosan nanoparticles are effective for oral insulin delivery. *Pharm. Res.* **2007**, *24*, 2198–2206.

Scomparin, A.; Salmaso, S.; Bersani, S.; Satchi-Fainaro, R.; Caliceti, P. Novel folated and nonfolatedpullulanbioconjugates for anticancer drug delivery. *Eur. J. Pharm. Sci.* **2011**, *42*, 547–558.

She, W.; Li, N.; Luo, K. Guo, C.; Wang, G.; Geng, Y.; Gu, Z. Dendronizedheparindoxorubicin conjugate-based nanoparticle as pH-responsive drug delivery system for cancer therapy. *Biomaterials.* **2013**, *34*, 2252–2264.

Shi, M. X.; Cai, Q. F.; Yao, L. M.; Mao, Y. B.; Ming, Y. L.; Ouyang, G. L. Antiproliferation and apoptosis induced by curcumin in human ovarian cancer cells. *Cell Biol. Int. Rep.* **2006**, *30*, 221–226.

Shingel, K. I. Current knowledge on biosynthesis, biological activity, and chemical modification of the exopolysaccharide, pullulan. *Carbohydr. Res.* **2004**, *339*, 447–460.

Shrivastava, S. K.; Jain, D. K.; Shrivastava, P. K.; Trivedi, P. Flurbiprofen- and suprofen-dextran conjugates: synthesis, characterization and biological evaluation. *Trop. J. Pharm. Res.* **2009**, *8*, 221–229.

Singh, R.; Lillard, J. W. Jr. Nanoparticle-based targeted drug delivery. *Exp. Mol. Pathol.* **2009**, *86*, 215–223.

Sinha, V. R.; Kumria, R. Colonic drug delivery: Prodrug approach. *Pharm. Res.* **2001**, *18*, 557–564.

Stubbs, S. S.; Morrell, R. M. Intravenous methylprednisolone sodium succinate: Adverse reactions reported in association with immunosuppressive therapy. *Transplant. Proc.* **1973**, *5*, 1145–1146.

Tanaka, T.; Hamano, S.; Fujishima, Y.; Kaneo, Y. Uptake of pullulan in cultured rat liver parenchymal cells. *Biol. Pharm. Bull.* **2005**, *28*, 560–562.

Tang, X.-H.; Xie, P.; Ding, Y.; Chu, L.-Y.; Hou, J.-P.; Yang, J. L.; Song, X.; Xie, Y.-M. Synthesis, characterization, and in vitro and in vivo evaluation of a novel pectin–adriamycin conjugate. *Bioorg. Med. Chem.* **2010**, *18*, 1599–1609.

Tao, Y.; He, J.; Zhang, M.; Hao, Y.; Liu, J.; Ni, P. Galactosylated biodegradable poly (ε-caprolactone-cophosphoester) random copolymer nanoparticles for potent hepatoma-targeting delivery of doxorubicin. *Polym. Chem.* **2014**, *5*, 3443–3452.

Thomsen, L. B.; Lichota, J.; Kim, K. S.; Moos, T. Gene delivery by pullulan derivatives in brain capillary endothelial cells for protein secretion. *J. Control. Release*, **2011**, *151*, 45–50.

Tong, R.; Cheng, J. Anticancer polymeric nanomedicines. *Polym. Rev.* **2007**, *47*, 345–381.

Torchilin, V. P. Targeted pharmaceutical nanocarriers for cancer therapy and imaging. *AAPS J.* **2007**, *9*, 128–147.

Tseng, C-L.; Su, W.-Y.; Yen, K.-C.; Yang, K.-C.; Lin, F.-H. The use of biotinylated-EGF modified gelatin nanoparticle carrier to enhance cisplatin accumulation in cancerous lungs via inhalation. *Biomaterials* **2009**, *30*, 3476–3485.

Uchino, H.; Matsumura, Y.; Negishi, T.; Koizumi, F.; Hayashi, T.; Honda, T.; Nishiyama, N.; Kataoka, K.; Naito, S.; Kakizoe, T. Cisplatin-incorporating polymeric micelles (NC-6004) can reduce nephrotoxicity and neurotoxicity of cisplatin in rats. *Br. J. Cancer* **2005**, *93*, 678–687.

Van, S.; Das, S. K.; Wang, X.; Feng, Z.; Jin, Y.; Hou, Z.; Chen, F.; Pham, A.; Jiang, N.; Howell, S. B.; Yu, L. Synthesis, characterization and biological evaluation of poly(L-γ-glutamyl-glutamine)-paclitaxel nanoconjugate. *Int. J. Nanomed.* **2010**, *5*, 825–837.

Varshosaz, J.; Emami, J.; Tavakoli, N.; Fassihi, A.; Minaiyan, M.; Ahmadi, F.; Dorkoosh, F. Synthesis and evaluation of dextran–budesonide conjugates as colon specific pro-drugs for treatment of ulcerative colitis. *Int. J. Pharm.* **2009**, *365*, 69–76.

Vercauteren, R.; Bruneel, D.; Schacht, E.; Duncan, R. Effect of the chemical modification of dextran on the degradation by dextranase. *J. Bioact. Compat. Polym.* **1990**, *5*, 4–15.

Verma, A. K.; Chanchal, A.; Chutani, K. Augmentation of antitumor activity of cisplatin by pectin nano-conjugates in B-16 mouse model: pharmacokinetics and in-vivo bio-distribution of radio-labeled, hydrophilic nano-conjugates. *Int. J. Nanotechnol.* **2012**, *9*, 872–886.

Verma, A. K.; Sachin, K. Novel hydrophilic drug polymer nano-conjugates of cisplatin showing long blood retention profile: Its release kinetics, cellular uptake and bio-distribution. *Curr. Drug Deliv.*. **2008**, *5*, 120–126.

Vittorio, O.; Voliani, V.; Faraci, P.; Karmakar, B.; Iemma, F.; Hampel, S.; Kavallaris, M.; Cirillo, G. Magnetic catechin-dextran conjugate as targeted therapeutic for pancreatic tumor cells. *J. Drug Target.* **2014**, *22*, 408–415.

Vyas, S.; Trivedi, P.; Chaturvedi, S. C. Ketorolac-dextran conjugates: Synthesis, in vitro and in vivo evaluation. *Acta Pharmaceutica*, **2007**, *57*, 441–450.

Wang, Q.-W.; Liu, X.-Y.; Liu, L.; Feng, J.; Li, Y.-H.; Guo, Z.-J.; Mei, Q.-B. Synthesis and evaluation of the 5-fluorouracil-pectin conjugate targeted at the colon. *Med. Chem. Res.* **2007**, *16*, 370–379.

Wang, W.; Yao, J.; Zhou, J. P.; Lu, Y.; Wang, Y.; Tao, L.; Li, Y. P. Urocanic acid-modified chitosan-mediated P53 gene delivery inducing apoptosis of human hepatocellular carcinoma cell line HepG2 is involved in its antitumor effect in vitro and in vivo. *Biochem. Biophys. Res. Commun.* **2008**, *377*, 567–572.

Wang, Y.; Chen, H.; Liu, Y.; Wu, J.; Zhou, P.; Wang, Y.; Li, R.; Yang, X.; Zhang, N. pH-sensitive pullulan-based nanoparticle carrier of methotrexate and combretastatin A4 for the combination therapy against hepatocellular carcinoma. *Biomaterials*, **2013**, *34*, 7181–7190.

Wang, Y.; Liu, P.; Qiu, L.; Sun, Y.; Zhu, M.; Gu, L.; Di, W.; Duan, Y. Toxicity and therapy of cisplatin-loaded EGF modified mPEG-PLGA-PLL nanoparticles for SKOV3 cancer in mice. *Biomaterials*, **2013**, *34*, 4068–4077.

Wang, Y.; Yang, X.; Yang, J.; Wang, Y.; Chen, R.; Wu, J.; Liu, Y.; Zhang, N. Self-assembled nanoparticles of methotrexate conjugated O-carboxymethyl chitosan: Preparation, characterization and drug release behavior in vitro. *Carbohydr. Polym.* **2011**, *86*, 1665–1670.

Wang, Y.; Zhou, J.; Qiu, L.; Wang, X.; Chen, L.; Liu, T.; Di, W. Cisplatin-alginate conjugate liposomes for targeted delivery to EGFR-positive ovarian cancer cells. *Biomaterials* **2014**, *35*, 4297–4309.

Wente, M. N.; Kleeff, J.; Büchler, M. W.; Wers, J.; Cheverton, P.; Langman, S.; Friess, H. DE-310, a macromolecular prodrug of the topoisomerase-I-inhibitor exatecan (DX-8951), in patients with operable solid tumors. *Invest. New Drugs*, **2005**, *23*, 339–347.

Willats, W. G. T.; McCartney, L.; Mackie, W.; Knox, J. P. Pectin: Cell biology and prospects for functional analysis. *Plant Mol. Biol.* **2001**, *47*, 9–27.

Woodley, J. F. Peptidase activity in the GI tract: Distribution between luminal contents and mucosal tissue. *Proc. Int. Symp. Control. Release Bioact. Mater.* **1991**, *18*, 337–338.

Xiao, H.; Qi, R.; Liu, S.; Hu, X.; Duan, T.; Zheng, Y.; Huang, Y.; Jing, X. Biodegradable polymer-cisplatin (IV) conjugate as a prodrug of cisplatin (II). *Biomaterials*, **2011**, *32*, 7732–7739.

Xin, Y.; Yuan, J. Schiff's base as a stimuli-responsive linker in polymer chemistry. *Polym. Chem.* **2012**, *3*, 3045–3055.

Xu, C.; He, W.; Lv, Y.; Qin, C.; Shen, L.; Yin, L. Self-assembled nanoparticles from hyaluronic acid–paclitaxel prodrugs for direct cytosolic delivery and enhanced antitumor activity. *Int. J. Pharm.* **2015**, *493*,172–181.

Xu, X.; Li, Y.; Shena, Y.; Guo, S. Synthesis and in vitro cellular evaluation of novel antitumor norcantharidin-conjugated chitosan derivatives. *Int. J. Biol. Macromol.* **2013**, *62*, 418–425.

Yang, R.; Meng, F.; Ma, S.; Huang, F.; Liu, H.; Zhong, Z. Galactose-decorated cross-linked biodegradable poly (ethylene glycol)-*b*-poly (ε-caprolactone) block copolymer micelles for enhanced hepatoma-targeting delivery of paclitaxel. *Biomacromolecules*, **2011**, *12*, 3047–3055.

Yang, R.; Zhang, S.; Kong, D.; Gao, X.; Zhao, Y.; Wang, Z. Biodegradable polymer-curcumin conjugate micelles enhance the loading and delivery of low-potencycurcumin. *Pharm. Res.* **2012**, *29*, 3512–3525.

Yura, H.; Yoshimura, N.; Hamashima, T.; Akamatsu, K.; Nishikawa, M.; Takakura, Y.; Hashida, M. Synthesis and pharmacokinetics of a novel macromolecular prodrug of tacrolimus (FK506), FK506-dextran conjugate. *J. Control. Release*, **1999**, *57*, 87–99.

Zhang, H.; Li, F.; Yi, J.; Gu, C.; Fan, L.; Qiao, Y.; Tao, Y.; Chenga, C.; Wu, H. Folate-decorated maleilatedpullulan–doxorubicin conjugate for active tumor-targeted drug delivery. *Eur. J. Pharm. Sci.* **2011**, *42*, 517–526.

Zheng, D.; Duan, C.; Zhang, D.; Jia, L.; Liu, G.; Liu, Y.; Wang, F.; Li, C.; Guo, H.; Zhang, Q. Galactosylated chitosan nanoparticles for hepatocyte-targeted delivery of oridonin. *Int. J. Pharm.* **2012**, *436*, 379–386.

Zhou, N.; Zan, X.; Wang, Z.; Wu, H.; Yin, D.; Liao, C.; Wan, Y. Galactosylatedchitosan–polycaprolactone nanoparticles for hepatocyte-targeted delivery of curcumin. *Carbohydr. Polym.* **2013**, *94*, 420–429.

Zhou, P.; Li, Z.; Chau, Y. Synthesis, characterization and in vivo evaluation of poly(ethyleneoxide-coglycidol)-platinate conjugate. *Eur. J. Pharm. Sci.* **2010**, *41*, 464–472.

CHAPTER 6

GLYCAN-BASED NANOCARRIERS IN DRUG DELIVERY

SONGUL YASAR YILDIZ,[1] MERVE ERGINER,[1] TUBA DEMIRCI,[2] JUERGEN HEMBERGER,[2] and EBRU TOKSOY ONER[1,*]

[1]IBSB-Industrial Biotechnology and Systems Biology Research Group, Bioengineering Department, Marmara University, Istanbul, Turkey, *E-mail: ebru.toksoy@marmara.edu.tr

[2]Institute for Biochemical Engineering and Analytics, University of Applied Sciences, Giessen, Germany

CONTENTS

ABSTRACT

Glycans are linear or branched polymer molecules composed of repeating sugar units adjoined by glycosidic bonds. Next to DNA and protein, glycans are considered as the third dimension in molecular biology and various properties of glycans such as biocompatibility, solubility, potential for modification, and innate bioactivity offer great potential for their use in drug delivery systems (DDS). Hence, recently, studies on glycans and their applications as nanoparticle drug delivery systems are accelerated. In this chapter, after a brief introduction to glycans and glycan-based DDS, various glycans including chitosan, hyaluronic acid, alginate, dextran, hydroxyethyl starch, levan and mannan that are used in drug delivery are summarized and finally their role as nanocarriers in DDS are discussed.

6.1   INTRODUCTION

Nanotechnology is useful for synthesis of self-assembly or fabricated material and devices in small and desired scale (Kanapathipillai et al., 2014). Nanoparticles in the context of drug delivery systems (DDS) are used as carriers to deliver drugs or biomolecules to the target site. These nano-sized materials with diameter 1–100 nm are called nanocarriers with features such as high surface area and high ligand density on the surface that carry drugs or imaging agents (Buzea et al., 2007). They also have the ability to pass barriers in order to reach targeted compartments in the body or cell (Peer et al., 2007; Sutherland, 2002).

Nanocarriers can be prepared in a variety of formulations such as nano-capsules, nanospheres, nanoliposomes, nanomicelles, etc. Use of nano-carrier systems in drug delivery has significant benefits. Nanocarriers can penetrate cells and tissue gaps as well as through the smallest capillary vessels without any difficulty to arrive at the target organs such as brain, spinal cord, lung, spleen, lymph and liver. Since rapid clearance of nanoparticles by phagocytes is avoided, their duration in blood stream is greatly prolonged. Because the responsiveness of glycan nanocarriers to ion, pH and/or temperature changes and their biodegradability, they could reveal properties of controlled release (Liu et al., 2008). They can induce positive effects like increases in drug efficacy and decreases in toxic side effects. As DDS, nanocarriers can contain drugs or biomolecules in their interior structures and/or drugs or biomolecules can be absorbed on their outer surfaces. At present, nanocarriers have been extensively used to deliver drugs, nucleic acids, proteins, polypeptides, genes, vaccines, etc. (Yasar Yildiz and Toksoy Oner, 2014).

Glycans are linear or branched polymeric molecules composed of repeating sugar units adjoined by glycosidic bonds. They are highly abundant in nature and generally have low processing costs. They can be of algal, plant, microbial or animal origin, and are widely used in formulation of both conventional and novel DDS. Many pharmacological properties of conventional drugs can be improved by the help of DDS. DDS are designed to modify the pharmacokinetics (PK) and biodistribution (BD) of their associated drugs, or to function as drug reservoirs, or both.

In recent years, a growing number of research activities are focused on glycans and their derivatives for their possible application as nanoparticle DDS (Lemarchand et al., 2004; Liu et al., 2008; Sinha et al., 2006) that will be reviewed in this chapter.

6.2 IMPORTANCE OF GLYCANS AND GLYCOTECHNOLOGY

Glycans can be defined as "compounds consisting of a large number of monosaccharides linked glycosidically" (Panitch et al., 2014). These polymeric molecules can be homopolysaccharides or heteropolysaccharides of monomeric sugar residues, and can be either branched or linear. Moreover, the term glycan can be used for the sugar part of a glycoconjugate (glycolipid, glycoprotein and proteoglycan), even though the carbohydrate is only an oligosaccharide (Dwek, 1996). In glycoproteins, there are two types of linkages between oligosasaccharides and proteins. The first type, called N-linked glycans, includes the binding of N-acetylglucosamine to the amide side chain of asparagine. The second type, O-linked glycans, contains the binding of C-1 of N-acetylgalactosamine to the hydroxyl function of serine or threonine (Gorelik et al., 2001).

Glycans have biological roles that contribute to physical/structural integrity, formation of extracellular matrix, protein folding, signal transduction, and data transport between cells. Most of the glycans are present on the outer surface of the cell or secreted to the extracellular environment. As a result, glycans and glycoconjugates take roles in various interactions between cell–cell, cell–matrix, and cell–molecule that are critical to the development and function of a complex multicellular organism (von der Lieth et al., 2009). Moreover, they can also be involved in the interactions between different organisms (Varki and Sharon, 2009). Glycans can function as adhesion molecules, recognition molecules and signaling molecules due to their variable complex structure on the cell surface (Ofek et al., 2003a, 2003b; Sharon and Lis, 1989). Likewise, cell surface glycans take roles in many physiologically significant functions such as embryonic development, cell differentiation, growth of organism, cell–cell recognition, cell signaling, host–pathogen interaction during infection, host immune response, development of disease, metastasis, intracellular trafficking and localization, rate of degradation and membrane rigidity (Ghazarian et al., 2011).

Even though glycans are one of the four major components of living cells, their functions are less understood compared to protein, lipid or DNA. Nonetheless, glycans are the most abundant and structurally varied biopolymers present in nature. Glycosylation (glycan biosynthesis) consist of the many pathways and enzymatic activities that generate the variability of the secretory pathway (Lowe and Marth, 2003).

With the development of new technologies for studying the structures and functions of glycans, the term "glycobiology" emerged and is defined as the science of the synthesis, structure, function, degradation, and evolution of sugars that are extensively distributed in nature, and the proteins that interact with them (Varki and Sharon, 2009).

In the natural sciences, one of the most rapidly growing fields is glycobiology. It is relevant to several diverse areas of biomedicine, basic research, and biotechnology and cover research areas including carbohydrate chemistry, glycan formation and degradation, enzymology, recognition of glycans by specific proteins (lectins and glycosaminoglycan-binding proteins), roles of glycans in complex biological systems and analysis or manipulation of glycans by a variety of methods. Research in glycobiology thus requires a background not only in chemical synthesis, structure, biosynthesis, and functions of glycans, but also in the general disciplines of protein chemistry, molecular genetics, physiology, developmental biology, cell biology, and medicine (Varki and Sharon, 2009).

Various properties of glycans such as biocompatibility, solubility, potential for modification, and innate bioactivity offer great potential for their use in DDS (Yasar Yildiz and Toksoy Oner, 2014).

Contrary to numerous synthetic polymers, toxicity levels of glycans are either zero or close to zero (Dang and Leong, 2006; Mizrahy and Peer,

2012; Ratner and Bryant, 2004). For instance, dextrans that are homopolymers of glucose, show high biocompatibility and low toxicity, that can be used for the production of microspheres with no inflammatory response following subcutaneous injection into rats (Cadée et al., 2001) and also for the production of biocompatible hydrogels for controlled prolonged therapeutic release (Coviello et al., 2007).

Glycans are naturally present in the body, as a result, most of them are degraded enzymatically. They can be broken down to their monomeric or oligomeric building blocks during the enzymatic catalysis, and recycled in order to reuse them for storage, structural support, or even cell signaling applications (Jain et al., 2007). Accordingly, mechanisms for release of therapeutics from glycan-based carrier systems are mostly by enzymatic degradation (Mehvar, 2003).

Glycans can be easily modified chemically or biochemically because of the presence of various groups on the molecular chain that could easily be activated and modified giving rise to several kinds of glycan derivatives. Chemical reactions such as oxidation, sulfation, esterification, amidation, or grafting methods can be used for the modification of glycans (Yang et al., 2011). The character of the glycans can also be altered by these modifications. As an example, oxidation of hydroxyl group increases biodegradability, whereas sulfonation creates a heparin-like glycan with improved blood compatibility (Kumar et al., 2004).

Many glycans have inherent bioactivities, in particular antimicrobial, anti-inflammatory and mucoadhesive properties. Through charge interactions, positively charged glycans are able to bind to the mucosal layers that are negatively charged (Bernkop-Schnürch and Dünnhaupt, 2012; Boddohi et al., 2008; Bonferoni et al., 2010). Hydrogen bonding of neutral or negatively charged glycans provides an alternative mechanism for mucoadhesion (Reddy et al., 2011). Nanoparticle carriers prepared along with bioadhesive glycans could extend the residence time and hence accelerate the absorbance of loaded drugs (Liu et al., 2008). Besides, most glycans in nature are antimicrobial, such as chitosan (Dai et al., 2011). Others have been identified as inflammation reducing biopolymers. Anti-inflammatory activity is believed to be caused by the binding of immune-related acute phase and complement proteins (Young, 2008), which reflects the well-known fact that glycans interact with a variety of proteins.

6.3 GLYCAN-BASED DRUG DELIVERY SYSTEMS

Approaches, formulations, technologies, and systems for the process of administering a pharmaceutical compound to achieve a therapeutic effect in an organism are described by the term "drug delivery" (Tiwari et al., 2012). In literature, many drug delivery methods such as oral-, parenteral-, topical- and enteral administrations are described, but these conventional methods have several disadvantages, for instance most drugs have poor specificity and high toxicity. By oral or parenteral administration, active components enter to systemic circulation with the consequence of metabolic degradation that results in insufficient therapy as well as in the need for higher doses with the consequence of increased side effects. Glycan-based DDS are an alternative method to overcome some of these problems (Sinha et al., 2006). Glycan-based DDS have unique features such as low toxicity, biocompatibility, biodegradability, stability, low cost, nonimmunogenicity, hydrophilic nature and availability of reactive sites for chemical modification (Mizrahy and Peer, 2012).

Modifications may change the hydrophobicity, solubility, physicochemical and biological characteristic of a glycan. This feature can be advantageous in drug delivery via drug conjugation to the glycans for sustained release, or via ligand attachment for targeting to a desired site. These benefits make them particularly suitable for medical drug application, including gene, protein, and antigen delivery, as well as diagnostic devices. Valuable properties of glycan-based DDS make them usable in different formulations, such as nanoparticles, micelles, and hydrogels. Protein-sugar interactions are essential for many cellular functions and these interactions can be used to overcome toxicity and immunogenicity of protein-based drugs. Protein-based drugs can be modified through conjugation with sugars. Foreign epitopes, which activate the immune response, can be hidden by the sugar residues. Therefore, serum half-life of unstable drugs can be extended (Zhang et al., 2015).

Drug delivery can be divided into two main groups, namely, passive and active targeting (Figure 6.1). Passive targeting uses permeability of tissue that results from alteration of barrier function of an organ during the disease condition such as ischemia that occur during a stroke and infectious diseases like multiple sclerosis, Alzheimer's, encephalopathy and cancer. Rapid

FIGURE 6.1   Types of drug delivery systems.

vascularization of tumor tissue provides further permeability to tumor tissue. The presence of cytokines and other growth factors increases the permeability of this tissue, which allows passive diffusion of the drugs into targeted tissue. In passive targeting one of the most important properties is the size of the drug and it needs to be smaller than the pore diameter of the dilated vasculature. The surface of the drug is another significant characteristic and it should be hydrophilic in order to circumvent clearance by macrophages. If the surface of the drug is hydrophilic, it can protect itself against adsorption of plasma protein to the surface. This hydrophilic surface protection can be obtained by using hydrophilic polymer coating, like polyethylene glycol (PEG), poloxamines, glycans, or poloxamers or by using block or branched amphiphilic copolymers (Moghimi and Hunter, 2000; Park et al., 2005a). Passive targeting system is subdivided into three groups: leaky vasculature, tumor microenvironment and local drug application.

For the leaky vasculature passive targeting, effect of polymer on improved permeation and retention to form nanoparticles was published for the first time in 1985 (Maeda and Matsumura, 1988). When compared with normal tissue endothelium, the capillary endothelium system of malignant tissue reveals more permeation to macromolecules which in turn provides permeability to circulating polymeric nanoparticles into the tumor. More drugs get accumulated in the tumor tissue due to the lack of lymphatic drainage in this tissue. The drug concentration may reach levels 10–100 times higher than that of free circulating concentration of drug if appropriate biodegradable polymer is used.

Tumor microenvironment type passive drug targeting uses the advantage of tumor environment. Chemotherapeutic agent is conjugated to a tumor-specific molecule and administered into the body in an inactive state. Tumor environment converts the drug-polymer conjugate into a more active form when it reaches its target site, which is known as tumor activated prodrug therapy (Sharma et al., 2013; Sinha et al., 2006).

In local drug application type passive targeting, direct application of the drug to tumor tissue avoids systemic toxicity and promotes accumulation of the drug at the tumor site. Intra-tumoral application is most common delivery form of passive targeting with increased concentration at the tumor and decreased toxicity in the body (Sinha et al., 2006).

Use of glycosylated nano-assemblies is a common method in passive targeting. This method can be used especially for tumors (Torchilin, 2010). In the progression of cancer, partial angiogenesis results in leaky vasculature and enlargement of gap junctions. Hence, macromolecular carriers can enter the interstitial space of tumor and remain for prolonged time periods due to enhanced permeability and retention (EPR) effect (Byrne et al., 2008). In a variety of research approaches, sugar-based polymers are used for local drug delivery by mimicking some physicochemical properties of proteins such as surface charge. Due to these physicochemical properties they can be involved in biological processes at target sites (Gu et al., 2014; York et al., 2012). Moreover, these systems may display natural bioactivity by competitive inhibition of the binding of ligands to their receptors and disturbing the normal ligand-activated pathways, for instance immune response, intracellular signaling, and other disease cascades. In many studies, the ability of ionic liposomes, micelles, or nanoparticles is used to favorably accumulate in macrophages and repress atherogenesis (Chono et al., 2005; Lewis et al., 2013; Petersen et al., 2014; Plourde et al., 2009; York et al., 2012). Anionic nanoparticles consisting of sugar-based amphiphilic polymers (SBAPs) with a carbohydrate backbone and a PEG tail can be used to target the lesion sites in cardiovascular research and additionally, these nanoparticles bind macrophage and smooth muscle cell scavenger receptors to limit cholesterol accumulation in atherosclerosis and neointima hyperplasia in restenosis, respectively (Lewis et al., 2013; Zhang et al., 2015). Moreover, sulfated ester nanoparticles with a lactose core can also be used for passive targeting due to their high selectin binding efficiency on leukocytes and platelets in blood. Interaction between dendritic β-lactose-PEG glycopolymers with terminal sulfate moieties and leukocyte-selectin receptors cause blockage of chemokines from binding to the epithelium. This competitive binding of dendritic β-lactose-PEG glycopolymers provides anti-inflammatory effects due to the multivalent properties of the glycopolymers (Rele et al., 2005; Weinhart et al., 2011).

In active targeting, targeting is achieved by the conjugation of the nanoparticles with a drug to a targeting site. Active targeting allows accumulation of the drug at the tumor tissue, within individual cancer cells, intracellular organelles, or specific molecules in cancer cells (Sinha et al., 2006). This method is used to direct nanoparticles to the cancer cell by the help of specific interactions, like ligand-receptor, antibody-antigen, and lectin-carbohydrate.

Active targeting can be achieved in three ways: receptor targeting, antibody targeting and carbohydrate-directed targeting.

Receptor targeting approach is based on the initiation of endocytosis through receptor activation. On human cancer cells, specific antigens and receptors are overexpressed. Through receptor-mediated endocytosis, drug-conjugate gets into the cells and drug is released inside the cells. A prerequisite for achieving high specificity with this approach is high and homogeneous expression of targeted receptors on tumor cells, but not on normal cells. In the mechanism of receptor-mediated endocytosis, targeting conjugates firstly bind to the receptors, after that; plasma membrane enclosures around the ligand–receptor complex to form an endosome. This endosome is transferred to specific organelles, where acidic pH or enzymes could release the drug (Sharma et al., 2013).

Antibody targeting approach is based on monoclonal antibody mechanism for targeting nanoparticles to solid tumor tissue in vivo. For instance, drug prepared via conjugation of the anti-HER-2 monoclonal antibody fragments with liposomal-grafted polyethylene glycol chain can be targeted towards the human epidermal growth factor receptor 2 (HER-2) on cancer cells. Cellular uptake of the drug is increased making antibody targeting a new opportunity for development of strategies for drug delivery system in breast cancer (Kirpotin et al., 2006). Abundant receptor and antigen on the surface of the tumor tissue and the ligand/antibody with high affinity for these cell surface molecules are necessary for optimum delivery of drugs. Unfortunately, many of the receptors useful for tumor targeting, for instance epidermal growth factor or low-density lipoprotein receptors are often also found on a wide variety of cell types (Sharma et al., 2013; Sinha et al., 2006).

In carbohydrate-directed targeting, nonimmunogenic protein lectins can recognize and bind to cell specific carbohydrate structures that are present on the surface of all cells in the body for cellular interaction. This specific reaction can be used for carbohydrate-directed targeting (Sharma et al., 2013). Tumor cells present different carbohydrate structures on their surface when compared to normal cells. Some lectins recognize the different patterns of cell surface carbohydrates on tumor cells. It has been reported that lectins affect adhesion to the extracellular matrix or endothelium, tumor vascularization, tumor cell survival and many other processes that are vital for metastatic spread and growth (Gorelik et al., 2001; Raz et al., 1986). Targeting can be achieved either by direct lectin targeting or reverse lectin targeting. Formation of nanoparticles with carbohydrate moieties that are directed to certain lectins is termed as *direct lectin targeting* and incorporation of lectins into nanoparticles that are directed to cell surface carbohydrates is termed *reverse lectin targeting*.

Up to now, DDS that have been developed established on this novel interaction between carbohydrates and lectins are targeted to whole tissue (Yamazaki et al., 2000), and hence it could be harmful to healthy tissues. In spite of some of their obstacles, lectins are continuing to be studied further for the development of "smart carrier" molecules for drug delivery and their unique affinity for sugar moieties on the surface of tumor tissue seems to be an attractive tool for further enhancement of nano-drug delivery.

6.3.1 GLYCANS USED IN DRUG DELIVERY SYSTEMS

Various glycans have been used as drugs in DDS due to their different applications. Main characteristics of commonly investigated glycans for drug delivery are discussed in the following subsections.

6.3.1.1 Chitosan

Chitosan is a natural, linear glycan composed of β-(1,4)-linked D-glucosamine and N-acetyl-D-glucosamine and it is obtained by deacetylation of chitin. Commercial chitosan is positively charged, with amino groups, in neutral and alkaline pH and has a molecular weight between 3.8–20 kDa. It is insoluble in water and organic solvents. However, protonation of the amine groups in acidic aqueous solution under pH 6.5 provide solubility of chitosan. Introduction of some hydrophilic units and removal of hydrogen atoms results in enhanced water solubility of chitosan (Sinha et al., 2004). Various chemical modification reactions are also available for chitosan. Degradation of chitosan in the body is mediated by enzymes such as lysozymes, chitotriodidase, di-N-acetylchitobiase and products of those degradations are nontoxic. Biological studies have indicated that, chitosan has antibacterial, antifungal, hemostatic, analgesic, and mucoadhesive features, which play an important role in drug delivery and tissue healing (Shelke et al., 2014). Several chitosan-based drug delivery applications like nasal, ocular, oral, parenteral, transdermal or usage as wound dressing material are possible due to low cytotoxicity and biocompatibility features of chitosan and the availability of chemical modification (Mizrahy and Peer, 2012).

Since chitosan is one of the rare positively charged glycans with mucosal surface, electrostatic interactions with negatively charged macromolecules such as DNA, RNA and protein are possible and can be used to target solid tumors (Fang et al., 2001; Morille et al., 2008). Enhancement of administration route, targeting capacity and controlled delivery can be maintained by chitosan-based carriers due to increased cell membrane permeability of hydrophobic drugs (Huang et al., 2015).

Chitosan and derivatives were found toxic to several bacteria, yeast, fungi and parasites. Chitosan-based hydrogels can be produced by cross-linking techniques including ionic gelation, covalent cross-linking, and *in situ* gelation. Studies with hydrogels of chitosan with 3D porous structure were used for wound healing and tissue engineering (Shelke et al., 2014).

6.3.1.2   Hyaluronic Acid

Hyaluronic acid (HA) is a mucoglycan that is composed of repeating disaccharide units of D-glucuronic acid and N-acetyl D-glucosamine linked by β-(1,3) and β-(1,4) glycosidic bonds which occurs naturally in all living organisms (Sirisha, 2015). HA is a natural bacterial capsule component. HA is neither sulfated nor linked to proteins. Other desirable properties are water solubility, biodegradability and biocompatibility (Mizrahy and Peer, 2012). Molecular weight of HA is above 10^6 Dalton and many of its biological activities such as immunosuppressivity, antiangiogenicity, anti-inflammatory feature and inhibition of cell proliferation are due to its high molecular weight. On the other hand, low molecular weight HA promotes angiogenesis and cell migration. Due to carboxylic and hydroxyl groups of HA, several chemical modifications are possible for chemical and mechanical transformations without loss of bioactivity (Shelke et al., 2014). HA has various applications in diverse fields of drug designing and in creation of new biomaterials. Recently, various administration routes, including nasal, parenteral, ophthalmic, pulmonary and topical has been studied for HA based drug delivery (Brown and Jones, 2005; Sirisha, 2015). Cell-receptor targeting is one of the major active drug targeting strategy. HA is the major ligand for CD44 and CD168 cell receptors, which are overexpressed on ovarian, colon, stomach, and many other types of cancer cells (Mizrahy and Peer, 2012). Hyaluronic acid specific chemotherapy can be performed to target overexpressed CD44 receptors (Huang et al., 2015). Another drug delivery method is the conjugation of peptide drugs on HA as a prodrug. Enzymatic degradation of the prodrug after administration leads to release of hyaluronic acid and peptide (Huang et al., 2015).

6.3.1.3   Alginate

Alginate is a linear anionic and hydrophilic glycan composed of alternating blocks of 1,4-linked β-D-mannuronic acid and α-L-glucuronic acid residues. Some parts of the polymer consist of homo-monomeric blocks and alternating mannuronic and glucuronic acid residues. Physical properties and molecular weight of this glycan depends on the monomeric composition. It is produced by bacteria and brown algae. Like other natural biopolymers, alginate is biocompatible, nontoxic, water-soluble and highly mucoadhesive. Commercial alginates have molecular weight ranging from 400–500 kDa. Various formulations including hydrogels, injectable solutions and ionic or covalent cross-linkers can be prepared for alginate-based drug delivery applications. Alginate is one of the most used glycan for regeneration and repair of the tissue or in a supporting matrix of delivery system (Mizrahy and Peer, 2012). Due to ability to interact with cationic polyelectrolytes and proteoglycans, alginate can be used in electrostatic delivery system of cationic drugs. Alginate has long-term stability under biological conditions and compatible with a wide variety of substances. As a result, it can be used for controlled and enhanced bioavailable drug-delivery systems.

6.3.1.4   Dextran

Dextran is a high molecular weight glycan and obtained from lactic acid bacteria cultures such as *Leuconostoc mesenteroides* NRRL B-512. It is a homopolymer of glucose residues with α-1,6 linkages, with occasional branching from α-1,2, α-1,3 and α-1,4 linkages to the backbone glucose units. It is a promising polymer for drug delivery applications due to features like solubility in water and various solvents, biodegradability, nonspecific cell adhesion, protein adsorption resistance, structure transformation convenience, and low cost (Huang et al., 2015; Mizrahy and Peer, 2012). There are various applications of dextran in DDS. Varshosaz et al. (Varshosaz et al., 2011) performed studies with budesonide (type of glucocorticosteroid) to provide the connection between dextran with different molecular masses and glutaric acid for prodrug preparation. Results of the study indicated that dextran with high molecular mass has a positive effect on the solubility of budesonide. Prodrugs prepared in this study were used for treatment of ulcerative colitis. In another study, drug-loaded block copolymer micelles were obtained with dextran and polycaprolactone via disulfide bonds by Sun et al. (Sun et al., 2010). These micelles with an average particle size of 60 nm were sensitive to a reductive environment. They also showed that sustainable and full release of the drug could be observed within 10 h in a reducing environment.

6.3.1.5 Hydroxyethyl Starch

Hydroxyethyl starch (HES) is a semisynthetic derivate of amylopectin. Starch is the reserve form of glucose in many plants and amylopectin is the highly branched component of starch. HES is obtained through hydrolysis and subsequent hydroxyethylation of amylopectin. The extent of hydroxyethylation (molar substitution) determines the degradation and elimination. HES is similar to human glycogen, because of its molecular structure. It is nontoxic, nonantigenic, nonmutagenic and has no hemolytic potential. Thereby it is commonly used for plasma expansion in emergency health situations. Polymer conjugation-technology plays important role in peptide-drug delivery and pharmaceutical chemistry (Hemberger and Orlando, 2006). Because of the immune defense of the body, most peptide-drugs have short plasma half-life. However, plasma half-life of drugs can be prolonged by using sugar-based polymers because of their chemically inert, low immunogenic features and availability of hydroxyl groups for coupling (Hemberger et al., 2009). Renal clearance time of peptides within the body is short, by polymer-conjugated molecule the half- life is prolonged 15–18 times. Moreover, glycan modification reduces both the antigenicity and the immunogenicity of proteins, by masking the antigenic epitopes of the native protein.

6.3.1.6 Levan

Levan is a homopolysaccharide composed of β-D-fructofuranose with β-(2–6) linkages between fructose rings. Levan is produced by the action of levansucrase enzyme by many bacteria, fungi and actinomycetes through conversion of sucrose (Han and Clarke, 1990). This naturally occurring polymer is strongly adhesive, biocompatible and soluble in oil and water which makes this polymer preferable in foods, feeds, cosmetics, pharmaceutical and chemical industry (Kang et al., 2009). Levan-based DDS has been reported to form nanoparticles by self-assembly (Renuart and Viney,

2000). Nanostructured adhesive thin films for biomedical applications of levan were produced by matrix-assisted pulsed laser evaporation (Sima et al., 2011, 2012). The incorporation of phosphorylated-levan leads to thin films with higher adherent capacity for cells. Microbial levan microparticles were used as encapsulating agent for vitamin E and magnetic levan microparticles were used for trypsin immobilization (Maciel et al., 2012; Nakapong et al., 2013). Sarilmiser and Toksoy Oner (2014) investigated anticancer activity of oxidized levan against different human cancer cell lines and found that increasing oxidation degree and dose increases anticancer activity. Ternary blend films of levan with polyethylene oxide (PEO) and chitosan were investigated for biological, morphological and structural features (Bostan et al., 2014). For biomedical and tissue engineering this films could be used as wound healing bandages or surgical sealants (Costa et al., 2013). In the study of Sezer et al. (2011) various biodegradable levan-based nanoparticle systems with different particular size, charge, and release profiles were investigated and it was shown that levan nanoparticles can be used effectively as drug carriers for peptide and proteins. In a recent study, it was found that cancer cells have good affinity to functionalized low molecular weight (LMW) levan encapsulated solid lipid nanoparticles (Kim et al., 2015).

6.3.1.7 Mannan

Mannan is one of the important members of the hemicellulose family. Mannan is present in four different forms, each having a β-1,4-linked backbone containing mannose (linear mannan) or a combination of glucose and mannose residues (glucomannan) and occasional side chains of α-1,6-linked galactose residues (galactomannan/galactoglucomannan) (Yasar Yildiz and Toksoy Oner, 2014). Its mucoadhesive properties and highly flexible conformation make mannan a highly preferred polymer for cancer targeting nano-based material (Jain et al., 2010). In many studies, mannans with various molecular weight and functional modifications have been used as a drug delivery agent (Apostolopoulos et al., 2006). It is known that like the other glycans, mannans are powerful anticancer agents as they are natural ligands for mannose receptors, which are widely present on dendritic cells (Martin and Jiang, 2010). Budzynska et al. (2007) revealed that mannan-methotrexate conjugate improved antitumor activity significantly when compared with free methotrexate in mouse model of leukemia. Nanoparticles coated with galactomannan were developed and investigated for endocytosis by macrophages, dendritic cells, and liver cells and the results indicated that liver and colon macrophages and mouse brain were targeted easily by mannosylated liposomes (Gupta et al., 2009; Park et al., 2005b).

6.4 GLYCAN-BASED NANOCARRIERS

Pharmacologically active drugs which are used in chemotherapy or conventional drug delivery target cancerous tissue with poor specificity. Main problem of those existing anticancer agents are systematic toxicity and adverse effects exerted on both healthy and tumor cells. Another problem is undesired toxicity due to rapid elimination or long distance circulation through lymphatic system which in turn requires the use of large quantities of a drug. Moreover, tissues like intestine, kidney, hearth and spleen with higher flow may also be adversely affected by high drug concentrations (Landesman-Milo et al., 2015).

Nanocarriers are mainly used for enhancement of lifetime and controlled release of encapsulated active agents, and targeting specifically selected single cell types with reduced risk of systemic toxicity in nontarget cells. They are protected against in vivo degradation, with the main advantage of these systems being the elimination of the carrier from body after the release of the drug. Particle size is an important parameter for nanoparticles delivery. It is known that particles larger than 200 nm may effect immunological reactions or can cause accumulation of drug in the injected area. In vivo activity of nanoparticles is mainly size-dependent and nanoparticles with size of 100–200 nm have faster clearance kinetics from blood.

Besides particle size, surface to volume ratio, charge and geometry of the surface and shape of the nanoparticle are also very important for designing a nanoparticle with desired features. Charge of the nanoparticles depends on functional groups and also play crucial role at targeting and activity of the nanoparticles (Spencer et al., 2015).

Different nanocarrier types are available with different physiochemical properties for disease treatment with drug delivery, imaging, photo

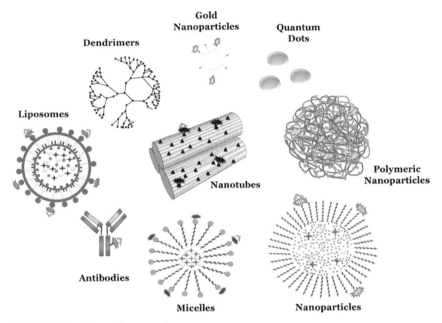

FIGURE 6.2 Types of nanocarriers.

thermal ablation of tumors or detection of apoptosis (Figure 6.2). Inorganic nanoparticles, polymeric nanoparticles, solid-liquid nanoparticles, liposomes, micelles (polymeric or lipid-based), dendrimers, carbon nanotubes, nanorods, gold nanoparticles, and nanocrystals are the most investigated nanocarriers (Duncan, 2006; Faraji and Wipf, 2009; Ferrari, 2005; Lavan et al., 2003; Peer et al., 2007).

Polymeric micelles are colloidal particles with hydrophobic tail and hydrophilic head and generally have size range between 5–100 nm. They are composed of amphiphiles and surface active agents. Polymeric micelles are good carriers for hydrophilic drugs. They are self-assembling with the increase of amphiphiles concentration in aqueous environment. Their functional modification, biocompatibility and biodegradability make them preferable for nanocarrier DDS (Nie et al., 2007; Sinha et al., 2006; Tanaka et al., 2009).

Dendrimers have tree like symmetrical branched structure in nanometer size and can be ideal for anticancer drug carriers with active core site. Their biocompatibility and pharmacokinetic parameters vary upon shape, size and branching type, length and degree. They can be functionalized chemically without difficulty by changing end groups of dendrimers with hydrophilic ones to increase water solubility that facilitate effective distribution of hydrophobic drugs. They can be used as carrier to treat different diseases like AIDS, different cancer types, malaria, etc. (Brinton et al., 2008; Sharma et al., 2013; Sinha et al., 2006).

Liposomes are self-assembled lipid bilayers with aqueous hydrophobic core and lipid bilayer colloidal membrane. Drugs can be loaded into both the core and the membrane compartment depending on the active pharmaceutical ingredient. They are highly biocompatible and amphiphilic. They provide long lasting drug release with reduced cytotoxicity and enhanced drug accumulation (Sharma et al., 2013).

Nanotubes are single or multi walled structures where atoms are arranged as tubes with inorganic or organic forms. Carbon nanotubes are cylindrical nano-shape structures of carbon allotropes. They serve multiple functions depending on their biocompatibility, water solubility and chemical modification resulting in not only good carriers for DDS but also good materials for cancer cell imaging. However, there are still some toxicity reports and oxidative-stress pathway dependent cell death that may contradict the usage of nanotubes (Faraji and Wipf, 2009; Lamprecht et al., 2009; Sahoo et al., 2011).

Nanorods are semiconducting materials or metals with nano-sized rod shaped structure with good biocompatibility and high surface area. They can be synthesized as chemically inert. For example nontoxic gold nanorods are used for imaging and DDS (Parab et al., 2009; Tong et al., 2009).

Inorganic nanoparticles are composed of metals, metal oxides, and ceramics and produced with different size and porosities to protect entrapped molecule from denaturation or degradation. Polymeric nanoparticles have many features like biodegradability, biocompatibility and chemical modification of the surface gives chance to influence the pharmacokinetics of the drugs (Faraji and Wipf, 2009).

Nanocrystals are aggregates of crystalline-formed molecules and generally have hydrophobic compounds that are surrounded by thin hydrophilic layer which determines biological activity and distribution. Drugs with low solubility can be distributed with nanocrystals for bioavailability enhancement (Sagadevan and Periasamy, 2014).

Hydrophilic groups of glycans such as hydroxyl, carboxyl and amino groups provide solubility in water and these groups form noncovalent bonds with mucosal membranes and biological tissues (Liu et al., 2008). Bioadhesion and mucoadhesion to the target site are provided by the hydrophilic properties of most of the glycan nanocarriers. Glycans also have the possibility for chemical modification of the macromolecules to bind drugs or targeting agents. The hydrophilic glycan-based nanocarriers also have the huge advantage of prolonged circulation in blood, which increases the chance of passive targeting of the nanoparticles into the tumor tissues (Mitra et al., 2001). In addition to these distinguished physical and chemical properties, glycans are low-cost materials and found in abundant sources, which make them good biomaterials to medical and even pharmaceutical applications (Lemarchand et al., 2005; Park et al., 2010; Rinaudo, 2008).

6.4.1 PREPARATION OF GLYCAN-BASED NANOCARRIERS

Glycan-based DDS can be prepared by different preparation methods, which can be divided into four main categories: covalent crosslinking, ionic crosslinking, polyelectrolyte complexation, and self-assembly of hydrophobically modified glycans (Weinhart et al., 2011).

6.4.1.1 Covalent Crosslinking

Covalent bonds are introduced between the chains of glycan for the preparation of nanoparticle by covalent crosslinking that allows preparation of relatively robust nanoparticles. Covalent crosslinking is often not preferred because of the toxicity of the crosslinker agents and probable undesired side-reactions with the active ingredient (Bhattarai et al., 2010; Dash et al., 2011). On the other hand, biocompatible crosslinkers like natural di- and tricarboxylic acids have been used to prepare biodegradable chitosan nanoparticles by the aid of the water-soluble carbodiimides as condensation agents (Bodnar et al., 2005; Liu et al., 2008). The condensation reaction involves the carboxylic groups of the natural acids and the amine groups of chitosan.

6.4.1.2 Ionic Crosslinking

Ionic crosslinking has some advantages over covalent crosslinking due to milder conditions and simpler experimental procedures. Nanoparticles can be prepared by crosslinking polyelectrolytes ionically with multivalent ions of opposite charge under suitable experimental conditions such as appropriate pH and dilute concentrations of precursor materials (Dash et al., 2011; Liu et al., 2008; Vauthier and Bouchemal, 2009). For instance, nanoparticles of alginate can be prepared by crosslinking Ca^{2+} ions in aqueous media with the ionized carboxyl groups on the alginate chains (Hamidi et al., 2008). Chitosan nanoparticles are the most widely studied nanoparticles prepared by this method. Crosslinks are introduced by the action of tripolyphosphate (TPP) (Calvo et al., 1997). TPP is a generally recognized as safe (GRAS) substance by the Unites States Food and Drug Administration (FDA) (Lin et al., 2008) and is negatively charged over a wide pH range. Hence, it can electrostatically interact with the protonated and positively charged amine groups of chitosan to form nanogels or nanoparticles. The crosslinking process can also be achieved by hydrogen bonding between hydroxyl groups of chitosan (Bhattarai et al., 2010). Parameters that potentially affect preparation of nanoparticles by ionic crosslinking includes the ionic strength of the solvent (Huang and Lapitsky, 2011), the type (Liu et al., 2008) and molecular weight (Janes and Alonso, 2003; Yang and Hon, 2009) of glycan, the glycan concentration (Gan et al., 2005; Liu and Gao, 2009), the cross-linker to glycan ratio (Opanasopit et al., 2008; Shah et al., 2009), the pH (Ma et al., 2002; Zhang et al., 2004), and the mixing conditions (Dong et al., 2013; Tsai et al., 2008). Considering differences in bond strength, ionically cross-linked nanoparticles are in general less robust than covalently cross-linked nanoparticles (Berger et al., 2004).

6.4.1.3 Polyelectrolyte Complexation

Intermolecular electrostatic interactions result in the formation of complexes between glycans that have polyelectrolyte character and oppositely charged polymers. Chitosan is one of the few positively charged glycans and because of that, it is widely used. Other glycans require chemical modification such as oligo- or polyamines to obtain extensive positive charge. A common example of polyelectrolyte complexation is the interaction of chitosan with negatively charged alginate (Liu et al., 2008). Chitosan can also form polyelectrolyte complexes directly with negatively charged macromolecular drugs, such as nucleic acids (Buschmann et al., 2013). The mechanism for particle formation involves noncovalent, electrostatic interactions. Nanoparticles prepared by this method may be more robust than ionically cross-linked particles due to the large portion of glycans involved in the complex formation (Bhattarai et al., 2010).

6.4.1.4   Self-Assembly

Glycans with an amphiphilic character can spontaneously form nanoparticles in the form of micelles or polymer aggregates in an aqueous environment. The micelle formation takes place at concentrations above the critical micelle concentration. The underlying mechanism consists of inter and intra molecular associations between hydrophobic parts on the glycan chains to minimize the interfacial free energy (Liu et al., 2008). The particles can exhibit a core-shell structure with a hydrophobic core and a hydrophilic shell. Glycans are generally hydrophilic, but can naturally possess some hydrophobic parts or be hydrophobically modified to facilitate self-assembly. Chitosan for instance has been modified with long-chained linolenic acid to increase the amphiphilic character of the glycan (Liu et al., 2005). There has been some debate about the suitability of such self-assembled nanoparticles for drug delivery purposes, due to the possible loss of particle integrity upon dilution after administration (Rapoport, 2007). The nanoparticle preparation may also be a mixture of the different methods. Nanoparticles can for instance be prepared by covalent crosslinking and subsequently coated with a glycan of opposite charge through polyelectrolyte complexation on the particle surface.

Glycan-based nanocarriers can be divided into two main groups: glycan-functionalized nanocarriers and glycan-constructed nanocarriers.

6.4.2 GLYCAN-FUNCTIONALIZED NANOCARRIERS

Nanocarriers may adsorb proteins on their surface due to the hydrophobic interactions and surface energy of in vivo systems. This protein adsorption is known as opsonization which results in phagocytosis of the nanocarrier by the immune system (Kang et al., 2015).

Once nanoparticles are designed, they can be functionalized to differentiate surface characteristics or dispersions. Poly (ethylene glycol) (PEG) modification (PEGylation) can prolong in vivo plasma half-life and reduce opsonization (Kang et al., 2015). PEG is biologically inert and the most preferred compound for the functionalization. It can be covalently linked onto nanoparticle surfaces to reduce immunological reactions (Calvo et al., 2001, 2002). Main limitations of those artificial polymer functionalized nanocarriers (such as PEGylated nanocarriers) are biodegradability, biocompatibility and toxic products of degradation thus natural materials gained importance on functionalization (Frenz et al., 2015; Kang et al., 2015).

Properties such as biocompatibility, biodegradability, high water solubility, diverse functionalities and their well-defined structure make glycans well suited for biomedical applications as nanocarriers and as a possible alternative for PEGylation due to higher biodegradability ratios (Kang et al., 2015).

Carbohydrates play a crucial role in the communication and signaling of organisms. Most of the glycans present on the cell surface and have various important functions for cellular recognition, interaction, adhesion, signal transmission, cellular growth regulation, inflammation, metastasis, and immunologic reactions like bacterial and viral attachment. Cell surface glycans are entry sites for drugs (Ahmad et al., 2015). There are three classes of proteins which serve as a receptor for carbohydrate ligands: enzymes, immunoglobulins and lectins (Yamazaki et al., 2000). Examples of ligand and cell specificity were reported for both mono- and polysaccharides. Mannose and galactose can bind to different type of cells such as dendritic cells, alveolar macrophages or hepatic tumor cells via C type lectin receptors, rhamnose targets human skin cells while functionalized dextran prefers vascular smooth muscle cells and human endothelial cells (Cansell et al., 1999; Chavez-Santoscoy et al., 2012; Chen et al., 2014; Letourneur et al., 2000; Martínez-Ávila et al., 2009).

Wu et al. (2013) used thiol-functionalized mannose to improve the mesoporous silica nanoparticle pore activity depending on low PH conditions to target cancer cells or high glucose conditions like high blood sugar. Ahire et al. (2013) also worked on mannose functionalized silica nanoparticles to target MCF-7 (human breast adenocarcinoma cells) cells.

Another study was performed with glycol nanoparticle-based nanosensors with monosaccharide ligands like mannose, galactose, fucose and sialic acid by El-Boubbou et al. (2010). Monosaccharides immobilized onto amine functionalized nanoparticles are used for not only detection and differentiation of cancer cells but also to profile quantitatively the carbohydrate binding abilities of these cells.

Farr et al. (2014) synthesized functionalized super paramagnetic nanoparticles with Sialyl Lewis[x] (SX) to image early inflammation. SX is an important antigen for blood groups and has binding affinity and specificity to endothelial inflammatory E-P selectin proteins. These proteins are expressed as a result of inflammation on activated vascular endothelial cells. After treatment, accumulation at the brain vasculature was observed using MRI due to relaxation time of nanoparticles. Lartigue et al. (2009) studied phosphonated rhamnose for functionalization of nanoparticles. These highly magnetic iron oxide nanoparticles were used to image targeting effects on human skin cells.

Thiol functionalized gold nanoparticles were studied by Martinez-Avila et al. (2009) to design a carbohydrate-based drug against HIV using different oligomannosides which have high binding affinity to C type lectin on dendritic cells.

Frigell et al. (2014) increased cross rate of gold nanoparticles through blood-brain barrier with glycosylated gold nanoparticles. Sun et al. (2014) investigated toxicity difference of chitosan functionalized silver nanoparticles with hypothermia agents of PEGylated gold nanorods. Chitosan functionalized silver nanoparticles showed higher anticancer activity. Another study with chitosan functionalized gold nanoparticles was performed by Boca et al. (2011) and found that ethylene-glycan modified chitosan gold nanoparticles had enhanced stability and tumor targeting ability. Valodkar et al. (2012) highlighted the high antibacterial potential of starch functionalized copper nanoparticles against Gram-negative and Gram-positive bacteria. Gum Arabic capped gold nanoparticles with encapsulated epirubicin were studied by Devi et al. (2015) Surface of the gold nanoparticles was also functionalized with folic acid. Results showed that these functionalized nanoparticles had enhanced cytotoxicity against A549 (Human Lung Adenocarcinoma) cells compared to free epirubicin.

Needham et al. (2012) studied surface modification of micelles with hyaluronic acid. These micelles showed a decreased cytotoxicity and increased transfection efficiency. Functionalized micelles with lactabionic acid on the surface were investigated to target asialoglycoprotein receptor (ASGP-R) of liver cancer cells. Another micelle functionalization study with hyaluronic acid was performed by Liu et al. (2014) with glycan functionalized nanoparticle loaded with lovastin. Hyaluronic acid was adsorbed electrostatically to the cationic lipid core and it is observed that nanoparticle had fewer accumulations in liver and better efficiency on atherosclerotic lesion targeting which led to suppressed advancement of atherosclerosis.

Wang et al. (2014) studied dopamine encapsulated gold nanocages. Hyaluronic acid layer was used to protect dopamine by sealing nanopores and also gave specific cellular internalization to nanocage with the interactions of the CD44 receptors on cancer cells. Yu et al. (Yu et al., 2014) studied azide functionalized and self-assembled nanorods with low cytotoxicity. Results showed that nanorods can be used as cell glues depending on their high affinity to *E.coli* carbohydrate receptors for bacterial agglutination. Ohyanagi et al. (2011) studied sialic acid functionalized quantum dots and results showed that sialic acid functionalization prolonged the half-life in in vivo compared to other monosaccharide functionalized quantum dots.

6.4.3 GLYCAN-CONSTRUCTED NANOCARRIERS

Glycans are not only used for functionalization of the nanocarriers but also used for construction of those nano-sized devices/materials for biomedical applications since glycans play numerous roles in organisms from immunogenity to cell recognition, communications and so on. Many glycans play a role in different parts of the homeostatic mechanism. For instance sialic acid, mainly bound to glycoproteins is essential for the communication and recognition with the immune system. It is known that erythrocytes without sialic acid on the surface are removed rapidly from the blood by the immune system. Another monosaccharide, mannose, plays a crucial role in protein glycosylation. Mannose binding C-type lectin proteins are important for cell surface recognition and communication. Studies with mannose generally focus on cell surface targeting. Galactose is essential for cell targeting or blood type detection due to antigen structure. Hyaluronic acid is a common glycan for vertebrate tissues but is mostly found in connective tissues and body fluids with many functions like lubrication, plasma protein regulation, filtration, homeostasis of the water. Furthermore rhamnose which is generally found in bacteria and higher organisms such as plants plays important roles in cell survival.

Pietrzak-Nguyen et al. (2014) studied mannose nanocapsules which allow encapsulation of hydrophilic materials in the core. Those nanocapsules more particularly are deposited in lungs. Another study with the use of nanocapsule was performed by Roux et al. (2012) Azide functionalized sucrose was preferred for nanocapsulation and core was filled with miglyol for hydrophobic molecule upload. Acrylated dextran nanocapsules were tested to observe enzyme degradation possibility and pH changes by Malzahn et al. (2013). Antibacterial agent silver nanoparticle loaded into water soluble potato starch nanocapsules with the possibility of shell functionalization was reported by Taheri et al. (2014).

Self-assembled trehalose nanoparticles were prepared to protect proteins from physical and chemical degradation during storage by Giri et al. (2011). Gel forming ability of the trehalose under dehydrating conditions protects cells internal organelles. Ray et al. (2013) worked with CD44 overexpressing cancer cell binding ability of hexadecylated glycan nanoparticles which are self-assembled in water. Use of these nanoparticles revealed higher therapeutic potential in presence of epigallocatechin-3-gallate, which is a green tea polyphenol. Accumulation of the paclitaxel in human liver HepG2 carcinoma cells via oral uptake was found in the research of Li et al. (2013) by using self-assembled mitotic inhibitor hyaluronic acid- paclitaxel nanoparticles. Multilayered glycan vesicles were also used with different glycan types at the core and shell. A study with starch and hyaluronic acid was performed by Kwag et al. (2014) who constructed starch-hyaluronic acid nanoparticles. Amine groups of functionalized starch were reacted with activated ester groups of hyaluronic acid and this hyaluronic acid-starch shell-core was self-assembled in phosphate buffered saline. These particles were treated with amylase and hollow inner core was obtained. These vesicles can be used for protein or peptide encapsulation and release of these materials can be accelerated by the enzymatic degradation of hyaluronic acid shell by hyaluronidase. Different nanoparticle studies with $CaCl_2$ and $FeCl_3$ cross-linked carboxy-methyl-chitin with antibacterial activity and controlled release abilities were reported by Dev et al. (2010).

6.5 CONCLUDING REMARKS

Nanocarriers as DDS are considered to improve the pharmacological and therapeutic properties of drugs. The integration of the drug into a nanocarrier can offer targeting possibilities, protection against degradation and provide controlled release. Since nanocarriers have small dimensions, they are able to cross biological barriers and function on cellular level. Nanocarrier-drug conjugates are more effective when compared with the traditional form of drugs. Nanocarriers accumulate drugs in target sites and hence, they can reduce the toxicity and other adverse side effects in healthy tissues. As a result, the required doses of drugs are lower.

Up to date, nanocarriers are mostly studied in terms of their physicochemical properties, in vitro toxicity and drug-loading ability. On the other hand, the more important topics, such as the specific interaction of these nanoparticles with human tissues, their effect on human metabolism and the wider application of these nanoparticles for drug delivery, etc. needs to be investigated in more detail. Moreover, new methods for the earlier diagnosis of diseases and more efficient therapies by a new generation of multifunctional nanostructured materials-based on glycans, modified glycans and glycan-based nanocarriers has gained enormous interest in recent years. Therefore, more glycan-based nanoparticles will emerge in the future, that will significantly enrich the versatility of nanoparticle carrier agents. With more investigation and understanding of polysaccharides and discovery of more functional polysaccharides, the sequences of glycans can be controlled more specifically by chemical and biosynthetic methods and the structure of glycans can be modified in order to improve their properties, such as hydrophobicity, hydrophilicity, ionic strength sensitivity, pH sensitivity, temperature sensitivity, and others. It will further increase the use of glycans in the drug delivery.

Moreover, there are numerous limitations and side effects related with the traditional cancer treatment methods. Hence, new approaches are being comprehensively studied and investigated for anticancer drug delivery to tumor cells. For the development of an anticancer drug delivery system, nanoparticles are promising candidates due to their various benefits over conventional chemotherapy such as improvement of the drug water solubility, protection from macrophage uptake, resulting in an enhancement of the blood circulation time, and finally an increased specificity and controlled release to the desired final target. Especially, glycan-based nanoparticles not only have importance in the ability of chemotherapeutic drugs to be effectively delivered to specific tumor cells but also have advantages such as water solubility, biocompatibility and biodegradability. Although major development on the glycan-based systems for the anticancer drug delivery has been made recently, an ideal carrier has not yet been achieved.

KEYWORDS

- drug delivery systems
- glycan
- glycotechnology
- nanocarriers
- nanoparticles
- polysaccharides

REFERENCES

Ahire, J. H.; Chambrier, I.; Mueller, A.; Bao, Y.; Chao, Y. Synthesis of D-mannose capped silicon nanoparticles and their interactions with MCF-7 human breast cancerous cells. *ACS Appl. Mater. Interfaces.* **2013,** *5*(15), 7384–7391.

Ahmad, M. U.; Ali, S. M.; Ahmad, A.; Sheikh, S.; Chen, P.; Ahmad, I. Carbohydrate mediated drug delivery: Synthesis and characterization of new lipid-conjugates. *Chem. Phys. Lipids.* **2015**, *186*, 30–38.

Apostolopoulos, V.; Pietersz, G. A.; Tsibanis, A.; Tsikkinis, A.; Drakaki, H.; Loveland, B. E.; Piddlesden, S. J.; Plebanski, M.; Pouniotis, D. S.; Alexis, M. N. Pilot phase III immunotherapy study in early stage breast cancer patients using oxidized mannan-MUC1 [ISRCTN71711835]. *Breast Cancer Res.* **2006**, *8*(3), R27.

Berger, J.; Reist, M.; Mayer, J. M.; Felt, O.; Peppas, N.; Gurny, R. Structure and interactions in covalently and ionically cross-linked chitosan hydrogels for biomedical applications. *Eur. J. Pharm. Biopharm.* **2004**, *57*(1), 19–34.

Bernkop-Schnürch, A.; Dünnhaupt, S. Chitosan-based drug delivery systems. *Eur. J. Pharm. Biopharm.* **2012**, *81*(3), 463–469.

Bhattarai, N.; Gunn, J.; Zhang, M. Chitosan-based hydrogels for controlled, localized drug delivery. *Adv Drug Deliv Rev.* **2010**, *62*(1), 83–99.

Boca, S. C.; Potara, M.; Gabudean, A.-M.; Juhem, A.; Baldeck, P. L.; Astilean, S. Chitosan-coated triangular silver nanoparticles as a novel class of biocompatible, highly effective photothermal transducers for in vitro cancer cell therapy. *Cancer Lett.* **2011**, *311*(2), 131–140.

Boddohi, S.; Killingsworth, C. E.; Kipper, M. J. Polyelectrolyte multilayer assembly as a function of pH and ionic strength using the polysaccharides chitosan and heparin. *Biomacromolecules.* **2008**, *9*(7), 2021–2028.

Bodnar, M.; Hartmann, J. F.; Borbely, J. Preparation and characterization of chitosan-based nanoparticles. *Biomacromolecules.* **2005**, *6*(5), 2521–2527.

Bonferoni, M.; Sandri, G.; Ferrari, F.; Rossi, S.; Larghi, V.; Zambito, Y.; Caramella, C. Comparison of different in vitro and ex vivo methods to evaluate mucoadhesion of glycol-palmitoyl chitosan micelles. *J. Drug Delivery Sci. Technol.* **2010**, *20*(6), 419–424.

Bostan, M. S.; Mutlu, E. C.; Kazak, H.; Keskin, S. S.; Toksoy Oner, E.; Eroglu, M. S. Comprehensive characterization of chitosan/PEO/levan ternary blend films. *Carbohydr. Polym.* **2014**, *102*, 993–1000.

Brinton, L. A.; Sherman, M. E.; Carreon, J. D.; Anderson, W. F. Recent trends in breast cancer among younger women in the United States. *J. Natl. Cancer Inst.* **2008**, *100*(22), 1643–1648.

Brown, M.; Jones, S. A. Hyaluronic acid: a unique topical vehicle for the localized delivery of drugs to the skin. *J Eur Acad Dermatol Venereol.* **2005**, *19*(3), 308–318.

Budzynska, R.; Nevozhay, D.; Kanska, U.; Jagiello, M.; Opolski, A.; Wietrzyk, J.; Boratynski, J. Antitumor activity of mannan–methotrexate conjugate in vitro and in vivo. *Onkol Res.* **2007**, *16*(9), 415–421.

Buschmann, M. D.; Merzouki, A.; Lavertu, M.; Thibault, M.; Jean, M.; Darras, V. Chitosans for delivery of nucleic acids. *Adv Drug Deliv Rev.* **2013**, *65*(9), 1234–1270.

Buzea, C.; Pacheco, I. I.; Robbie, K. Nanomaterials and nanoparticles: sources and toxicity. *Biointerphases.* **2007**, *2*(4), MR17-MR71.

Byrne, J. D.; Betancourt, T.; Brannon-Peppas, L. Active targeting schemes for nanoparticle systems in cancer therapeutics. *Adv Drug Deliv Rev.* **2008**, *60*(15), 1615–1626.

Cadée, J.; Brouwer, L.; Den Otter, W.; Hennink, W.; Van Luyn, M. A comparative biocompatibility study of microspheres-based on cross-linked dextran or poly (lactic-co-glycolic) acid after subcutaneous injection in rats. *J. Biomed. Mater. Res.* **2001**, *56*(4), 600–609.

Calvo, P.; Gouritin, B.; Chacun, H.; Desmaële, D.; D'Angelo, J.; Noel, J.-P.; Georgin, D.; Fattal, E.; Andreux, J. P.; Couvreur, P. Long-circulating PEGylated polycyanoacrylate nanoparticles as new drug carrier for brain delivery. *Pharm. Res.* **2001**, *18*(8), 1157–1166.

Calvo, P.; Gouritin, B.; Villarroya, H.; Eclancher, F.; Giannavola, C.; Klein, C.; Andreux, J. P.; Couvreur, P. Quantification and localization of PEGylated polycyanoacrylate nanoparticles in brain and spinal cord during experimental allergic encephalomyelitis in the rat. *Eur. J. Neurosci.* **2002**, *15*(8), 1317–1326.

Calvo, P.; Remunan-Lopez, C.; Vila-Jato, J.; Alonso, M. Novel hydrophilic chitosan-poly-ethylene oxide nanoparticles as protein carriers. *J. Appl. Polym. Sci.* **1997**, *63*(1), 125–132.

Cansell, M.; Parisel, C.; Jozefonvicz, J.; Letourneur, D. Liposomes coated with chemically modified dextran interact with human endothelial cells. *J. Biomed. Mater. Res.* **1999**, *44*(2), 140–148.

Chavez-Santoscoy, A. V.; Roychoudhury, R.; Pohl, N. L.; Wannemuehler, M. J.; Narasimhan, B.; Ramer-Tait, A. E. Tailoring the immune response by targeting C-type lectin receptors on alveolar macrophages using "pathogen-like" amphiphilic polyanhydride nanoparticles. *Biomaterials.* **2012**, *33*(18), 4762–4772.

Chen, W.; Zou, Y.; Meng, F.; Cheng, R.; Deng, C.; Feijen, J.; Zhong, Z. Glyconanoparticles with sheddable saccharide shells: a unique and potent platform for hepatoma-targeting delivery of anticancer drugs. *Biomacromolecules.* **2014**, *15*(3), 900–907.

Chono, S.; Tauchi, Y.; Deguchi, Y.; Morimoto, K. Efficient drug delivery to atherosclerotic lesions and the antiatherosclerotic effect by dexamethasone incorporated into liposomes in atherogenic mice. *J. Drug Targeting.* **2005**, *13*(4), 267–276.

Costa, R. R.; Neto, A. I.; Calgeris, I.; Correia, C. R.; Pinho, A. C.; Fonseca, J.; Öner, E. T.; Mano, J. F. Adhesive nanostructured multilayer films using a bacterial exopolysaccharide for biomedical applications. *J. Mater. Chem. B* **2013**, *1*(18), 2367–2374.

Coviello, T.; Matricardi, P.; Marianecci, C.; Alhaique, F. Polysaccharide hydrogels for modified release formulations. *J. Controlled Release.* **2007**, *119*(1), 5–24.

Dai, T.; Tanaka, M.; Huang, Y.-Y.; Hamblin, M. R. Chitosan preparations for wounds and burns: antimicrobial and wound-healing effects. *Expert Rev Anti Infect Ther.* **2011** Jul; *9*(7), 857–879.

Dang, J. M.; Leong, K. W. Natural polymers for gene delivery and tissue engineering. *Adv. Drug Deliv. Rev.* **2006**, *58*(4), 487–99.

Dash, M.; Chiellini, F.; Ottenbrite, R. M.; Chiellini, E. Chitosan—A versatile semi-synthetic polymer in biomedical applications. *Prog. Polym. Sci.* **2011**, *36*(8), 981–1014.

Dev, A.; Mohan, J. C.; Sreeja, V.; Tamura, H.; Patzke, G.; Hussain, F.; Weyeneth, S.; Nair, S.; Jayakumar, R. Novel carboxymethyl chitin nanoparticles for cancer drug delivery applications. *Carbohydr. Polym.* **2010**, *79*(4), 1073–1079.

Devi, P. R.; Kumar, C. S.; Selvamani, P.; Subramanian, N.; Ruckmani, K. Synthesis and characterization of Arabic gum capped gold nanoparticles for tumor-targeted drug delivery. *Materials Lett.* **2015**, *139*, 241–244.

Dong, Y.; Ng, W. K.; Shen, S.; Kim, S.; Tan, R. B. Scalable ionic gelation synthesis of chitosan nanoparticles for drug delivery in static mixers. *Carbohydr. Polym.* **2013,** *94*(2), 940–945.

Duncan, R. Polymer conjugates as anticancer nanomedicines. *Nat. Rev. Cancer.* **2006,** *6*(9), 688–701.

Dwek, R. A. Glycobiology: toward understanding the function of sugars. *Chem. Rev.* **1996,** *96*(2), 683–720.

El-Boubbou, K.; Zhu, D. C.; Vasileiou, C.; Borhan, B.; Prosperi, D.; Li, W.; Huang, X. Magnetic glyconanoparticles: a tool to detect, differentiate, and unlock the glyco-codes of cancer via magnetic resonance imaging. *J. Am. Chem. Soc.* **2010,** *132*(12), 4490–4499.

Fang, N.; Chan, V.; Mao, H.-Q.; Leong, K. W. Interactions of phospholipid bilayer with chitosan: effect of molecular weight and pH. *Biomacromolecules.* **2001,** *2*(4), 1161–1168.

Faraji, A. H.; Wipf, P. Nanoparticles in cellular drug delivery. *Bioorg. Med. Chem.* **2009,** *17*(8), 2950–2962.

Farr, T. D.; Lai, C.-H.; Grünstein, D.; Orts-Gil, G.; Wang, C.-C.; Boehm-Sturm, P.; Seeberger, P. H.; Harms, C. Imaging early endothelial inflammation following stroke by core shell silica superparamagnetic glyconanoparticles that target selectin. *Nano Lett.* **2014,** *14*(4), 2130–2134.

Ferrari, M. Cancer nanotechnology: opportunities and challenges. *Nat. Rev. Cancer.* **2005,** *5*(3), 161–171.

Frenz, T.; Grabski, E.; Durán, V.; Hozsa, C.; Stępczyńska, A.; Furch, M.; Gieseler, R. K.; Kalinke, U. Antigen presenting cell-selective drug delivery by glycan-decorated nanocarriers. *Eur. J. Pharm. Biopharm.* **2015,** *95,* 13–7.

Gan, Q.; Wang, T.; Cochrane, C.; McCarron, P. Modulation of surface charge, particle size and morphological properties of chitosan–TPP nanoparticles intended for gene delivery. *Colloids Surf., B.* **2005,** *44*(2), 65–73.

Ghazarian, H.; Idoni, B.; Oppenheimer, S. B. A glycobiology review: carbohydrates, lectins and implications in cancer therapeutics. *Acta Histochem.* **2011,** 113(3), 236–247.

Giri, J.; Li, W. J.; Tuan, R. S.; Cicerone, M. T. Stabilization of proteins by nanoencapsulation in sugar–glass for tissue engineering and drug delivery applications. *Adv. Mater.* **2011,** *23*(42), 4861–4867.

Gorelik, E.; Galili, U.; Raz, A. On the role of cell surface carbohydrates and their binding proteins (lectins) in tumor metastasis. *Canc. Metastasis Rev.* **2001,** *20*(3–4), 245–277.

Gu, L.; Faig, A.; Abdelhamid, D.; Uhrich, K. Sugar-based amphiphilic polymers for biomedical applications: from nanocarriers to therapeutics. *Acc. Chem. Res.* **2014,** *47*(10), 2867–2877.

Gupta, A.; Gupta, R. K.; Gupta, G. Targeting cells for drug and gene delivery: emerging applications of mannans and mannan binding lectins. *J. Sci. Ind. Res.* **2009,** *68*(6), 465–483.

Hamidi, M.; Azadi, A.; Rafiei, P. Hydrogel nanoparticles in drug delivery. *Adv Drug Deliv Rev.* **2008,** *60*(15), 1638–1649.

Han, Y. W.; Clarke, M. A. Production and characterization of microbial levan. *J. Agric. Food. Chem.* **1990,** *38*(2), 393–396.

Hemberger, J.; Orlando, M. Use of Starch Derivatives to Improve Protein Drug Solubility and Efficacy. *Chem. Ing. Tech.* **2006**, *78*(9), 1188–1188.

Hemberger, J.; Orlando, M.; Sommermeyer, K.; Eichner, W.; Frie, S.; Lutterbeck, K.; Jungheinrich, C.; Scharpf, R. Coupling proteins to a modified polysaccharide, **2009**, Google Patents.

Huang, G.; Mei, X.; Xiao, F.; Chen, X.; Tang, Q.; Peng, D. Applications of Important Polysaccharides in Drug Delivery. *Curr. Pharm. Des.* **2015**, *5*, 3692–3695.

Huang, Y.; Lapitsky, Y. Monovalent salt enhances colloidal stability during the formation of chitosan/tripolyphosphate microgels. *Langmuir.* **2011**, *27*(17), 10392–10399.

Jain, A.; Agarwal, A.; Majumder, S.; Lariya, N.; Khaya, A.; Agrawal, H.; Majumdar, S.; Agrawal, G. P. Mannosylated solid lipid nanoparticles as vectors for site-specific delivery of an anticancer drug. *J. Controlled Release.* **2010**, *148*(3), 359–367.

Jain, A.; Gupta, Y.; Jain, S. K. Perspectives of biodegradable natural polysaccharides for site-specific drug delivery to the colon. *J Pharm Pharm Sci.* **2007**, *10*(1), 86–128.

Janes, K.; Alonso, M. Depolymerized chitosan nanoparticles for protein delivery: preparation and characterization. *J. Appl. Polym. Sci.* **2003**, *88*(12), 2769–2776.

Kanapathipillai, M.; Brock, A.; Ingber, D. E. Nanoparticle targeting of anticancer drugs that alter intracellular signaling or influence the tumor microenvironment. *Adv Drug Deliv Rev.* **2014**, *79*, 107–118.

Kang, B.; Opatz, T.; Landfester, K.; Wurm, F. R. Carbohydrate nanocarriers in biomedical applications: functionalization and construction. *Chem. Soc. Rev.* **2015**, *21*(44–22), 8301–8325.

Kang, S. A.; Jang, K.-H.; Seo, J.-W.; Kim, K. H.; Kim, Y. H.; Rairakhwada, D.; Seo, M. Y.; Lee, J. O.; Ha, S.; Kim, C. Levan: applications and perspectives. Microbial production of biopolymers and polymer precursors. *Caister Academic Press*, Norfolk, UK **2009**, 145–161.

Kazak Sarilmiser, H.; Toksoy Oner, E. Investigation of anticancer activity of linear and aldehyde-activated levan from *Halomonas smyrnensis* AAD6 T. *Biochem. Eng. J.* **2014**, *92*, 28–34.

Kim, S.-J.; Bae, P. K.; Chung, B. H. Self-assembled levan nanoparticles for targeted breast cancer imaging. *Chem. Commun.* **2015**, *51*(1), 107–110.

Kirpotin, D. B.; Drummond, D. C.; Shao, Y.; Shalaby, M. R.; Hong, K.; Nielsen, U. B.; Marks, J. D.; Benz, C. C.; Park, J. W. Antibody targeting of long-circulating lipidic nanoparticles does not increase tumor localization but does increase internalization in animal models. *Cancer Res.* **2006**, *66*(13), 6732–6740.

Kumar, M. R.; Muzzarelli, R. A.; Muzzarelli, C.; Sashiwa, H.; Domb, A. Chitosan chemistry and pharmaceutical perspectives. *Chem. Rev.* **2004**, *104*(12), 6017–6084.

Kwag, D. S.; Oh, K. T.; Lee, E. S. Facile synthesis of multilayered polysaccharidic vesicles. *J. Controlled Release.* **2014**, *187*, 83–90.

Lamprecht, C.; Liashkovich, I.; Neves, V.; Danzberger, J.; Heister, E.; Rangl, M.; Coley, H. M.; McFadden, J.; Flahaut, E.; Gruber, H. J. AFM imaging of functionalized carbon nanotubes on biological membranes. *Nanotechnology.* **2009**, *20*, 434001, 1–7.

Landesman-Milo, D.; Ramishetti, S.; Peer, D. Nanomedicine as an emerging platform for metastatic lung cancer therapy. *Canc. Metastasis Rev.* **2015**, *20*(43), 291–301.

Lartigue, L.; Oumzil, K.; Guari, Y.; Larionova, J.; Guérin, C.; Montero, J.-L.; Barragan-Montero, V.; Sangregorio, C.; Caneschi, A.; Innocenti, C. Water-soluble rhamnose-coated Fe3O4 nanoparticles. *Organic Lett.* **2009**, *11*(14), 2992–2995.

Lavan, D. A.; McGuire, T.; Langer, R. Small-scale systems for in vivo drug delivery. *Nat. Biotechnol.* **2003**, *21*(10), 1184–1191.

Lemarchand, C.; Gref, R.; Couvreur, P. Polysaccharide-decorated nanoparticles. *Eur. J. Pharm. Biopharm.* **2004**, *58*(2), 327–341.

Lemarchand, C.; Gref, R.; Lesieur, S.; Hommel, H.; Vacher, B.; Besheer, A.; Maeder, K.; Couvreur, P. Physicochemical characterization of polysaccharide-coated nanoparticles. *J. Controlled Release.* **2005**, *108*(1), 97–111.

Letourneur, D.; Parisel, C.; Prigent-Richard, S.; Cansell, M. Interactions of functionalized dextran-coated liposomes with vascular smooth muscle cells. *J. Controlled Release.* **2000**, *65*(1), 83–91.

Lewis, D. R.; Kholodovych, V.; Tomasini, M. D.; Abdelhamid, D.; Petersen, L. K.; Welsh, W. J.; Uhrich, K. E.; Moghe, P. V. In silico design of antiatherogenic biomaterials. *Biomaterials.* **2013**, *34*(32), 7950–7959.

Li, J.; Huang, P.; Chang, L.; Long, X.; Dong, A.; Liu, J.; Chu, L.; Hu, F.; Liu, J.; Deng, L. Tumor targeting and pH-responsive polyelectrolyte complex nanoparticles-based on hyaluronic acid-paclitaxel conjugates and chitosan for oral delivery of paclitaxel. *Macromol. Res.* **2013**, *21*(12), 1331–1337.

Lin, Y.-H.; Sonaje, K.; Lin, K. M.; Juang, J.-H.; Mi, F.-L.; Yang, H.-W.; Sung, H.-W. Multi-ion-cross-linked nanoparticles with pH-responsive characteristics for oral delivery of protein drugs. *J. Controlled Release.* **2008**, *132*(2), 141–149.

Liu, C.-G.; Desai, K. G. H.; Chen, X.-G.; Park, H.-J. Linolenic acid-modified chitosan for formation of self-assembled nanoparticles. *J. Agric. Food. Chem.* **2005**, *53*(2), 437–441.

Liu, H.; Gao, C. Preparation and properties of ionically cross-linked chitosan nanoparticles. *Polym. Adv. Technol.* **2009**, *20*(7), 613–619.

Liu, L.; He, H.; Zhang, M.; Zhang, S.; Zhang, W.; Liu, J. Hyaluronic acid-decorated reconstituted high density lipoprotein targeting atherosclerotic lesions. *Biomaterials.* **2014**, *35*(27), 8002–8014.

Liu, Z.; Jiao, Y.; Wang, Y.; Zhou, C.; Zhang, Z. Polysaccharides-based nanoparticles as drug delivery systems. *Adv Drug Deliv Rev.* **2008**, *60*(15), 1650–1662.

Lowe, J. B.; Marth, J. D. A genetic approach to mammalian glycan function. *Annu. Rev. Biochem.* **2003**, *72*(1), 643–691.

Ma, Z.; Yeoh, H. H.; Lim, L. Y. Formulation pH modulates the interaction of insulin with chitosan nanoparticles. *J. Pharm. Sci.* **2002**, *91*(6), 1396–1404.

Maciel, J.; Andrad, P.; Neri, D.; Carvalho, L.; Cardoso, C.; Calazans, G.; Aguiar, J. A.; Silva, M. Preparation and characterization of magnetic levan particles as matrix for trypsin immobilization. *J. Magn. Magn. Mater.* **2012**, *324*(7), 1312–1316.

Maeda, H.; Matsumura, Y. Tumoritropic and lymphotropic principles of macromolecular drugs. *Crit. Rev. Ther. Drug Carrier Syst.* **1988**, *6*(3), 193–210.

Malzahn, K.; Marsico, F.; Koynov, K.; Landfester, K.; Weiss, C. K.; Wurm, F. R. Selective interfacial olefin cross metathesis for the preparation of hollow nanocapsules. *ACS Macro Lett.* **2013**, *3*(1), 40–43.

Martin, T.; Jiang, W. Anti-Cancer agents in medicinal chemistry (Formerly current medicinal chemistry-anti-cancer agents). *Anti-Cancer Agents Med. Chem.* **2010**, *10*(1), 1.

Martínez-Ávila, O.; Hijazi, K.; Marradi, M.; Clavel, C.; Campion, C.; Kelly, C.; Penadés, S. Gold Manno-Glyconanoparticles: Multivalent Systems to Block HIV-1 gp120 Binding to the Lectin DC-SIGN. *Chem. Eur. J.* **2009**, *15*(38), 9874–9888.

Mehvar, R. Recent trends in the use of polysaccharides for improved delivery of therapeutic agents: pharmacokinetic and pharmacodynamic perspectives. *Current Pharm. Biotechnol.* **2003**, *4*(5), 283–302.

Mitra, S.; Gaur, U.; Ghosh, P.; Maitra, A. Tumour targeted delivery of encapsulated dextran–doxorubicin conjugate using chitosan nanoparticles as carrier. *J. Controlled Release.* **2001**, *74*(1), 317–323.

Mizrahy, S.; Peer, D. Polysaccharides as building blocks for nanotherapeutics. *Chem. Soc. Rev.* **2012**, *41*(7), 2623–2640.

Moghimi, S. M.; Hunter, A. C. Poloxamers and poloxamines in nanoparticle engineering and experimental medicine. *Trends Biotechnol.* **2000**, *18*(10), 412–420.

Morille, M.; Passirani, C.; Vonarborg, A.; Clavreul, A.; Benoit, J.-P. Progress in developing cationic vectors for nonviral systemic gene therapy against cancer. *Biomaterials.* **2008**, *29*(24), 3477–3496.

Nakapong, S.; Pichyangkura, R.; Ito, K.; Iizuka, M.; Pongsawasdi, P. High expression level of levansucrase from Bacillus licheniformis RN-01 and synthesis of levan nanoparticles. *Int. J. Biol. Macromol.* **2013**, *54*, 30–36.

Needham, C. J.; Williams, A. K.; Chew, S. A.; Kasper, F. K.; Mikos, A. G. Engineering a polymeric gene delivery vector-based on poly (ethylenimine) and hyaluronic acid. *Biomacromolecules.* **2012**, *13*(5), 1429–1437.

Nie, S.; Xing, Y.; Kim, G. J.; Simons, J. W. Nanotechnology applications in cancer. *Annu. Rev. Biomed. Eng.* **2007**, *9*, 257–288.

Ofek, I.; Hasty, D. L.; Doyle, R. J. Bacterial adhesion to animal cells and tissues. American Society of Microbiology, USA, **2003a**.

Ofek, I.; Hasty, D. L.; Sharon, N. Anti-adhesion therapy of bacterial diseases: prospects and problems. *FEMS Immunol. Med. Microbiol.* **2003b**, *38*(3), 181–191.

Ohyanagi, T.; Nagahori, N.; Shimawaki, K.; Hinou, H.; Yamashita, T.; Sasaki, A.; Jin, T.; Iwanaga, T.; Kinjo, M.; Nishimura, S.-I. Importance of sialic acid residues illuminated by live animal imaging using phosphorylcholine self-assembled monolayer-coated quantum dots. *J. Am. Chem. Soc.* **2011**, *133*(32), 12507–12517.

Opanasopit, P.; Apirakaramwong, A.; Ngawhirunpat, T.; Rojanarata, T.; Ruktanonchai, U. Development and characterization of pectinate micro/nanoparticles for gene delivery. *AAPS PharmSciTech.* **2008**, *9*(1), 67–74.

Panitch, A.; Paderi, J. E.; Sharma, S.; Stuart, K. A.; Vazquez-Portalatin, N. M. Extracellular matrix-binding synthetic peptidoglycans, **2014**, Google Patents.

Parab, H. J.; Chen, H. M.; Lai, T.-C.; Huang, J. H.; Chen, P. H.; Liu, R.-S.; Hsiao, M.; Chen, C.-H.; Tsai, D.-P.; Hwu, Y.-K. Biosensing, cytotoxicity, and cellular uptake studies of surface-modified gold nanorods. *J. Phys. Chem. C.* **2009**, *113*(18), 7574–7578.

Park, E. K.; Lee, S. B.; Lee, Y. M. Preparation and characterization of methoxy poly (ethylene glycol)/poly (ε-caprolactone) amphiphilic block copolymeric nanospheres for tumor-specific folate-mediated targeting of anticancer drugs. *Biomaterials.* **2005a,** *26*(9), 1053–1061.

Park, J. H.; Saravanakumar, G.; Kim, K.; Kwon, I. C. Targeted delivery of low molecular drugs using chitosan and its derivatives. *Adv Drug Deliv Rev.* **2010,** *62*(1), 28–41.

Park, K.-H.; Sung, W. J.; Kim, S.; Kim, D. H.; Akaike, T.; Chung, H.-M. Specific interaction of mannosylated glycopolymers with macrophage cells mediated by mannose receptor. *J. Biosci. Bioeng.* **2005,** *99*(3), 285–289.

Peer, D.; Karp, J. M.; Hong, S.; Farokhzad, O. C.; Margalit, R.; Langer, R. Nanocarriers as an emerging platform for cancer therapy. *Nat. Nanotechnol.* **2007,** *2*(12), 751–760.

Petersen, L. K.; York, A. W.; Lewis, D. R.; Ahuja, S.; Uhrich, K. E.; Prud'homme, R. K.; Moghe, P. V. Amphiphilic nanoparticles repress macrophage atherogenesis: Novel core/shell designs for scavenger receptor targeting and down-regulation. *Mol. Pharmaceutics.* **2014,** *11*(8), 2815–2824.

Pietrzak-Nguyen, A.; Fichter, M.; Dedters, M.; Pretsch, L.; Gregory, S. H.; Meyer, C.; Doganci, A.; Diken, M.; Landfester, K.; Baier, G. Enhanced in vivo targeting of murine nonparenchymal liver cells with monophosphoryl lipid A functionalized microcapsules. *Biomacromolecules.* **2014,** *15*(7), 2378–2388.

Plourde, N. M.; Kortagere, S.; Welsh, W.; Moghe, P. V. Structure– Activity Relations of Nanolipoblockers with the Atherogenic Domain of Human Macrophage Scavenger Receptor A. *Biomacromolecules.* **2009,** *10*(6), 1381–1391.

Rapoport, N. Physical stimuli-responsive polymeric micelles for anticancer drug delivery. *Prog. Polym. Sci.* **2007,** *32*(8), 962–990.

Ratner, B. D.; Bryant, S. J. Biomaterials: where we have been and where we are going. *Annu Rev Biomed Eng.* **2004,** *6*, 41–75.

Ray, L.; Kumar, P.; Gupta, K. C. The activity against Ehrlich's ascites tumors of doxorubicin contained in self assembled, cell receptor targeted nanoparticle with simultaneous oral delivery of the green tea polyphenol epigallocatechin-3-gallate. *Biomaterials.* **2013,** *34*(12), 3064–3076.

Raz, A.; Meromsky, L.; Lotan, R. Differential expression of endogenous lectins on the surface of nontumorigenic, tumorigenic, and metastatic cells. *Cancer Res.* **1986,** *46*(7), 3667–3672.

Reddy, P. C.; Chaitanya, K.; Rao, Y. M. A review on bioadhesive buccal drug delivery systems: current status of formulation and evaluation methods. *DARU.* **2011,** *19*(6), 385.

Rele, S. M.; Cui, W.; Wang, L.; Hou, S.; Barr-Zarse, G.; Tatton, D.; Gnanou, Y.; Esko, J. D.; Chaikof, E. L. Dendrimer-like PEO glycopolymers exhibit antiinflammatory properties. *J. Am. Chem. Soc.* **2005,** *127*(29), 10132–10133.

Renuart, E.; Viney, C. Biological fibrous materials: self-assembled structures and optimized properties. *Pergamon Mat. Ser.* **2000,** *4*, 223–267.

Rinaudo, M. Main properties and current applications of some polysaccharides as biomaterials. *Polym. Int.* **2008,** *57*(3), 397–430.

Roux, R.m.; Sallet, L.; Alcouffe, P.; Chambert, S.p.; Sintes-Zydowicz, N.; Fleury, E.; Bernard, J. Facile and rapid access to glyconanocapsules by CuAAC interfacial polyaddition in miniemulsion conditions. *ACS Macro Lett.* **2012,** *1*(8), 1074–1078.

Sagadevan, S.; Periasamy, M. A review on role of nanostructures in drug delivery system. *Rev. Adv. Mater. Sci.* **2014,** *36*(2), 112–117.

Sahoo, N. G.; Bao, H.; Pan, Y.; Pal, M.; Kakran, M.; Cheng, H. K. F.; Li, L.; Tan, L. P. Functionalized carbon nanomaterials as nanocarriers for loading and delivery of a poorly water-soluble anticancer drug: a comparative study. *Chem. Commun.* **2011**, *47*(18), 5235–5237.

Sezer, A. D.; Kazak, H.; Ö ner, E. T.; Akbuğa, J. Levan-based nanocarrier system for peptide and protein drug delivery: optimization and influence of experimental parameters on the nanoparticle characteristics. *Carbohydr. Polym.* **2011**, *84*(1), 358–363.

Shah, S.; Pal, A.; Kaushik, V.; Devi, S. Preparation and characterization of venlafaxine hydrochloride-loaded chitosan nanoparticles and in vitro release of drug. *J. Appl. Polym. Sci.* **2009**, *112*(5), 2876–2887.

Sharma, A.; Jain, N.; Sareen, R. Nanocarriers for diagnosis and targeting of breast cancer. *Biomed Res. Int.* **2013**.

Sharon, N.; Lis, H. Lectins as cell recognition molecules. *Science.* **1989**, *246*(4927), 227–234.

Shelke, N. B.; James, R.; Laurencin, C. T.; Kumbar, S. G. Polysaccharide biomaterials for drug delivery and regenerative engineering. *Polym. Adv. Technol.* **2014**, *25*(5), 448–460.

Sima, F.; Axente, E.; Sima, L.; Tuyel, U.; Eroglu, M.; Serban, N.; Ristoscu, C.; Petrescu, S.; Toksoy Oner, E.; Mihailescu, I. Combinatorial matrix-assisted pulsed laser evaporation: Single-step synthesis of biopolymer compositional gradient thin film assemblies. *Appl. Phys. Lett.* **2012**, *101*(23), 233705.

Sima, F.; Mutlu, E. C.; Eroglu, M. S.; Sima, L. E.; Serban, N.; Ristoscu, C.; Petrescu, S. M.; Toksoy Oner, E.; Mihailescu, I. N. Levan nanostructured thin films by MAPLE assembling. *Biomacromolecules.* **2011**, *12*(6), 2251–2256.

Sinha, R.; Kim, G. J.; Nie, S.; Shin, D. M. Nanotechnology in cancer therapeutics: bioconjugated nanoparticles for drug delivery. *Mol. Canc. Therapeut.* **2006**, *5*(8), 1909–1917.

Sinha, V.; Singla, A.; Wadhawan, S.; Kaushik, R.; Kumria, R.; Bansal, K.; Dhawan, S. Chitosan microspheres as a potential carrier for drugs. *Int. J. Pharm.* **2004**, *274*(1), 1–33.

Sirisha, V. Polysaccharide nanoparticles: preparation and their potential application as drug delivery systems. *IJRANSS* **2015**, *3*(5), 69–94.

Spencer, D. S.; Puranik, A. S.; Peppas, N. A. Intelligent nanoparticles for advanced drug delivery in cancer treatment. *Curr. Opin. Chem. Eng.* **2015**, *7*, 84–92.

Sun, H.; Guo, B.; Li, X.; Cheng, R.; Meng, F.; Liu, H.; Zhong, Z. Shell-sheddable micelles-based on dextran-SS-poly (ε-caprolactone) diblock copolymer for efficient intracellular release of doxorubicin. *Biomacromolecules.* **2010**, *11*(4), 848–854.

Sun, I.-C.; Na, J. H.; Jeong, S. Y.; Kim, D.-E.; Kwon, I. C.; Choi, K.; Ahn, C.-H.; Kim, K. Biocompatible glycol chitosan-coated gold nanoparticles for tumor-targeting CT imaging. *Pharm. Res.* **2014**, *31*(6), 1418–1425.

Sutherland, A. J. Quantum dots as luminescent probes in biological systems. *Curr. Opin. Solid State Mater. Sci.* **2002**, *6*(4), 365–370.

Taheri, S.; Baier, G.; Majewski, P.; Barton, M.; Förch, R.; Landfester, K.; Vasilev, K. Synthesis and antibacterial properties of a hybrid of silver–potato starch nanocapsules by miniemulsion/polyaddition polymerization. *J. Mater. Chem. B.* **2014**, *2*(13), 1838–1845.

Tanaka, T.; Decuzzi, P.; Cristofanilli, M.; Sakamoto, J. H.; Tasciotti, E.; Robertson, F. M.; Ferrari, M. Nanotechnology for breast cancer therapy. *Biomed. Microdevices.* **2009**, *11*(1), 49–63.

Tiwari, G.; Tiwari, R.; Sriwastawa, B.; Bhati, L.; Pandey, S.; Pandey, P.; Bannerjee, S. K. Drug delivery systems: An updated review. *Int J Pharm Investig.* **2012**, *2*(1), 2–11.

Tong, L.; Wei, Q.; Wei, A.; Cheng, J. X. Gold nanorods as contrast agents for biological imaging: optical properties, surface conjugation and photothermal effects. *J. Photochem. Photobiol.* **2009**, *85*(1), 21–32.

Torchilin, V. P. Passive and active drug targeting: drug delivery to tumors as an example. In *Drug Delivery. Springer*, Germany. **2010**, *197*, 3–53.

Tsai, M. L.; Bai, S. W.; Chen, R. H. Cavitation effects versus stretch effects resulted in different size and polydispersity of ionotropic gelation chitosan–sodium tripolyphosphate nanoparticle. *Carbohydr. Polym.* **2008**, *71*(3), 448–457.

Valodkar, M.; Rathore, P. S.; Jadeja, R. N.; Thounaojam, M.; Devkar, R. V.; Thakore, S. Cytotoxicity evaluation and antimicrobial studies of starch capped water soluble copper nanoparticles. *J. Hazard. Mater.* **2012**, *201*, 244–249.

Varki, A., Cummings, R. D., Esko, J. D. et al., Eds. Essentials of Glycobiology. 2nd edition. Cold Spring Harbor (NY): Cold Spring Harbor Laboratory Press, New York, USA, 2009.

Varshosaz, J.; Emami, J.; Ahmadi, F.; Tavakoli, N.; Minaiyan, M.; Fassihi, A.; Mahzouni, P.; Dorkoosh, F. Preparation of budesonide–dextran conjugates using glutarate spacer as a colon-targeted drug delivery system: in vitro/in vivo evaluation in induced ulcerative colitis. *J. Drug Targeting.* **2011**, *19*(2), 140–153.

Vauthier, C.; Bouchemal, K. Methods for the preparation and manufacture of polymeric nanoparticles. *Pharm. Res.* **2009**, *26*(5), 1025–1058.

von der Lieth, C.-W.; Lütteke, T.; Frank, M. Bioinformatics for glycobiology and glycomics: an Introduction. John Wiley & Sons, USA, **2009**.

Wang, Z.; Chen, Z.; Liu, Z.; Shi, P.; Dong, K.; Ju, E.; Ren, J.; Qu, X. A multistimuli responsive gold nanocage–hyaluronic platform for targeted photothermal and chemotherapy. *Biomaterials.* **2014**, *35*(36), 9678–9688.

Weinhart, M.; Gröger, D.; Enders, S.; Riese, S. B.; Dernedde, J.; Kainthan, R. K.; Brooks, D. E.; Haag, R. The Role of Dimension in Multivalent Binding Events: Structure–Activity Relationship of Dendritic Polyglycerol Sulfate Binding to L-Selectin in Correlation with Size and Surface Charge Density. *Macromol. Biosci.* **2011**, *11*(8), 1088–1098.

Wu, S.; Huang, X.; Du, X. Glucose and pH Responsive Controlled Release of Cargo from Protein Gated Carbohydrate Functionalized Mesoporous Silica Nanocontainers. *Angew. Chem. Int. Ed.* **2013**, *125*(21), 5690–5694.

Yamazaki, N.; Kojima, S.; Bovin, N.; Andre, S.; Gabius, S.; Gabius, H.-J. Endogenous lectins as targets for drug delivery. *Adv Drug Deliv Rev.* **2000**, *43*(2), 225–244.

Yang, H.-C.; Hon, M.-H. The effect of the molecular weight of chitosan nanoparticles and its application on drug delivery. *Microchem. J.* **2009**, *92*(1), 87–91.

Yang, J.-S.; Xie, Y.-J.; He, W. Research progress on chemical modification of alginate: A review. *Carbohydr. Polym.* **2011**, *84*(1), 33–39.

Yasar Yildiz, S.; Toksoy Oner, E. Mannan as a promising bioactive material for drug nanocarrier systems. Application of Nanotechnology in Drug Delivery, Intechopen, Croatia. **2014**, *9*, 311–342.

York, A. W.; Zablocki, K. R.; Lewis, D. R.; Gu, L.; Uhrich, K. E.; Prud'homme, R. K.; Moghe, P. V. Kinetically Assembled Nanoparticles of Bioactive Macromolecules Exhibit Enhanced Stability and Cell-Targeted Biological Efficacy. *Adv. Mater.* **2012,** *24*(6), 733–739.

Young, E. The antiinflammatory effects of heparin and related compounds. *Thromb. Res.* **2008,** *122*(6), 743–752.

Yu, J.; Jin, J.; Cheng, B.; Jaroniec, M. A noble metal-free reduced graphene oxide–CdS nanorod composite for the enhanced visible-light photocatalytic reduction of CO_2 to solar fuel. *J. Mater. Chem. A.* **2014,** *2*(10), 3407–3416.

Zhang, H.; Oh, M.; Allen, C.; Kumacheva, E. Monodisperse chitosan nanoparticles for mucosal drug delivery. *Biomacromolecules.* **2004,** *5*(6), 2461–2468.

Zhang, Y.; Chan, J. W.; Moretti, A.; Uhrich, K. E. Designing polymers with sugar-based advantages for bioactive delivery applications. *J. Controlled Release.* **2015,** *10*(219), 355–68.

PHYSIOLOGICAL AND CLINICAL CONSIDERATIONS OF DRUG DELIVERY SYSTEMS CONTAINING CARBON NANOTUBES AND GRAPHENE

SAMI MAKHARZA,[1,2,*] GIUSEPPE CIRILLO,[3] ORAZIO VITTORIO,[4] and SILKE HAMPEL[2]

[1]Department of Chemistry, Faculty of Science & Technology, Hebron University, Hebron, Palestine, *E-mail: makharza.sami@gmail.com

[2]Leibniz Institute for Solid State and Materials Research Dresden, Dresden, Germany

[3]Department of Pharmacy, Health and Nutritional Sciences, University of Calabria, Rende (CS), Italy

[4]Children's Cancer Institute Australia, Lowy Cancer Research Centre, University of New South Wales, Sydney, Australia

CONTENTS

ABSTRACT

Nanotechnology is a rapidly growing field of research in biomaterials, medicine, and pharmacy for creating novel drug delivery systems. The understanding of nanotechnology in biological applications is still insufficient. Many years ago, the nanotechnology entered medicine and made a massive progress especially in drug delivery; the principle depends on ultra-small particles that have the ability to break into the cellular compartments. These nanoparticles such as carbon nanotubes (CNT, (1D)), graphene or graphene oxide (2D), and Fullerene (0D) used as vehicles to produce efficient therapy by means of carrying drug molecules due to their unique properties, including, noticeable cell membrane permeation, high drug loading capacity, and mostly less to no systemic toxicity. The intrinsic cytotoxicity of free drug molecules affect their pharmacological profile and their therapeutic efficacy, due to an act of changing in physical location, therefore diminishing their pharmacological progression.

This chapter focuses on understanding the physiological and clinical issues concerning the shape, surface chemistry, and dimensionality of Nano-sized particles. Moreover, the chapter will discuss the cellular paradigm (in vitro and in vivo) in order to have a complete overview of the applicability of these engineered nanoparticles.

7.1 INTRODUCTION

In 1959, Dr. Richard Feynman (Laureate, 1964, Physics) said a visionary sentence in stellar speech "There's Plenty of Rooms at the Bottom" and "that it would have an immense number of applications" in which he predicted that it would be possible to put 24 volumes of the *Encyclopedia Britannica* on the head of a pin (Joachim and Plévert, 2009). Indeed, what was fancied in that time is now reality (Nano Era) in twenty-first century. The first use of Nanotechnology term was in 1974 introduced by Norio Taniguchi (Taniguchi, 1974). The first book in Nanotechnology written in 1986 by Drexler, the term "Nano" comes from the Greek word "Nanos" meaning "extremely small or dwarf," it was found that the subdivision of an object for example 1 cm^3 to 1 nm^3 will produce 10^{21} cubes of 1 nm size and the surface area will be 6x, 10^{21}, while the total volume remains no change (Drexler, 1987). The surface area to volume ratio is a way of expressing the change in size to create an extremely number of surfaces that are useful in many fields, as for instance, in drug delivery applications. This feature opened the gate for incorporation of nanostructured materials as platforms or vehicles to carry the drug molecules and deliver it to a certain places in tissue organisms.

7.2 PROPERTIES OF DRUG DELIVERY SYSTEMS

The performance of astute drug delivery system depends on increasing the therapeutic activities on the targeted site and decreasing the side effects to normal tissues. Over the past decays, great efforts have been developed to use the nanoparticles such as CNT, graphene, Fullerene, and liposomes as nanocarrier drug delivery systems.

7.2.1 CHARACTERISTICS OF PRISTINE CNTS IN DRUG DELIVERY

Structurally, pristine single walled carbon nanotube (SWCNT) can be considered as rolled-up of sp^2 bonded carbon atoms in graphene sheet. Multi-walled carbon nanotubes consist of two or more rolled up cylinders of graphene sheets (Figure 7.1). Typically both types of CNTs have diameter ranging from 1 upto 50 nm, however, the lengths are thousands more than diameter, this length to diameter ratio candidates CNTs to be used as platforms in drug delivery application (Aboofazeli, 2010; Tasis et al., 2006; Wang et al., 2005).

Since the discovery of CNTs by Iijima in 1991, an enormous number of articles have been manifesting the great interest in their applicability in targeted delivery of drug molecules (Aboofazeli, 2010; Cirillo et al., 2013; Hua He et al., 2013; Ilbasmiş-Tamer et al., 2010; Shao et al., 2013; Tian et al., 2006; Vashist et al., 2011; Yongbin Zhang et al., 2010; Zhang et al., 2011). The substantial method of CNTs production is chemical vapor deposition (CVD), within the preparation process a tiny metals contaminant the CNT containers (Zhang and Zhu, 2002). Currently, intensive efforts have been made to improve the quality of produced CNTs, as well as enhance the purity. It has been found that the presence of trace amount

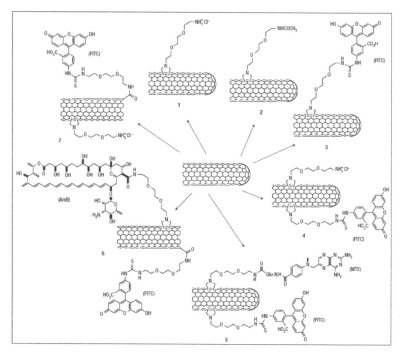

FIGURE 7.1 The chemical structure of monolayer and multilayer graphene, carbon nanotube and buckyball. (Reprinted from Lloyd-Hughes, J.; Jeon, T. I. A Review of the Terahertz Conductivity of Bulk and Nano-Materials. *J. Infrared, Millimeter, Terahertz Waves* **2012**, *33*(9), 871–925. With permission of Springer.)

of metal impurities affects the scope of using CNTs in physiological and clinical purposes, as it has been contributing remarkable to toxicity (Magrez et al., 2006). On the other hand, CNTs considered as hydrophobic regime which make them insoluble and less dispersible in physiological conditions, So that, it is essential to have good solubilized and dispersed nanoparticles without aggregation prior enforcing them in therapeutic purposes. Therefore, the solubility of CNTs is a great challenge to overcome the hydrophobicity and improve their biocompatibility. The dispersion of CNTs in cellular medium is also considered as a prerequisite to render good enough uniformity and stability. The main task assign to researchers is the ability to overcome these drawbacks in solubility and dispersibility of CNTs in water or physiological solution (Widenkvist, 2010). There are four dominant approaches for enhancing the solubility and stability of CNTs, including functionalization of the outer shells of CNTs by using biocompatible molecules (Widenkvist, 2010; Cirillo et al., 2013), solvent dispersion technique (Kharissova et al., 2013), biomolecular dispersion, and application of surfactants adsorbed on the sidewalls of CNTs (Haggenmueller et al., 2008). Among these interesting approaches, functionalization revealed higher stability and dispensability, because functionalized CNTs modify the surface for being more hydrophilic character and can easily increase the binding capacity with drug molecules.

7.2.1.1 Covalent Functionalization of CNTs for Targeted Drug Delivery

CNTs exhibit no sign aggregation or phase separation after effective functionalization. For physiological and clinical considerations, CNTs must be soluble in cell culture media to avoid cytotoxicity (Aboofazeli, 2010; Shao et al., 2013). Thus, covalent modification of CNTs is necessary and more widely used for better stability and dispersibility. As nanocarrier, CNTs designed in order to hold anticancer drugs, protein or DNA and internalize or excrete at the cellular level (Haggenmueller et al., 2008; He et al., 2013). It seems that the type of functionalization and functionalized materials determine the fate of CNTs in cellular conditions. Different mechanisms have been developed to elucidate the effect of CNTs functionalization, as well as the way of endo- and exocytosis. CNT is needle-like structure and can easily penetrate the cell membrane as a passive mechanism (Ilbasmiş-Tamer et al., 2010). Shi Kam et al. (2004) suggested an endocytosis pathway for acid – functionalization of CNTs. The type of functionalization depends on the chemical nature of the biomaterials bind to CNTs, as well as the reaction conditions. In covalent functionalization as shown in, the biomolecules linked the CNTs by formation of strong chemical bond for special applications to ensure high stability of drug delivery system as shown in Figure 7.2 (Toma, 2009).

7.2.1.2 Non-Covalent Functionalization of CNTs for Targeted Drug Delivery

In noncovalent approach, the interaction between CNTs and biomaterial is considered as physioadsorption, for instance, π–π interaction, Van der Waals forces or hydrogen bonds (Cirillo et al., 2013). On the other hand,

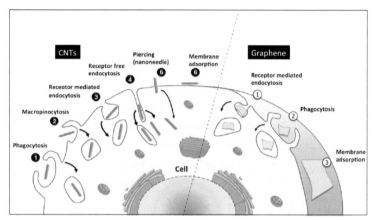

FIGURE 7.2 Chemical structures covalently functionalized SWCNTs, 1–7: Ammonium-functionalized CNT; Acetamido functionalized CNT; CNT functionalized with fluorescein isothiocyanate (FITC); CNT bifunctionalized with ammonium groups and FITC; CNT bifunctionalized with methotrexate (MTX) and FITC; shortened CNT bifunctionalized with amphotericin B (AmB) and FITC; shortened CNT bifunctionalized with ammonium groups and FITC (through an amide linkage), respectively. (Reprinted by permission from Macmillan Publishers Ltd: Kostarelos, K.; Lacerda, L.; Pastorin, G.; Wu, W.; Wieckowski, S.; Luangsivilay, J.; Godefroy, S.; Pantarotto, D.; Briand, J.-P.; Muller, S.; et al. Cellular Uptake of Functionalized Carbon Nanotubes Is Independent of Functional Group and Cell Type. *Nat. Nanotechnol.* **2007,** 2(2), 108–113.)

noncovalent CNTs protein interaction could influence the CNT-cell membrane interfaces, most studies reported that the endocytosis and the exocytosis of CNTs functionalized protein have the same rate and revoke by kidneys (Liu et al., 2009; Zhang et al., 2009).

Prior functionalization, both covalent and noncovalent, CNTs should proceed oxidation treatment in order to build up more functional groups like epoxy, carboxyl, hydroxyl, etc., distributed on the outer shells of CNTs, these groups play an important role in formation of small solubilizing clusters dispersed homogenously in cell culture media. A wide variety of materials have been experienced as functional biomolecules for drug delivery systems, such as protein, DNA, biopolymers, phospholipids, etc.

7.2.2 CHARACTERISTICS OF PRISTINE GRAPHENE AND ITS DERIVATIVES IN DRUG DELIVERY

Graphene is a one layer in graphite regime (Figure 7.1, (Geim and Novoselov, 2007). It has tremendous extraordinary physical and chemical properties for a wide variety of applications, especially after an outstanding milestone in graphene-based research by two scientists, they awarded Nobel Prize in physics in 2010 (Mattevi et al., 2009; Lee et al., 2013). The first report about graphene published in 2004, opened up the gate for new innovative physiochemical materials in pharmacology, medicine, and biology. To date, graphene and its derivate graphene oxide (GO) and reduced graphene oxide (rGO) classified as well-studied nanoparticles in medicine in comparison with carbon nanotubes, quantum dots, magnetic nanoparticles, and others. The cytotoxicity profile of graphene nanoparticles is not well elucidated because of the limited research since the initial report by Sun et al. (2008). As CNTs, graphene is also poorly soluble in water and physiological conditions due to the hydrophobic character of carbon atoms that build up the structure of graphene, in the same context, the dispersibility of graphene in physiological media is embarrassing (Widenkvist, 2010). In addition, the size distribution (lateral width and thickness) of graphene nano-sheets is crucial and must be controlled to have a proper uniformity, which has been considered as a challenge in physiological and clinical purposes (Akhavan et al., 2012; Makharza et al., 2013b). On the other hand, the toxicity of graphene nanoparticles is concentration dependent in cellular studies leading to cellular damage and inducing cell apoptosis through more ingress into various cellular compartments. Figure 7.3, demonstrates different mechanisms between graphene nanoparticles and cellular membrane in comparison with CNTs. The chemical nature of graphene and CNTs surfaces determine the way of interaction with lipid bilayer. In conclusion, pristine graphene has a lot of drawbacks in biological applications concerning the solubility, size, and dispersibility, in order to overcome these problems; the research community recommends modifying the surface of graphene by using biocompatible molecules in order to render stability and compatibility in biological solutions.

7.2.2.1 Functionalization of Graphene

Graphite (multilayers of graphene) can be oxidized to form graphite oxide by using three main oxidation reactions (Brodie, 1859; Hummers and Offeman, 1958; Staudenmaier, 1898). One of the advantages of graphite oxide is well dispersed in water or physiological environments due to

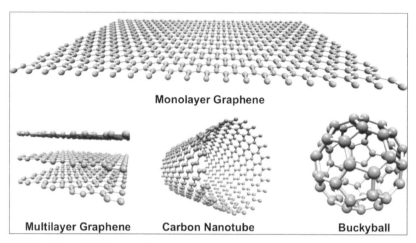

FIGURE 7.3 Cellular uptake mechanisms of CNT (left) and graphene (right) nanoparticles. (Reprinted with permission from Cyrill Bussy, Hanene Ali-Boucettâ, and Kostas Kostarelos. Safety Considerations for Graphene: Lessons Learnt from Carbon Nanotubes, *Acc. Chem. Res.,* **2013**, *46* (3), pp 692–701. © 2013 American Chemical Society.)

various hydrophilic functional groups, such as carbonyl, epoxide, and carboxylic groups distributed randomly on the basal planes and edges of graphene sheets (Makharza et al., 2013a). These groups possess the ability to coordinate with other molecules by covalent or noncovalent interactions. Many studies revealed that graphene oxide demonstrated less to no cytotoxicity with various cell lines such as HeLa cells, L929 cells, A549 human lung cancer cells, and human fibroblasts, this behavior of GO depends on several parameters such as concentration, shape, size, oxidation level, incubation time in cell culture lines, etc., and can determine the fate of cells in physiological and clinical environments (Akhavan et al., 2012; Makharza et al., 2014). In particular, surface functionalization of GO (covalent or noncovalent) by using biomaterials produces high biocompatibility in cell culture media as well as exhibits good stability and dispersibility, since it can reduce the hydrophobic character of GO and cell membrane. MTT proliferation assay revealed that Chitosan functionalized GO showed significant biocompatibility to L929 cell line for long of incubation period. Moreover, functionalization can reduce the reactivity of oxygen groups that mediate apoptosis through caspase-3 activation (Yang et al., 2013). Dai *et al.* reported polyethylene glycol covalently functionalized nanographene oxide (PEG-NGO) for delivering water insoluble anticancer therapeutic drugs (SN38; a camptothecin (CPT) analog). The report also confirmed that NGO-PEG system possesses high biocompatibility without sign of toxicity, as well as increase the therapeutic drug efficiency. In addition, they found that NGO particles functionalized with an antibody molecule loaded Doxorubicin as anticancer drug exhibited high killing tendency of selective damaging of abnormal cells (Liu et al., 2008). After this report, many studies affirmed the interaction between NGO and functionalized materials to render good stability and biocompatiblity.

Most of the available techniques for delivering therapeutics by NGO is-based on the fabrication of polymer nanohybrids. The synthesis of GO nanohybrids can be achieved by two main approaches, namely the "grafting from" and "grafting to" (Spizzirri et al., 2015).

The "grafting to" approach involves preformed polymer chains reacting with the surface of either pristine or prefunctionalized GO. Here, the structure of GO helps in understanding the reaction mechanism. The readymade polymers with reactive end groups can react with the functional groups on the GO surfaces with the advantage that polymers with controlled molecular weight and polydispersity can be used. The main limitation of this technique is that initial binding of polymer chains sterically hinders the diffusion of additional macromolecules to the GO surface, leading to a low grafting density. Also, only polymers containing reactive functional groups can be used (Cirillo et al., 2012). The "grafting from" approach involves the polymerization of monomers from surface-derived initiators on GO. These initiators are covalently attached using the various functionalization reactions developed for small molecules. The advantage of "grafting from" approach is that the polymer growth is not limited by steric hindrance, allowing high molecular weight polymers to be efficiently grafted. In addition, nanocomposites with quite high grafting density can be prepared. However, this method requires a strict control of the amounts of initiator and substrate as well as accurate control of polymerization conditions. Moreover, the continuous π-electronic properties of GO would be destroyed by the acid oxidation (required for initiator linkage). As a result, compared with the "grafting from," the "grafting to" has much less alteration of the structure of GO (Sakellariou et al., 2013).

7.2.2.2 Graphene and Its Derivate Graphene Oxide As Targeted Drug Delivery Platforms

As mentioned above, GO and its derivatives possess superior physicochemical and functional properties for use as innovative carriers for several different therapeutics including small (e.g., anticancer) or large (e.g., antibodies and protein) drug molecules or genes (e.g., DNA, RNA) (Parveen et al., 2012). Nevertheless, some critical drawbacks are still to be overcome for reaching an effective clinical applicability of GO-based carriers, and they are mainly due to limited blood–brain transport efficiency and renal clearance (Ali-Boucetta et al., 2013). As reported in literature (Wu et al., 2015), the effectiveness of a GO-based delivery vehicle depends on three factors, namely: (i) an optimal loading capacity; (ii) the high biocompatibility and the absence of any trace of toxicity; (iii) the possibility to control and target the release of the loaded therapeutics. The latter can be achieved by the conjugation of target units, including folic acid, transferrin and folic acid on the GO surface (Nasongkla et al., 2004; Daniels et al., 2012; Wu et al., 2015). Over the last decades, several different research groups have developed different approach to effectively deliver therapeutics agents (mainly anticancer) by exploiting the chemical features of GO. One of the simplest methods is the use of un-modified GO. For this purpose, a key requirements of the selected drug is an effective and strong interaction with GO surface, and this was reached by employing doxorubicin (DOX), which is effectively released from the carrier in the acidic environment of cancer cells (Depan et al., 2011; Mendes et al., 2013; Yang et al., 2008). A direct upgrade of this concept is the introduction of pH-sensitive units (polymers in this case) with the obtainment of hybrid hydrogels allowing a better control on the drug release. As pH-sensitive materials polymacrylic acid (Wang et al., 2014b; Zhao et al., 2015) and polyvinylalcohol (Bai et al., 2010). Other researchers developed and effective methods for inserting GO into Konjac glucomannan and Sodium alginate hydrogels (Wang et al., 2014d).

In the latter case GO nanosheets were found to significantly influence the micromorphology, swelling ability, and drug loading of the composite hydrogel. Interestingly, in vitro release studies showed that the burst release phenomenon could be avoided and excellent pH sensitive system could be achieved. Efficient GO-hydrogel hybrids have been synthesized by the noncovalent coating of GO with cyclic RGD-modified chitosan. The nanohybrid was proposed as drug delivery system for the targeted delivery of Doxorubicin to hepatocellular carcinoma, with the GO counterpart conferring efficient loading properties, while the GO-chitosan hydrogen bonding interactions allow the selective drug release in the acidic tumor environment due to the pH-responsive behavior. Furthermore, the nanocomposite is able to recognize hepatoma cells and promote drug uptake by the cells, especially for those overexpressing integrins as confirmed by cellular uptake and proliferation studies (Wang et al., 2014a). In Zo et al. (2013), a valuable approach for the preparation of hybrid hydrogels composed of GO and GO derivatives is the noncovalent inclusion into cyclodextrins to confer high elastic behavior at elevated temperature and effectively deliver the DOX to cancer tissues (Hu et al., 2014). In addition to internal cellular changes in pH, ion concentration, and temperature, there are other external stimuli such as ultrasound, magnetic, and electric fields that can be used to trigger the release of a drug from GO-based carrier.

An example of using a magnetic field to control the drug delivery is reported in (Zhou et al., 2010) where GO/Fe_3O_4 nanocomposite have been developed and characterized. In a different approach, a noncovalent approach was employed for the preparation of a flexible, electrically conducting hydrogel-based on commercial superabsorbent polymer and hyperbranched polymer. Here, the incorporation of reduced graphene oxide (rGO) resulted in excellent electrical self-healing properties and water-absorption reusability (Peng et al., 2014). Hybrid hydrogel membranes composed of reduced graphene oxide nanosheets and a poly (vinyl alcohol) matrix were proposed as an electrically responsive drug release system for the anesthetic drug lidocaine hydrochloride by Liu et al. (2012). The presence of rGO into the nanohybrids was found to act as a physical barrier to inhibit the drug release, while the exposure to an electrical stimulus highly enhances the release. He and Gao (2010) immobilized different kinds of functional moieties or polymers onto GP sheets via nitrene chemistry to obtain highly engineered composite materials able to delivery therapeutics upon the application of an external stimulus. An electro-responsive delivery vehicles is proposed in (Rana et al., 2011). Here, the chemistry of Chitosan is explored for the GO conjugation via amide bond formation between the carboxylic acid groups of GO and NH_2 groups of Chitosan.

The delivery of model drugs (Ibuprofen and 5-Fluorouracil in this case) is closely dependent on the ionization properties of the two drugs, with the release reaching a faster rate when the drugs are in the ionized form. External stimulation was also at the basis of the release of the antiinflammatory drug Dexamethasone from electrospinning-based polypyrrole nanohybrids developed in Weaver et al. (2014).

An electro-conductive hydrogel system was prepared by incorporation of rGO into Jeffamine polyetheramine and polyethylene glycol diglycidyl ether by one-step polymerization method (MacKenna et al., 2015). The obtained hybrid hydrogels showed enhanced mechanical and electrical properties depending on rGO content, and were investigated as delivery device for methyl orange. A significant reduction in passive release of the dye was observed by incorporating rGO, while upon electrical stimulation, the release rate and dosage could be tuned through variation of the percentage w/w of rGO, as well as of the polarity and amplitude of the applied electric potential. In a different approach, Pickering emulsion polymerization method was proposed for the fabrication of core-shell structured polystyrene microspheres containing GO units (Kim et al., 2014).

Here, GO shows the double activity of coating the polymer microspheres and imparting electro-responsive properties to the polymer composite suspension under an applied electric field.

In a study by Li et al. (2013) a hyaluronic acid-NGO nanohybrids was proposed as a cancer cell target and photoactivity switchable nanoplatform for photodynamic therapy. The photoactivity of photosensitizer adsorbed on nanocarriers was mostly quenched in aqueous solution to ensure biocompatibility, but was quickly recovered after the release from nanocarriers upon cellular uptake. The photodynamic therapy was remarkably improved resulting in efficiency 10 times higher than that of the free photosensitizer. Another key external stimulus which can be employed for controlling the delivery of drug molecules from NGO nanohybrids is the temperature.

Thermo-sensitive releasing behavior by GO nanohybrids was also reached by the covalent immobilization of thermo-responsive polymer nanoparticles on functionalized GO nanosheets. In the synthetic procedure, thermo-sensitive nanoparticles were first synthesized by free radical polymerization and GO nanosheets were noncovalently modified with a bi-functional linker able to provide reactive sites for the subsequent binding to the polymer. Then, Adriamycin was loaded with high efficiency and the whole system tested in suitable mouse models with interesting results (Wang et al., 2014c). Systems able to simultaneously respond to pH and temperature variations have been developed in Kavitha et al. (2014). Here, the loading properties of GO surface via π-π stacking and hydrophobic interactions were also exploited for the preparation of pH-sensitive drug delivery devices with poly N-vinyl caprolactam acting as the polymer counterpart conferring high solubility and stability in water and physiological solutions to GO. GO was inserted via atom transfer radical polymerization and the resulting composite was proposed for the intracellular delivery of the anticancer drug Camptothecin with high releasing efficiency. The composite showed high loading ratio, and the loaded drug release is fast released at a reduced pH (typical of the tumor microenvironment). In addition, the presence of poly N-vinyl caprolactam allows a targeted and temperature-dependent delivery rate. Authors demonstrated that the pure nanocargo alone was almost nontoxic, whereas the nanohybrid showed strong potency against cancer cells, due to a cellular-uptake mechanism of energy-dependent endocytosis.

NGO carriers are also effectively employed in the so-called combination therapy, where two or more active or preactive pharmacological agents are simultaneously employed. This concept was firstly introduced in Gautschi et al. (2007), where both DOX and camptothecin (CPT) were loaded onto a folic acid GO carrier. Authors stated that the codelivery of both drugs had a better target efficacy and higher cytotoxicity than GO loaded with either DOX or CPT alone. Another valuable example is the work reported in Zhi et al. (2013). Here, a multifunctional nanocomplex composed of polyethylenimine (PEI)/poly(sodium 4-styrene sulfonate) (PSS)/GO termed PPG was used to evaluate the reversal effects of PPG as a carrier for adriamycin (ADR) along with miR-21 targeted small-interfering RNA (siRNA) (anti-miR-21) in cancer drug resistance. Cell experiments showed that PPG significantly enhanced the accumulation of ADR in MCF-7/ADR cells (an ADR-resistant breast cancer cell line) and exhibited much higher cytotoxicity than free ADR, suggesting that PPG could effectively reverse ADR resistance of MCF-7/ADR. It should be also mentioned that the binding properties of GO surface can be also explored for the preparation of gene delivery vector.

It has been shown that GO derivatives can improve the penetration of siRNA or plasmid DNA (pDNA) into cells protecting DNA from enzyme cleavage (Zhang et al., 2011a). Li et al. (2014) managed to pattern preconcentrated PEI/pDNA on absorbent GO mediating highly localized and efficient gene delivery. According to the authors, the patterned substrates exhibited excellent biocompatibility and enabled effective gene transfection for various cell lines including stem cells. The distinguishing property of PEI-GO compared to other vehicles is its ability to condense DNA at a low mass ratio (+49 mV) and effectively transport pDNA through the cytoplasm to the nucleus. Other carbon vectors such as GO/chitosan, GO-PEG (Kim et al., 2013; Liu et al., 2014; Tian et al., 2011; Tripathi et al., 2013; Zhang et al., 2013). In another approach (Kim et al., 2011), low-molecular weight branched polyethylenimine was employed as a cationic gene carrier, with GO acting as fluorescence reagent probe. According to the authors' hypothesis, the excellent photoluminescence activities of the GO-based nanoconstruct will definitely merit further attention in the development of more sophisticated carrier systems that could serve both as a gene delivery vector and bioimaging tool.

7.3   CONCLUSION

We have reviewed the most and rapid advances nanostructures used as functional tools for drug delivery, by focusing on CNTs and functionalized graphene. Some key examples of NGO-based carriers have been elucidated to highlight the tremendous potential impact of such delivery vehicles on the clinical treatment of different pathologies including cancer, infections and degenerative diseases. Further efforts have to be done in optimizing the efficiency and long-terms stability of such materials, but it is clear that by a combination of the knowledge of material scientists, nanotechnologists, chemists, biologists it is possible the development of multidisciplinary research panels for a bench to clinic applicability of such systems.

KEYWORDS

- carbon nanotube
- cellular uptake
- graphene
- nanocarrier
- nanoparticle
- therapeutic activity

REFERENCES

Aboofazeli, R. Carbon Nanotubes: A Promising Approach for Drug Delivery. *Iran. J. Pharm. Res. IJPR* **2010**, *9*(1), 1–3.

Akhavan, O.; Ghaderi, E.; Akhavan, A. Size-Dependent Genotoxicity of Graphene Nanoplatelets in Human Stem Cells. *Biomaterials* **2012**, *33*(32), 8017–8025.

Ali-Boucetta, H.; Bitounis, D.; Raveendran-Nair, R.; Servant, A.; Van den Bossche, J.; Kostarelos, K. Purified Graphene Oxide Dispersions Lack In Vitro Cytotoxicity and In Vivo Pathogenicity. *Adv. Healthc. Mater.* **2013**, *2*(3), 433–441.

Bai, H.; Li, C.; Wang, X.; Shi, G. A pH-Sensitive Graphene Oxide Composite Hydrogel. *Chem. Commun. (Camb).* **2010**, *46*(14), 2376–2378.

Brodie, B. C. On the Atomic Weight of Graphite. *Philos. Trans. R. Soc. London* **1859**, *149*(12), 249–259.

Cirillo, G.; Hampel, S.; Puoci, F.; Haase, D.; Ritschel, M.; Leonhardt, A.; Iemma, F.; Picci, N. Carbon Nanotubes – Imprinted Polymers : Hybrid Materials for Analytical Applications. In *Hybrid Materials for Analytical Applications, Materials Science and Technology*; Sabar Hutagalung, Ed.; InTech, 2012.

Cirillo, G.; Vittorio, O.; Hampel, S.; Spizzirri, U. G.; Picci, N.; Iemma, F. Incorporation of Carbon Nanotubes into a Gelatin-Catechin Conjugate: Innovative Approach for the Preparation of Anticancer Materials. *Int. J. Pharm.* **2013**, *446*(1–2), 176–182.

Daniels, T. R.; Bernabeu, E.; Rodríguez, J. A.; Patel, S.; Kozman, M.; Chiappetta, D. A.; Holler, E.; Ljubimova, J. Y.; Helguera, G.; Penichet, M. L. The Transferrin Receptor and the Targeted Delivery of Therapeutic Agents against Cancer. *Biochim. Biophys. Acta* **2012**, *1820*(3), 291–317.

Depan, D.; Shah, J.; Misra, R. D. K. Controlled Release of Drug from Folate-Decorated and Graphene Mediated Drug Delivery System: Synthesis, Loading Efficiency, and Drug Release Response. *Mater. Sci. Eng. C* **2011**, *31*(7), 1305–1312.

Eric Drexler. *Engines of Creation: The Coming Era of Nanotechnology*; Anchor, 1987.

Gautschi, O. P.; Frey, S. P.; Zellweger, R. Bone Morphogenetic Proteins in Clinical Applications. *ANZ J. Surg.* **2007**, *77*(8), 626–631.

Geim, A. K.; Novoselov, K. S. The Rise of Graphene. *Nat. Mater.* **2007**, *6*(3), 183–191.

Haggenmueller, R.; Rahatekar, S. S.; Fagan, J. A.; Chun, J.; Becker, M. L.; Naik, R. R.; Krauss, T.; Carlson, L.; Kadla, J. F.; Trulove, P. C.; et al. Comparison of the Quality of Aqueous Dispersions of Single Wall Carbon Nanotubes Using Surfactants and Biomolecules. *Langmuir* **2008**, *24*(9), 5070–5078.

He, H.; Gao, C. General Approach to Individually Dispersed, Highly Soluble, and Conductive Graphene Nanosheets Functionalized by Nitrene Chemistry. *Chem. Mater.* **2010**, *22*(17), 5054–5064.

He, H.; Pham-Huy, L. A.; Dramou, P.; Xiao, D.; Zuo, P.; Pham-Huy, C. Carbon Nanotubes: Applications in Pharmacy and Medicine. *Biomed Res. Int.* **2013**, 578290.

Hu, X.; Li, D.; Tan, H.; Pan, C.; Chen, X. Injectable Graphene Oxide/Graphene Composite Supramolecular Hydrogel for Delivery of Anti-Cancer Drugs. *J. Macromol. Sci. Part A* **2014**, *51*(4), 378–384.

Hummers, W. S.; Offeman, R. E. Preparation of Graphitic Oxide. *J. Am. Chem. Soc.* **1958**, *80*(6), 1339–1339.

Iijima, S. Helical Microtubules of Graphitic Carbon. *Nature* **1991**, *354*, 56–58.

Ilbasmiş-Tamer, S.; Yilmaz, Ş.; Banoğlu, E.; Değim, I. T. Carbon Nanotubes to Deliver Drug Molecules. *J. Biomed. Nanotechnol.* **2010**, *6*(1), 20–27.

Joachim, C.; Plévert, L. *Nanosciences: The Invisible Revolution*; World Scientific Publishing Co. Pte. Ltd., 2009.

Kavitha, T.; Kang, I.-K.; Park, S.-Y. Poly(N-Vinyl Caprolactam) Grown on Nanographene Oxide as an Effective Nanocargo for Drug Delivery. *Colloids Surf. B. Biointerfaces* **2014**, *115*, 37–45.

Kharissova, O. V.; Kharisov, B. I.; de Casas Ortiz, E. G. Dispersion of Carbon Nanotubes in Water and Non-Aqueous Solvents. *RSC Adv.* **2013**, *3*(47), 24812.

Kim, H.; Lee, D.; Kim, J.; Kim, T. Il; Kim, W. J. Photothermally Triggered Cytosolic Drug Delivery via Endosome Disruption Using a Functionalized Reduced Graphene Oxide. *ACS Nano* **2013**, *7*(8), 6735–6746.

Kim, H.; Namgung, R.; Singha, K.; Oh, I. K.; Kim, W. J. Graphene Oxide-Polyethylenimine Nanoconstruct as a Gene Delivery Vector and Bioimaging Tool. *Bioconjug. Chem.* **2011**, *22*(12), 2558–2567.

Kim, S. D.; Zhang, W. L.; Choi, H. J. Pickering Emulsion-Fabricated Polystyrene–graphene Oxide Microspheres and Their Electrorheology. *J. Mater. Chem. C* **2014**, *2*, 7541.

Kostarelos, K.; Lacerda, L.; Pastorin, G.; Wu, W.; Wieckowski, S.; Luangsivilay, J.; Godefroy, S.; Pantarotto, D.; Briand, J.-P.; Muller, S.; et al. Cellular Uptake of Functionalized Carbon Nanotubes Is Independent of Functional Group and Cell Type. *Nat. Nanotechnol.* **2007**, *2*(2), 108–113.

Lee, S. K.; Kim, H.; Shim, B. S. Graphene: An Emerging Material for Biological Tissue Engineering. *Carbon Lett.* **2013**, *14*(2), 63–75.

Li, F.; Park, S.-J.; Ling, D.; Park, W.; Han, J. Y.; Na, K.; Char, K. Hyaluronic Acid-Conjugated Graphene Oxide/Photosensitizer Nanohybrids for Cancer Targeted Photodynamic Therapy. *J. Mater. Chem. B* **2013**, *1*(12), 1678.

Li, K.; Feng, L.; Shen, J.; Zhang, Q.; Liu, Z.; Lee, S.-T.; Liu, J. Patterned Substrates of Nano-Graphene Oxide Mediating Highly Localized and Efficient Gene Delivery. *ACS Appl. Mater. Interfaces* **2014**, *6*(8), 5900–5907.

Liu, H.-W.; Hu, S.-H.; Chen, Y.-W.; Chen, S.-Y. Characterization and Drug Release Behavior of Highly Responsive Chip-like Electrically Modulated Reduced Graphene Oxide–poly(vinyl Alcohol) Membranes. *J. Mater. Chem.* **2012**, *22*(33), 17311.

Liu, X.; Ma, D.; Tang, H.; Tan, L.; Xie, Q.; Zhang, Y.; Ma, M.; Yao, S. Polyamidoamine Dendrimer and Oleic Acid-Functionalized Graphene as Biocompatible and Efficient Gene Delivery Vectors. *ACS Appl. Mater. Interfaces* **2014**, *6*, 8173–8183.

Liu, Z.; Fan, A. C.; Rakhra, K.; Sherlock, S.; Goodwin, A.; Chen, X.; Yang, Q.; Felsher, D. W.; Dai, H. Supramolecular Stacking of Doxorubicin on Carbon Nanotubes for In Vivo Cancer Therapy. *Angew. Chemie Int. ed.* **2009**, *48*(41), 7668–7672.

Liu, Z.; Robinson, J. T.; Sun, X.; Dai, H. PEGylated Nano-Graphene Oxide for Delivery of Water Insoluble Cancer Drugs. *J. Am. Chem. Soc.* **2008**, *130*(33), 10876–10877.

Lloyd-Hughes, J.; Jeon, T. I. A Review of the Terahertz Conductivity of Bulk and Nano-Materials. *J. Infrared, Millimeter, Terahertz Waves* **2012**, *33*(9), 871–925.

MacKenna, M.; Calvert, P.; Morrin, A.; Wallace, G.; Moulton, S. Electro-Stimulated Release from a Reduced Graphene Oxide Composite Hydrogel. *J. Mater. Chem. B* **2015**, *3*, 2530–2537.

Magrez, A.; Kasas, S.; Salicio, V.; Pasquier, N.; Seo, J. W.; Celio, M.; Catsicas, S.; Schwaller, B.; Forró, L. Cellular Toxicity of Carbon-Based Nanomaterials. *Nano Lett.* **2006**, *6*(6), 1121–1125.

Makharza, S.; Cirillo, G.; Bachmatiuk, A.; Ibrahim, I.; Ioannides, N.; Trzebicka, B.; Hampel, S.; Rümmeli, M. H. Graphene Oxide-Based Drug Delivery Vehicles: Functionalization, Characterization, and Cytotoxicity Evaluation. *J. Nanoparticle Res.* **2013a**, *15*(12), 2099–2124.

Makharza, S.; Cirillo, G.; Bachmatiuk, A.; Vittorio, O.; Mendes, R. G.; Oswald, S.; Hampel, S.; Rümmeli, M. H. Size-Dependent Nanographene Oxide as a Platform for Efficient Carboplatin Release. *J. Mater. Chem. B* **2013b**, *1*(44), 6107.

Makharza, S.; Vittorio, O.; Cirillo, G.; Oswald, S.; Hinde, E.; Kavallaris, M.; Buchner, B.; Mertig, M.; Hampel, S. Graphene Oxide – Gelatin Nanohybrids as Functional Tools for Enhanced Carboplatin Activity in Neuroblastoma Cells. *Pharm Res* **2014**, *32*(6), 2132–2143.

Mattevi, C.; Eda, G.; Agnoli, S.; Miller, S.; Mkhoyan, K. A.; Celik, O.; Mastrogiovanni, D.; Granozzi, G.; Garfunkel, E.; Chhowalla, M. Evolution of Electrical, Chemical, and Structural Properties of Transparent and Conducting Chemically Derived Graphene Thin Films. *Adv. Funct. Mater.* **2009**, *19*(16), 2577–2583.

Mendes, R. G.; Bachmatiuk, A.; Buechner, B.; Cuniberti, G.; Rümmeli, M. Carbon Nanostructures as Multi-Functional Drug Delivery Platforms. *J. Mater. Chem. B* **2013**, 401–428.

Nasongkla, N.; Shuai, X.; Ai, H.; Weinberg, B. D.; Pink, J.; Boothman, D. A.; Gao, J. cRGD-Functionalized Polymer Micelles for Targeted Doxorubicin Delivery. *Angew. Chemie – Int. Ed.* **2004**, *43*(46), 6323–6327.

Parveen, S.; Misra, R.; Sahoo, S. K. Nanoparticles: A Boon to Drug Delivery, Therapeutics, Diagnostics and Imaging. *Nanomedicine Nanotechnology, Biol. Med.* **2012**, *8*(2), 147–166.

Peng, R.; Yu, Y.; Chen, S.; Yang, Y.; Tang, Y. Conductive Nanocomposite Hydrogels with Self-Healing Property. *Rsc Adv.* **2014**, *4*(66), 35149–35155.

Rana, V. K.; Choi, M. C.; Kong, J. Y.; Kim, G. Y.; Kim, M. J.; Kim, S. H.; Mishra, S.; Singh, R. P.; Ha, C. S. Synthesis and Drug-Delivery Behavior of Chitosan-Functionalized Graphene Oxide Hybrid Nanosheets. *Macromol. Mater. Eng.* **2011**, *296*(2), 131–140.

Sakellariou, G.; Priftis, D.; Baskaran, D. Surface-Initiated Polymerization from Carbon Nanotubes: Strategies and Perspectives. *Chem. Soc. Rev.* **2013**, 677–704.

Shao, W.; Arghya, P.; Yiyong, M.; Rodes, L.; Prakash, S. Carbon Nanotubes for Use in Medicine : Potentials and Limitations. In *Syntheses and Applications of Carbon Nanotubes and Their Composites*; InTech, 2013; p. 28.

Shi Kam, N. W.; Jessop, T. C.; Wender, P. a; Dai, H. Nanotube Molecular Transporters: Internalization of Carbon Nanotube-Protein Conjugates into Mammalian Cells. *J. Am. Chem. Soc.* **2004**, *126*(22), 6850–6851.

Spizzirri, U.; Curcio, M.; Cirillo, G.; Spataro, T.; Vittorio, O.; Picci, N.; Hampel, S.; Iemma, F.; Nicoletta, F. Recent Advances in the Synthesis and Biomedical Applications of Nanocomposite Hydrogels. *Pharmaceutics* **2015**, *7*(4), 413–437.

Staudenmaier, L. Verfahren Zur Darstellung Der Graphitslure. *Berichte der Dtsch. Chem. Gesellschaft* **1898**, *31*(2), 1481–1487.

Sun, X.; Liu, Z.; Welsher, K.; Robinson, J. T.; Goodwin, A.; Zaric, S.; Dai, H. Nano-Graphene Oxide for Cellular Imaging and Drug Delivery. *Nano Res.* **2008**, *1*(3), 203–212.

Taniguchi, N. On the Basic Concept of NanoTechnology. In *Proc. Intl. Conf. Part II, Japan Society of Precision Engineering,* 1974; pp. 18–23.

Tasis, D.; Tagmatarchis, N.; Bianco, A.; Prato, M. Chemistry of Carbon Nanotubes. *Chem. Rev.* **2006**, *106*(3), 1105–1136.

Tian, B.; Wang, C.; Zhang, S.; Feng, L.; Liu, Z. Photothermally Enhanced Photodynamic Therapy Delivered by Nano-Graphene Oxide. *ACS Nano* **2011**, *5*(9), 7000–7009.

Tian, F.; Cui, D.; Schwarz, H.; Estrada, G. G.; Kobayashi, H. Cytotoxicity of Single-Wall Carbon Nanotubes on Human Fibroblasts. *Toxicol. Vitr.* **2006**, *20*(7), 1202–1212.

Toma, F. M. Covalently Functionalized Carbon Nanotubes and Their Biological Applications, International School For Advanced Studies, 2009.

Tripathi, S. K.; Goyal, R.; Gupta, K. C.; Kumar, P. Functionalized Graphene Oxide Mediated Nucleic Acid Delivery. *Carbon N. Y.* **2013**, *51*(1), 224–235.

Vashist, S. K.; Zheng, D.; Pastorin, G.; Al-Rubeaan, K.; Luong, J. H. T.; Sheu, F.-S. Delivery of Drugs and Biomolecules Using Carbon Nanotubes. *Carbon N.Y.* **2011**, *49*(13), 4077–4097.

Wang, C.; Chen, B.; Zou, M.; Cheng, G. Cyclic RGD-Modified Chitosan/graphene Oxide Polymers for Drug Delivery and Cellular Imaging. *Colloids Surfaces B Biointerfaces* **2014a**, *122*, 332–340.

Wang, H.; Gu, W.; Xiao, N.; Ye, L.; Xu, Q. Chlorotoxin-Conjugated Graphene Oxide for Targeted Delivery of an Anticancer Drug. *Int. J. Nanomedicine* **2014b**, *9*(1), 1433–1442.

Wang, H.; Sun, D.; Zhao, N.; Yang, X.; Shi, Y.; Li, J.; Su, Z.; Wei, G. Thermo-Sensitive Graphene Oxide–polymer Nanoparticle Hybrids: Synthesis, Characterization, Biocompatibility and Drug Delivery. *J. Mater. Chem. B* **2014c**, *2*(10), 1362.

Wang, J.; Liu, C.; Shuai, Y.; Cui, X.; Nie, L. Controlled Release of Anticancer Drug Using Graphene Oxide as a Drug-Binding Effector in Konjac Glucomannan/sodium Alginate Hydrogels. *Colloids Surfaces B Biointerfaces* **2014d**, *113*, 223–229.

Wang, Y.; Kim, M. J.; Shan, H.; Kittrell, C.; Fan, H.; Ericson, L. M.; Hwang, W.-F.; Arepalli, S.; Hauge, R. H.; Smalley, R. E. Continued Growth of Single-Walled Carbon Nanotubes. *Nano Lett.* **2005**, *5*(6), 997–1002.

Weaver, C. L.; LaRosa, J. M.; Luo, X.; Cui, X. T. Electrically Controlled Drug Delivery from Graphene Oxide Nanocomposite Films. *ACS Nano* **2014**, *8*(2), 1834–1843.

Widenkvist, E. Fabrication and Functionalization of Graphene and Other Carbon Nanomaterials in Solution, PhD Thesis, Uppsala University, 2010.

Wu, S.-Y.; An, S. S. A.; Hulme, J. Current Applications of Graphene Oxide in Nanomedicine. *Int. J. Nanomedicine* **2015**, *10*, 9–24.

Yang, K.; Li, Y.; Tan, X.; Peng, R.; Liu, Z. Behavior and Toxicity of Graphene and Its Functionalized Derivatives in Biological Systems. *Small* **2013**, *9*(9–10), 1492–1503.

Yang, X.; Zhang, X.; Liu, Z.; Ma, Y.; Huang, Y.; Chen, Y. High-Efficiency Loading and Controlled Release of Doxorubicin Hydrochloride on Graphene Oxide. *J. Phys. Chem. C* **2008**, *112*(45), 17554–17558.

Zhang, L.; Lu, Z.; Zhao, Q.; Huang, J.; Shen, H.; Zhang, Z. Enhanced Chemotherapy Efficacy by Sequential Delivery of siRNA and Anticancer Drugs Using PEI-Grafted Graphene Oxide. *Small* **2011a**, *7*(4), 460–464.

Zhang, L.; Wang, Z.; Lu, Z.; Shen, H.; Huang, J.; Zhao, Q.; Liu, M.; He, N.; Zhang, Z. PEGylated Reduced Graphene Oxide as a Superior srRNA Delivery System. *J. Mater. Chem. B* **2013**, *1*(6), 749.

Zhang, W.; Zhang, Z.; Zhang, Y. The Application of Carbon Nanotubes in Target Drug Delivery Systems for Cancer Therapies. *Nanoscale Res. Lett.* **2011b**, *6*(1), 555.

Zhang, X.; Meng, L.; Lu, Q.; Fei, Z.; Dyson, P. J. Targeted Delivery and Controlled Release of Doxorubicin to Cancer Cells Using Modified Single Wall Carbon Nanotubes. *Biomaterials* **2009**, *30*(30), 6041–6047.

Zhang, Y.; Ali, S. F.; Dervishi, E.; Xu, Y.; Li, Z.; Casciano, D.; Biris, A. S. Cytotoxicity Effects of Graphene and Single-Wall Carbon Nanotubes in Neural Phaeochromocytoma-Derived PC12 Cells. *ACS Nano* **2010**, *4*(6), 3181–3186.

Zhang, Y.; Zhu, J. Synthesis and Characterization of Several One-Dimensional Nanomaterials. *Micron* **2002**, *33*(6), 523–534.

Zhao, X.; Yang, L.; Li, X.; Jia, X.; Liu, L.; Zeng, J.; Guo, J.; Liu, P. Functionalized Graphene Oxide Nanoparticles for Cancer Cell-Specific Delivery of Antitumor Drug. *Bioconjug. Chem.* **2015**, *26*(1), 128–136.

Zhi, F.; Dong, H.; Jia, X.; Guo, W.; Lu, H.; Yang, Y.; Ju, H.; Zhang, X.; Hu, Y. Functionalized Graphene Oxide Mediated Adriamycin Delivery and miR-21 Gene Silencing to Overcome Tumor Multidrug Resistance In Vitro. *PLoS One* **2013**, *8*(3), 1–9.

Zhou, K.; Zhu, Y.; Yang, X.; Li, C. One-Pot Preparation of graphene/Fe_3O_4 Composites by a Solvothermal Reaction. *New J. Chem.* **2010**, *34*(12), 2950.

Zo, H. J.; Lee, J.-S.; Song, K.-W.; Kim, M.; Lee, G.; Park, J. S. Incorporation of Graphene Oxide into Cyclodextrin-Dye Supramolecular Hydrogel. *J. Incl. Phenom. Macrocycl. Chem.* **2013**, *79*(3–4), 357–363.

CHAPTER 8

NANOFIBERS: GENERAL ASPECTS AND APPLICATIONS

RAGHAVENDRA RAMALINGAM and
KANTHA DEIVI ARUNACHALAM*

Center for Environmental Nuclear Research, SRM University,
Kattankulathur, Kancheepuram – 603203, Tamilnadu, India,
*E-mail: kanthad.arunachalam@gmail.com

CONTENTS

ABSTRACT

Nanostructures are objects or structures which has at least one of its dimensions within nano-scale range. Nanofibers geometrically fall into the category of 1D nano-scale elements. Nanofiber can be considered either as a nanomaterial in view of its diameter or as a nanostructured material when filled with nanoparticles. Nanofibers gives rise to novel functions and find its applications in various fields such as catalysis, drug delivery, filter technology, functionalization of textiles, tissue engineering, cellular matrices, ultra-strong composites, bioreactors, electronics and energy application. In this chapter, a brief introduction to the different fabrication techniques for synthesizing nanofibers; in particular electrospinning, its working principle, the polymers and solvents used for nanofiber fabrication, parameters influencing the fiber formation and the applications of electrospun nanofibers in drug delivery are reviewed and discussed.

8.1 INTRODUCTION

Nanotechnology is a rapidly emerging multidisciplinary platform that influences both synthetic and naturally occurring materials in nanoscale dimensions (1–1000 nm). The influence of nanotechnology on the healthcare industry is substantial, particularly in the areas of disease diagnosis and treatment. Among various disciplines in nanotechnology, recent investigations and advancement in the field of drug delivery and tissue engineering have delivered high-impact contributions in translational research, with associated pharmaceutical products and applications (Sridhar et al., 2015). Based on the end-use products and applications, the structure and the composition of nanomaterial can be varied and custom-designed by applying the appropriate synthetic method (Gates et al., 2005; Zhang, 2003).

The concept of drug delivery system was formulated in the early 1970's and the first reliable system demonstrated was polymer lactic acid-based. Drug delivery is the process of administering a pharmaceutical compound to achieve a therapeutic effect in humans or animals (Ravikumar, 2008). An optimal drug delivery system assures availability of the active drug molecule at the site of action for appropriate time and duration. The concentration of the drug at the target site must always remain above the minimal effective concentration (MEC) and below the minimal toxic concentration (MTC). This concentration interval is known as the therapeutic window (Joshi et al., 2014). Even though it was introduced decades ago, extensive research have been reported in the last five to ten years espe-cially after the universalization of nanotechnology. Modern drug delivery systems are targeted to carry drugs, growth factors, genes and biomol-ecules for treating diseases (Balaji et al., 2015).

The prime objective of the novel drug delivery system is to deliver drug to the action site, enhancing its therapeutic effects and minimizing its toxic effects by increasing the drug concentration at the target sites and reducing the drug exposure to nontarget sites (Malik et al., 2015). Controlled drug systems enhance the safety, efficiency and reliability of drug therapy. It also release drug on demand in appropriate amount and over prolonged periods (Garg et al., 2012; Joshi et al., 2014). Most pre-ferred routes of drug delivery include oral, injection, trans mucosal, inha-lation and topical routes (Aulton and Taylor, 2007; Carrier, 2005; Florence and Attwood, 2008).

With the advancement in nanotechnology, various nanoforms such as liposomes, quantum dots, polymeric nanoparticles, dendrimers and nano-fibers have been attempted as drug delivery systems. Liposomes finds its application as nanocarriers with great promise for gene delivery and can-cer therapy (Torchilin, 2005), quantum dots are widely used as contrasting agents for imaging the living systems for the detection and diagnosis of the disease in vivo, Polymeric nanoparticles can be easily surface functional-ized for active targeting, dendrimers are hyper branched structures which can deliver both hydrophobic and hydrophilic drugs at the site of action, nanofibers as carriers of drug for cell/tissue regeneration, as scaffolds in wound dressing (Banyal et al., 2013; Wang et al., 2010).

Nanofibers can be produced by self-assembly, phase separation, elec-trospinning and other mechanical spinning techniques (Jayaraman et al., 2004; Ma et al., 2005; Sill and von Recum, 2008; Smith and Ma, 2004). Self-assembly produces nanofibers of 5–8 nm in diameter, but the com-plexity and low productivity of this technique limits its application. Further, these matrices tend to be relatively soft or nonstructured com-pared to the other processing techniques (Zhang et al., 2009). Phase sepa-ration is a simple technique that does not require specialized equipment, and generates fibers with diameters from 50–500 nm, similar to natural collagen fibers in ECM. This method also has its own limitations; it is effective only with a selected number of polymers and not feasible for scale-up (Jayaraman et al., 2004). Electrospinning surpasses the other methods and prevails as unique and most efficacious technique to fabricate nanofibers; since it is simple, inexpensive, and can be scaled up (Sridhar et al., 2015). Moreover, a diverse set of polymers can be used to produce fibers from a few micrometers down to the tens of nanometers in diameter with desired geometries and properties by adjusting the processing parameters (Sill and von Recum, 2008; Zhang et al., 2009). In this chapter, nanofibers, evolution of electrospinning, processing parameters, application of nanofibers as drug delivering systems will be discussed in detail.

8.2 NANOFIBERS

Nanofibers are defined as fibers (versatile class of one dimensional nanomaterial) with diameter less than 1000 nm. In current scenario, nanofiber find its applications in the healthcare systems as a tool for drug and cell delivery against various chronic diseases (Chaudhary et al., 2014). The unique properties of nanofibers like large surface area to volume ratio, superior stiffness as well as tensile strength, high porosity and flexibility in surface functionalities shows a significant role in transporting the bioactive molecules effectively to the appropriate site in the body (Caracciolo et al., 2013; Garg et al., 2012). Natural polymers like collagen and alginate; synthetic polymers like poly ε-caprolactone (PCL) and poly(vinylpyrrolidone) (PVP), and composite polymers (combination of natural–natural, natural–synthetic, and synthetic–synthetic) are successfully used for fabrication of nanofibers. Two important techniques namely adsorption and dipping are used for active and passive drug loadings in nanofibers. Nanofibers finds its application in different fields such as tissue engineering, drug delivery, cosmetics, filter media, protective clothing, wound dressing, homeostatic, and sensor devices. In future, the main research will be focused on delivery of plasmid DNA, large protein drugs, genetic materials, and autologous stem-cell to the target site (Garg et al., 2014).

8.2.1 CHARACTERISTICS OF NANOFIBERS SCAFFOLD

The unique characteristics of nanofibers such as biocompatibility, biodegradability, excellent mechanical property, sterility and controlled release pattern makes it an ideal candidate for drug as well as cell delivery. Nanofibers also supports cell adhesion due to their excellent porosity and pore interconnectivity, so it is widely used in tissue regeneration (Garg et al., 2014; Utreja et al., 2010).

8.2.1.1 Biocompatibility

Biocompatibility is related to the property of being biologically compatible to achieve proper host response without producing any adverse side effects (toxic, injurious or immune response) in living tissue (Farhana et al., 2010). Nanofiber scaffolds formulations exhibit excellent biocompatibility.

8.2.1.2 Biodegradability

Biodegradability- the relative ease of the material to degrade by any biological means such as hormones, acids, and body fluids (Girlich and Scholmerich, 2012). Nanofibers possess acceptable biodegradability profile and their degradation products have nontoxic nature as well as can be eliminated easily from implantation site of the body.

8.2.1.3 Porosity

Porosity – the ratio of the volume of interstices of a material to the volume of its mass or state of being porous (Abdelkader and Alany, 2012). Nanofiber has an adequate porosity, pore size distribution, and excellent interconnectivity between pores which help in cell in-growth as well as vascularization (Garg et al., 2014).

8.2.1.4 Targetability

Targetability is the ability of the carrier system to reach their predetermined site and release their loaded therapeutic agents on to the target site (Maira et al., 2012). Nanofiber formulations have excellent ability to deliver their encapsulated therapeutic agents to the target site. Due to their targetability, the dose and the frequency of drugs are significantly reduced (Garg et al., 2014).

8.2.1.5 Entrapment and Loading Efficiency

Entrapment efficiency is the ratio of weight of drug entrapped into a carrier system to the total drug added whereas loading efficiency is the ratio of drug to the weight of the total carrier system (all excipients taken together) (Parhi et al., 2012). Nanofibers have maximum entrapment as well as loading capacity so that the drug is released continuously for longer duration of period after introducing them into the body.

8.2.1.6 Toxicity

Toxicity is a grade to which a drug or formulation can harm an organism or affect an organism substructures (cells and organ) (Basile et al., 2012). Due to the biocompatible, biodegradable nature of nanofibers, most of them do not show toxicity or negligible toxicity in the body.

8.2.1.7 Stability and Sterility

Stability is defined as resistance to chemical change or to physical disintegration. Drugs need to maintain their structure and activity over a prolonged period of time (Sokolsky-Papkov et al., 2007). Nanofibers scaffold formulations show excellent stability at physiological temperature and their activity is maintained for longer duration of period.

Sterility is the state of being free from pathogenic microorganisms. Ultraviolet radiation (UVR), antimicrobial solutions (AMS), ethanol, gamma radiation have been successfully used for sterilization of nanofibers. However, these techniques affect the physicochemical properties of prepared formulations. The sterilization techniques caused changes in the nanofiber morphology, dimensions, and a greater reduction in polymeric molecular weight, respectively. Among these sterilization techniques, AMS sterilized nanofibers showed superior cellular adhesion (Garg et al., 2014).

8.2.1.8 Mechanical Property

Nanofibers possess good mechanical properties (tensile strength, young's modulus); their strength can be matched with that of implantation site tissues and sufficient enough to shield the active therapeutic agents from destructive compressive or tensile forces (Garg et al., 2014).

8.2.1.9 Controlled Release Behavior

Controlled release is a term denoting to the release or discharge of drug compounds into the desired site of body in response to stimuli or time (Sokolsky-Papkov et al., 2007). Nanofiber formulations shows excellent controlled as well as sustained release behavior in the body and it can also be processed accordingly to have burst release-based on the requirement.

8.2.1.10 Binding Affinity

Binding affinity is defined as how tightly the drug binds to the scaffold. To achieve better drug release profile, the binding affinity of the drug towards the scaffold must be sufficiently low (Sokolsky-Papkov et al., 2007).

8.3 FABRICATION TECHNIQUES FOR SYNTHESIZING NANOFIBERS

Several mechanical fiber spinning techniques have been used for processing natural and synthetic polymers into different types of porous nanofiber scaffolds. Some of the important techniques are described in the following subsections (Garg et al., 2014).

8.3.1 DRAWING TECHNIQUE

Drawing technique can produce nanofibers directly from viscous polymer liquids. It create very long single nanofibers analogous to dry spinning in fiber-forming industry. Figure 8.1 represents the schematic illustration of drawing

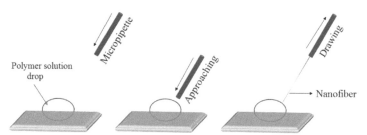

FIGURE 8.1 Nanofibers fabrication by drawing technique. (Modified from Xing, X.; Yu, H.; Zhu, D.; Zheng, J.; Chen, H.; Chen, W.; Cai, J. Sub-wavelength and Nanometer Diameter Optical Polymer Fibers as Building Blocks for Miniaturized Photonics Integration, Optical Communication; Dr. Narottam Das (Ed.), InTech, 2012. DOI: 10.5772/47822. https://creativecommons.org/licenses/by/3.0/).

technique. This technique mainly involves three steps; (i) a drop of polymer solution is applied over the substrate material, (ii) followed by a micropipette moved down towards the edge of the drop, and (iii) once the contact made, by retrieving the micropipette, the fiber is drawn out of the polymer droplet at certain rate (Ramakrishna et al., 2005). Fibers obtained from this technique has diameter ranging from 2 to 100 nm up to several μm; length from 10 mm to several cm. The quality of nanofiber mainly depends on material composition, drawing velocity, and speed at which the solvent evaporates. Various polymers such as PCL, PVA, blend of hyaluronic acid (HA), gelatin, PEO, poly(vinyl butyral) (PVB), and poly(methyl methacrylate) (PMMA) are successfully drawn into nanofibers using this technique. This technique is simple, inexpensive, easily modified, and longer fibers can be obtained. However, its major limitations includes time-consuming, not applicable for all polymers, lower productivity and very long single fiber with nonuniform fiber size. Xing et al. (Xing et al., 2008) fabricated flexible and elastic poly tri-methylene terephthalate (PTT) nanofibers by drawing process.

8.3.2 TEMPLATE SYNTHESIS TECHNIQUE

Polymer nanofibers can be fabricated using templates made from special materials such as self-ordered porous alumina. Alumina network templates with pore diameters from 25 to 400 nm, and pore depths from around 100 nm to several 100 μm have been fabricated. Schematic representation of the nanofiber synthesis by template synthesis is shown in Figure 8.2 (Xing et al., 2012). (a) The alumina hard template is attached to an underlying aluminum substrate. (b) Resin is infiltrated into the nanopores

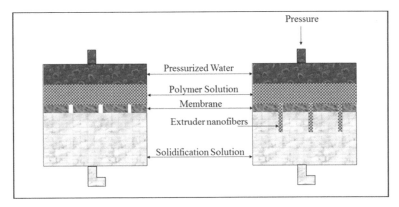

FIGURE 8.2 Nanofibers production by template synthesis technique. (From Xing, X.; Yu, H.; Zhu, D.; Zheng, J.; Chen, H.; Chen, W.; Cai, J. Sub-wavelength and Nanometer Diameter Optical Polymer Fibers as Building Blocks for Miniaturized Photonics Integration, Optical Communication; Dr. Narottam Das (Ed.), InTech, 2012. DOI: 10.5772/47822. https://creativecommons.org/licenses/by/3.0/)

by a pressure impregnation technique. (c) And then UV cross-linked. (d) The obtained array of cross-linked nanofibers is extracted from the alumina hard template by mechanical extraction and the alumina template can be reused further (Grimm et al., 2008). However, it is a time consuming process, cannot be scaled up, laborious and cannot make continuous nanofiber (Garg et al., 2014).

8.3.3 PHASE SEPARATION

In phase-separation, gel of a polymer is prepared by storing the homogeneous solution of the polymer at the required concentration in a refrigerator at the gelation temperature (Nayak et al., 2011). The gel is then immersed in distilled water for solvent exchange, followed by the removal from the distilled water, blotting with filter paper and finally transferring to a freeze-drying vessel leading to a nanofiber formation. The mechanism involved in this process is the separation of phases due to physical incompatibility. The schematic illustration of phase separation is shown in Figure 8.3. The type of polymers, type of solvents, gelation temperature, gelation duration, and thermal treatment affects the nanofibers morphology. This technique is simple, inexpensive and continuous nanofibers can be made. However, it suffers from limitations

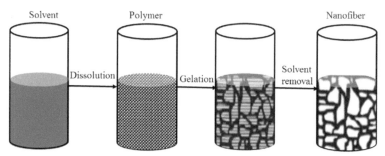

FIGURE 8.3   Nanofibers production by phase separation technique. (Modified from Xing, X.; Yu, H.; Zhu, D.; Zheng, J.; Chen, H.; Chen, W.; Cai, J. Sub-wavelength and Nanometer Diameter Optical Polymer Fibers as Building Blocks for Miniaturized Photonics Integration, Optical Communication; Dr. Narottam Das (Ed.), InTech, 2012. DOI: 10.5772/47822. https://creativecommons.org/licenses/by/3.0/)

such as time consuming process, laboratory scale production, lack of structural stability, difficult to maintain porosity, and not applicable for all polymers.

8.3.4 SELF-ASSEMBLY

Self-assembly is a formulating method where small molecules are used as basic building blocks which add-up to give nanofibers (Hartgerink et al., 2001). Through this technique, very fine fibers (7–100 nm in diameter; hundreds of nanometers in length) are produced. Liu et al. (2014), Dhinakaran et al. (2014), Nalluri et al. (2014) have prepared nanofibers using self-assembly. In self-assembly the final (desired) structure is 'encoded' in the shape of the small blocks, thus referred to as 'bottom-up' manufacturing technique. The synthesis of molecules for self-assembly often involves a chemical process called convergent synthesis. Figure 8.4 is a simple schematic representation on self-assembly for obtaining nanofibers. The main mechanism for a generic self-assembly is the intermolecular forces that bring the smaller units together and the shape of the smaller subunits which determine the overall shape of the macromolecular nanofiber. The main advantages are easy processability, easy to get smaller fibers, reproducible but it's a complex process, only lab scale production, inability to control dimensions of nanofibers, and difficult to maintain its porosity for longer duration of period (Endres et al., 2012).

8.3.5 ROTARY JET SPINNING

Rotary jet spinning (RJS) is a process to fabricate three-dimensional aligned nanofibers by exploiting a high-speed rotating nozzle to form a polymer jet which undergoes stretching before solidification. In this process fiber diameter, fiber morphology and web porosity can be controlled by varying rotational speed, nozzle geometry and solution properties. Schematic representation of RJS system is shown in Figure 8.5. The RJS system consists of a reservoir (with two side wall orifices) attached to the shaft of a motor with controllable rotation speed. The polymer solution is continuously fed

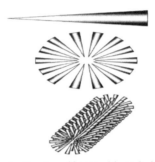

FIGURE 8.4 Nanofibers production by self-assembly technique. (From Xing, X.; Yu, H.; Zhu, D.; Zheng, J.; Chen, H.; Chen, W.; Cai, J. Sub-wavelength and Nanometer Diameter Optical Polymer Fibers as Building Blocks for Miniaturized Photonics Integration, Optical Communication; Dr. Narottam Das (Ed.), InTech, 2012. DOI: 10.5772/47822. https://creativecommons.org/licenses/by/3.0/)

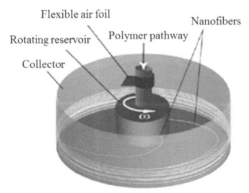

FIGURE 8.5 Schematic representation of Rotary jet spinning system. (Reprinted with permission from Badrossamay, M.R.; McIlwee, H.A.; Goss, J.A.; Parker, K.K. Nanofiber assembly by rotary jet-spinning. Nano Lett.2010, 10, 2257–2261. © 2010 American Chemical Society.)

to the reservoir at a suitable rate to maintain a constant hydrostatic pressure and continuous flow. The fibers are collected either on a stationary surrounding cylindrical collector or on coverslips held against the collector wall. The RJS technique has several advantages over electrospinning, such as: (a) no requirement of high voltage, (b) fiber fabrication is independent of solution conductivity, (c) it is applicable to polymeric emulsions and suspensions, and (d) higher productivity (Badrossamay et al., 2010).

8.3.6 MELT BLOWING

Melt blowing (MB) is a simple, versatile and one step process for the production of materials in micrometer and nanometer range. A typical MB process involves following components: extruder, metering pumps, die assembly, web formation, and winding. A molten polymer is extruded through the orifice of a die. Figure 8.6 represents the schematic illustration of melt blowing system. The fibers are formed by the elongation of the polymer streams coming out of the orifice by air-drag and are collected on the surface of a suitable collector in the form of a web. The average fiber diameter mainly depends on the throughput rate, melt viscosity, melt temperature, air temperature and air velocity (Nayak et al., 2011). The difficulty in fabricating nanofibers in melt blowing is due to the inability to design sufficiently small orifice in the die and the high viscosity of the polymeric melt. Nanofibers can be fabricated by

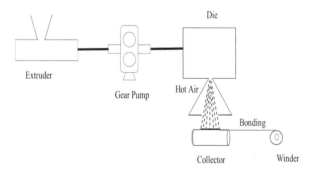

FIGURE 8.6   Nanofibers production by melt blowing technique. Modified from Hiremath and Bhat (2015).

special die designs with a small orifice, reducing the viscosity of the polymeric melt and suitable modification of the melt blowing setup. For example, Ellison et al. (2007) produced melt blown nanofibers of different polymers by a special designed single-hole die with small orifice. The use of small orifices made by an electric discharge machine for the production of super-hydrophobic nanofibers and microfibers has been reported.

The process is suitable for many melt-spinnable commercial polymers, copolymers and their blends such as polyesters, poly olefins, poly alcohols (PA), nylons, polyurethane (PU), poly (vinyl chloride) (PVC), poly(vinyl alcohol) (PVA) and ethylene vinyl acetate. In melt blowing, the sudden cooling of the fiber as it leaves the die can prevent the formation of nanofibers. This can be improved by providing hot air flow in the same direction of the polymer around the die. The hot air stream flowing along the filaments helps in attenuating them to smaller diameter. The viscosity of polymeric melt can be lowered by increasing the temperature, but there is a risk of thermal degradation at high temperature (Nayak et al., 2011).

8.3.7 ELECTROSPINNING

Electrospinning is a process that produces nanofibers through an electrically charged jet of polymer solution or polymer melt. Figure 8.7 shows the schematic representation of an electrospinning unit, which basically consists of three major components: a high voltage power supply, a spinneret and a grounded collector. It uses a high voltage source to inject charge of a certain polarity into a polymer solution or melt, which is then accelerated towards a collector of opposite polarity (Agarwal et al., 2008; Bhardwaj and Kundu, 2010; Liang et al., 2007; Vasita and Katti, 2006).

8.4 ELECTROSPINNING

Electrospinning has gained much attention in the last decade due to its versatility in spinning a wide variety of polymeric fibers and its ability to consistently produce fibers in the submicron range (Theron

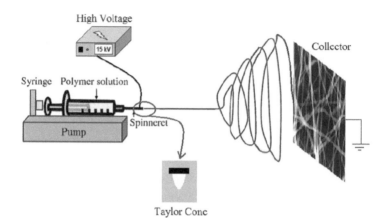

FIGURE 8.7 Schematic diagram of electrospinning apparatus (horizontal setup). (Reprinted from Ning Zhu; Xiongbiao Chen. Biofabrication of Tissue Scaffolds, Advances in Biomaterials Science and Biomedical Applications, Prof. Rosario Pignatello (Ed.), InTech, 2013. doi: 10.5772/54125. http://www.intechopen.com/books/advances-in-biomaterials-science-and-biomedical-applications/biofabrication-of tissue scaffolds. https://creativecommons.org/licenses/by/3.0/)

et al., 2005; Tsai et al., 2002). Electrospun nanofibers have several advantages over other mechanical fiber spinning techniques such as high porosity, extremely high surface-to-volume ratio, tunable porosity and the ability to control the nanofiber composition to achieve the desired results from its properties and functionality. Currently electrospinning is used in various applications such as nanocatalysis, tissue engineering scaffolds, protective clothing, filtration, biomedical, pharmaceutical, optical electronics, healthcare, biotechnology, defense and security, and environmental engineering (Bhardwaj and Kundu, 2010; Luu et al., 2003; Ramakrishna et al., 2006; Subbiah et al., 2005; Welle et al., 2007).

8.4.1 EVOLUTION OF ELECTROSPINNING

The process of using electrostatic forces to form synthetic fibers has been known for over 100 years (Sill and von Recum, 2008). Electrospinning has originated from electro spraying, where an electric charge is supplied to a conducting liquid which produces a jet which splits into fine particles that resemble a spray. It was first observed in 1897 by Rayleigh, studied in detail by Zeleny (1914). In 1934, Formhals was the first to patent electrospinning technique, describing an experimental setup for the production of polymer filaments using an electrostatic force (Bhardwaj and Kundu, 2010). Following the work of Formhals the focus turned towards better understanding of the electrospinning process; however, it took nearly 30 years before Taylor published work regarding the jet forming process (Sill and von Recum, 2008). In 1969, Taylor published his work regarding the jet forming process, in which for the first time he described the phenomena of Taylor cone, which develops from the pendant droplet when the electrostatic forces are balanced by surface tension. Taylor also observed the emission of a fiber jet from the apex of the cone, which explained the generation of fibers with significantly smaller diameters compared to the spinneret (Zhang et al., 2009).

Shortly after Taylor's work on the jet forming process was published, the focus of electrospinning turned towards understanding the relationships between individual processing parameters and the structural properties of electrospun fibers. In 1971, Baumgarten began to investigate the effect of varying certain solution and processing parameters (solution viscosity, flow rate, applied voltage, etc.) on the structural properties of electrospun fibers (Sill and von Recum, 2008; Zhang et al., 2009). Concurrently, other researchers began to examine the potential of electrospun fibrous matrices for other applications. In 1978, Annis and Bornat evaluated the feasibility of electrospun polyurethane mats as vascular prostheses. By 1985, Fisher and Annis were examining the long-term in vivo performance of an electrospun arterial prosthesis (Murugan and Ramakrishna, 2007). Electrospinning regained attention in the 1990's partly due to interest in nanotechnology, as the process allowed easy fabrication of fibers with nanoscale diameters from various polymers (Huang et al., 2003). Over the years, more than 200 polymers have been electrospun for various applications and the number is still increasing gradually with time.

8.4.2 PRINCIPLE OF ELECTROSPINNING

During electrospinning, electric potential is applied to a liquid droplet suspended from the spinneret tip, the liquid becomes charged (Chen et al., 2014; Hohman et al., 2001a, 2001b). Increase in electric potential leads to the elongation of the hemispherical surface of the droplet to form a conical shape structure known as Taylor cone. A further increase causes the electric potential to reach a critical value, at which it overcomes the surface tension forces of liquid to cause the formation of a jet that is ejected from the tip of the Taylor cone (Zhang et al., 2009). The charged jet undergoes instabilities and gradually thins in air primarily due to elongation and solvent evaporation forming nanofibers. Two different models were proposed to explain the nanofiber drawing process: single filament elongation or splitting of a fiber into several smaller fibers (Deitzel et al., 2001). According to single fiber theory, the fiber jet undergoes a chaotic whipping and bending trajectory while accelerating towards the collector due to repulsive interactions among like charges in the polymer jet (Reneker et al., 2000; Yarin et al., 2001). Doshi and Reneker hypothesized a different model involving the splitting of a fiber jet caused by the increase in charge repulsion during elongation (Doshi and Reneker, 1995). However, recent studies have imaged the unstable zone of fiber jet with the aid of high speed photography and revealed that a whipping instability leads the single fiber to bend and turn rapidly, thereby resulting in the incorrect perception of fiber splitting (Shin et al., 2001).

8.4.3 POLYMERS USED FOR NANOFIBER SCAFFOLD FABRICATION

A wide range of polymers has been used in electrospinning to obtain nanofibers. The selection of polymers plays a major role in the design of nanofiber systems with predetermined physicochemical parameters as well as release profiles for healthcare applications. By proper selection of polymers, the range of formulations, routes, and site of delivery can be broadened. Various properties such as biocompatibility, biodegradability, nontoxicity, hydrophilicity and proper mechanical strength are mainly considered at the time of selection of polymers. Natural, synthetic, and composite polymers have been used for nanofiber fabrication (Garg et al., 2014).

8.4.3.1 Natural Polymers

Some of the natural polymers which have been electrospun were chitosan (Beachley and Wen, 2010; Bhattarai et al., 2005; Geng et al., 2005; Jiang et al., 2014b; Rosic et al., 2012b), collagen (Badylak et al., 2009; Beachley and Wen, 2010; Bhardwaj and Kundu, 2010; Khadka and Haynie, 2012), elastin (Beachley and Wen, 2010; Khadka and Haynie, 2012), hyaluronic acid (Beachley and Wen, 2010; Li et al., 2012; Uppal et al., 2011), fibrinogen (McManus et al., 2007; Ravichandran et al., 2013; Wnek et al., 2003), gelatin (Maleknia and Majdi, 2014), etc. Advantages of naturally occurring polymers includes better biocompatibility, they carry specific protein sequences, such as RGD which enhances cell binding, similar to macromolecular substances present in the human body, more closely mimic the natural extracellular matrix of tissues, and remarkable physicochemical properties. In spite of its advantages they show batch-to-batch variation, limited availability, expensive, presence of immunogenic/pathogen moieties (Garg et al., 2014).

8.4.3.2 Synthetic Polymers

Some of the synthetic polymers include poly(ethylene oxide) (PEO) (Son et al., 2004; Kriegel et al., 2009; Fortunato et al., 2014), poly ε-caprolactone (PCL) (Zamani et al., 2010; Ganesh et al., 2012; Shalumon et al., 2013), poly(lactide-coglycolide) (PLGA) (Jang et al., 2009; Meng et al., 2011), poly(vinyl alcohol) (PVA) (Janković et al., 2013; Rošic et al., 2013), polylactic acid (PLA) (Jang et al., 2009; Casasola et al., 2014), poly(vinyl pyrrolidone) (PVP) (Dai et al., 2012b; Yu et al., 2012), etc. Advantages of synthetic polymer are reliable, often easily electrospinnable, well-defined structure, easily controlled physicochemical properties, fine tunable degradation kinetic as well as mechanical properties and consistently supplied in large quantities. However, limitations such as poor cell-recognition sites and poor affinity for cell attachment are associated with the use of synthetic polymers (Garg et al., 2014).

8.4.3.3 Composite Polymers

Researchers have tried combinations of different polymers (composite polymers) to prepare nanofibrillar scaffolds that do not trigger the host immune response, to preserve structural integrity, to tailor mechanical, degradation, biological properties and to ensure a cell-friendly microenvironment (Kriegel et al., 2009). Some of the composite polymers used are hyaluronic acid (HA), silk fibroin, and poly ε-caprolactone (PCL) (Pelipenko et al., 2015); cellulose acetate (CA) and poly(vinylpyrrolidone) (PVP); poly L-lactide (PLLA) fibers and nylon sutures; PLLA, PVP, PCL and poly(ethylene oxide) (PEO) blend (Garg et al., 2014).

8.4.4   SOLVENTS USED FOR ELECTROSPINNING

Once a polymer with desirable characteristics was selected, the next step is to prepare the polymer dispersion for electrospinning (Bhardwaj and Kundu, 2010; Khadka and Haynie, 2012). Basically, a solvent performs two crucial roles: firstly to dissolve the polymer molecules for forming the electrified jet and secondly to carry the dissolved polymer molecules towards the collector (Ohkawa et al., 2004).

Solvents should not chemically interact with the dissolved polymer (Bhardwaj and Kundu, 2010; Casasola et al., 2014; Jarusuwannapoom et al., 2005). Solvents decisively affect the surface tension and conductivity of the polymer dispersion, which is also affected by the polymer nature and concentration (Rosic et al., 2013). Due to its safety and biocompatibility, water is the most desirable solvent; however, its use is limited to hydrophilic polymers (Bhardwaj and Kundu, 2010), the solubility of polymers in water is often low (Bhattarai et al., 2005).

The commonly used organic solvents in electrospinning are acetone, dichloromethane, methanol, ethanol, acetic acid, dimethylformamide, ethyl acetate, trifluoroethanol, tetrahydrofuran, and formic acid (Bhardwaj and Kundu, 2010; Khadka and Haynie, 2012). In order to achieve optimal solution viscosity, surface tension, and solvent volatility, a combination of two or more solvents are often used (Li et al., 2006).

8.5   EFFECT OF VARIOUS PARAMETERS ON NANOFIBER FORMATION

The electrospinning process is affected by different parameters, making it challenging to handle. The parameters affecting electrospinning are classified broadly into 3 categories: solution parameters, process parameters, and ambient parameters (Kriegel et al., 2009; Liu et al., 2013a; Pelipenko et al., 2012; Tripatanasuwan et al., 2007). Despite some allegedly known correlations, the morphology of the electrospun product obtained is often different from expected one, indicating the interconnectivity of numerous known parameters and probably some unknown ones. Each of these parameters significantly affect the fiber morphology and by proper manipulation of these parameters we can obtain nanofibers of desired morphology and size (Chong et al., 2007).

8.5.1 SOLUTION PARAMETERS

Solution parameters such as polymer type and concentration, molecular weight, viscosity, conductivity, surface tension are the most frequently investigated (Pelipenko et al., 2015).

8.5.1.1 Polymer Type and Concentration

Polymer type has a significant effect on inter and intramolecular interactions reflected in the physical properties of solution (Gupta et al., 2005). Linear polymers are preferred instead of nonlinear polymers, because the latter form very viscous solutions or even gels at low polymer concentrations itself (McKee et al., 2004; Rosic et al., 2012a). Polymers with a polyelectrolyte nature are very difficult to spun (Rosic et al., 2012a) and thus uncharged polymers are preferable. Polyelectrolytes are also susceptible to intensive swelling, which leads to highly viscous solutions even at low concentrations. In addition, the polyelectrolyte nature of polymers causes intense repulsive forces, which result in jet instability while electrospinning (Pelipenko et al., 2015).

At low concentrations beads are formed instead of fibers and at very high concentrations the formation of continuous fibers are prohibited because of the inability to maintain the flow of the solution at the tip of the needle resulting in the formation of larger fibers (Haghi and Akbari, 2007; McKee et al., 2004; Sukigara et al., 2003). So, there should be an optimum solution concentration to obtain continuous nanofibers (Bhardwaj and Kundu, 2010).

8.5.1.2 Polymer Molecular Weight

Molecular weight of the polymer has a significant effect on rheological and electrical properties such as viscosity, surface tension, conductivity and dielectric strength (Haghi and Akbari, 2007). Molecular weight of the polymer reflects the number of entanglements of polymer chains in a solution. Solutions of low molecular weight polymers tend to form beads rather than fibers. Increasing the molecular weight of a polymer results in a reduction in the number of beads in the electrospun fibers (Geng et al., 2005; Gupta et al., 2005; Haghi and Akbari, 2007). Polymers with high molecular weights are preferable in order to enable a sufficient number of intermolecular entanglements. But, it has been observed that high molecular weight polymers are not always essential for nanofiber formation if the entanglements between polymers are replaced by sufficient intermolecular interactions (McKee et al., 2006).

8.5.1.3 Viscosity

Solution viscosity plays an important role in determining the fiber size and morphology during spinning of polymeric fibers. It has been found that with very low viscosity there is no continuous fiber formation and with very high viscosity there is difficulty in the ejection of jets from polymer solution, thus there is a requirement of optimal viscosity for electrospinning. Viscosity, polymer concentration and molecular weight of polymer are interconnected to each other (Bhardwaj and Kundu, 2010; Pelipenko et al., 2015).

8.5.1.4 Conductivity

Polymers are mostly conductive, with a few exceptions of dielectric materials, and the charged ions in the polymer solution are highly influential in jet formation. Solution conductivity is mainly determined by the polymer type, solvent used (Bhardwaj and Kundu, 2010). Polymer solutions with low conductivity cannot be electrospun due to the absence of a surface charge on the fluid droplet, which is required for Taylor cone formation. On the other hand, very high conductivity leads to a depleted tangential electric field along the surface of the fluid droplet, preventing Taylor cone formation (Angammana and Jayaram, 2011). In the case of uncharged polymers, the problem of low conductivity can be solved by adding salts, which can be also employed for nanofiber diameter manipulation. It has been reported that higher solution conductivities generally result in thinner nanofibers (Angammana and Jayaram, 2011; Cramariuc et al., 2013). In addition, the high solution conductivity enables the use of lower applied voltage; however, highly conductive solutions can be very unstable in the applied electric field (Bhardwaj and Kundu, 2010).

8.5.1.5 Surface Tension

Surface tension interferes with the electrospinning process, because it is the main force acting against Taylor cone formation and further jet elongation. The effect of surface tension on the morphology of nanofibers has been investigated by numerous authors (Hohman et al., 2001a; Zuo et al., 2005). Surface tension, more likely to be a function of solvent composition. Different solvents may contribute different surface tensions. Generally, low surface tension values result in the formation of fibers without beads and low voltages can be applied in electrospinning. High surface tension of a solution inhibits the electrospinning process because of instability of the jets and the generation of sprayed droplets (Bhardwaj and Kundu, 2010).

8.5.2 PROCESS PARAMETER

Process parameters include applied voltage, tip-to-collector distance, flow rate of polymer solution, nozzle design, and collector type (Bhardwaj and Kundu, 2010).

8.5.2.1 Applied Voltage

In electrospinning process the applied voltage to the solution plays a critical role. Only after attainment of threshold voltage, fiber formation occurs. Solutions with low conductivity, high surface tension, and/or high viscosity require higher voltages, and vice versa. A high voltage causes increased repulsive electrostatic forces, which lead to more extensive stretching of the electrospun jet, resulting in thinner fibers (Cramariuc et al., 2013; Pham et al., 2006). When applied voltage is too high the probability of bead formation in the electrospun product is much greater due to Taylor cone instability (Megelski et al., 2002; Pham et al., 2006). Thus, voltage influences fiber diameter, but the level of significance varies with the polymer solution concentration and on the distance between the tip and the collector (Yordem et al., 2008).

8.5.2.2   Flow Rate

The flow rate of the polymer influences the jet velocity and the material transfer rate. A lower feed rate is more desirable as the solvent will get enough time for evaporation (Yuan et al., 2004). Control over the solution flow rate depends mainly on the volatility of the solvent used. High flow rates may result in the generation of beads (Bhardwaj and Kundu, 2010) or deposition of undried nanofibers (Pham et al., 2006). The bead formation is mainly due to unavailability of proper drying time prior to reaching the collector (Wannatong et al., 2004; Yuan et al., 2004; Zuo et al., 2005).

8.5.2.3   Tip-To-Collector Distance

The distance between needle tip and collector has been examined to control the fiber diameter. It has been found that distance between the needle tip and collector should provide sufficient time for electrospun jet to dry before reaching the collector. If the distance is too short, the jet does not solidify before it reaches the collector, which results in nanofiber fusion and polymer film formation. Increasing the distance results in production of thinner nanofibers (Bhardwaj and Kundu, 2010); only if it is accompanied by increased flow rate and applied voltage (Pelipenko et al., 2015). The distances that are either too close or too far, beads have been observed (Bhardwaj and Kundu, 2010).

8.5.2.4   Nozzle Design

Many modifications of spinning nozzles have been made for producing different kinds of nanofibers (Jiang et al., 2014a; Yu et al., 2013). A single-channel nozzle allows formation of uniform nanofibers, whereas a coaxial nozzle enables formation of core–shell or even multilayered nanofibers (Maleki et al., 2013). In general, smaller nozzle diameter generally results in thinner nanofibers and vice versa.

8.5.2.5   Collector Types

The type of collector determines the orientation and morphology of the electrospun product. Randomly oriented nanofiber mats can be produced when a static planar collector is used, but aligned nanofibers can be obtained only if the collector is suitably modified (using dynamic collectors) (Vasita and Katti, 2006). Many researchers have tried with different types of collector in the place of standard collectors such as rotating cylinder, or a wheel-like disk (Hu et al., 2014), split electrodes (Jalili et al., 2006), stainless steel wires (Chuangcote and Supaphol, 2006), rotating drum (Subramanian et al., 2012), hollow metallic cylinder (Khamforoush and Mahjob, 2011) to obtain aligned nanofibers. Rotation of the collector during electrospinning allows production of nanofiber mats with uniform thickness. Zussman et al. (2003) have reported that too high rotation speed can damage nanofiber morphology due to stretching or disturb the pathway of the electrospun jet due to the generation of airflow by the high rotation speed of the collector.

8.5.3   AMBIENT PARAMETERS

Ambient parameters are not commonly considered among the parameters affecting electrospinning, though they play an important role. Studies have been conducted to examine the effects of ambient parameters (i.e., temperature and humidity) on the electrospinning process (Casper et al., 2004; De Vrieze et al., 2008; Medeiros et al., 2008; Mit-uppatham et al., 2004; Su et al., 2011). Higher environmental temperatures result in higher solvent evaporation rate and the formation of thicker nanofibers, and vice versa. At low relative humidity values rapid solvent evaporation causes the polymer solution to solidify soon as it comes out of the nozzle and it is subjected to voltage-induced stretching for a shorter time period, which results in thicker nanofibers. In the case of higher relative humidity, the solidification process is decelerated and the liquid in the electrospun jet has more time to flow and be stretched under the applied electric field resulting in the formation of thinner nanofibers at first, followed by gradual occurrence of bead on electrospun product morphology, and finally resulting in formation of a polymer film as a consequence of wet electrospun product deposition and its fusion on the collector (Pelipenko et al., 2015).

8.6 ELECTROSPUN NANOFIBER BASED DRUG DELIVERY SYSTEMS

At present Electrospun nanofibers are in the commercial arena, the reason for the interest in nanofibers as a drug delivery vehicle is due to its high surface area, unique pore structure, excellent mechanical properties in proportion to weight, flexibility for surface functionalization, possibility to incorporate different additives. These drug delivery system relies on the principle that dissolution rate of a particulate drug increases with increasing surface area of both the drug and the corresponding carrier (Verreck et al., 2003). Apart from other delivery systems nanofibers showed controlled release of the drug at a release rate as constant as possible over a longer period of time, and the release rate can be attuned depending on the application.

Controlled drug delivery from the carrier system is an essential feature to improve drug bioavailability with narrow absorption window (Malik et al., 2015). The recent advancement in electrospinning technology, such as coaxial electrospinning allows developing reservoir type of systems whereas the traditional methods of electrospinning are suitable for matrix type of controlled release formulations. Further, electrospinning technology can be successfully used for multiple coating of core materials, where the coating thickness can be precisely controlled ranging from nanometer to micrometer scale (Yang et al., 2007; Zeng et al., 2005).

Scientists dealing with nanofiber successfully fabricated control release nano drug delivery systems using biodegradable polymers (Demirci et al., 2014; Paaver et al., 2015). Non-woven, porous and multilayered component of nanofibers makes them a potential candidate to design controlled drug delivery systems (Carbone et al., 2014; Sultana et al., 2013). For example, a nanoscale controlled and multidrug delivery system-based on micelle-enriched electrospun nanofibers was developed. First, hydro-phobic curcumin encapsulated micelles (assembled from biodegradable PEG-PCL copolymer) and hydrophilic doxorubicin in PVA solution were prepared separately. Then both the solutions were mixed together and electrospun. A time-programmed release of the two drugs achieved due to the different domains of the two drugs within the nanofibers; moreover, the release can be temporally and spatially regulated (Yang et al., 2014).

8.6.1 WOUND DRESSING AND TRANSDERMAL DRUG DELIVERY

Wound dressing is a material applied onto a wound with or without medication, to give protection and also assist in faster healing. An ideal wound dressing should have the ability to absorb exudates from wound site, maintain sufficient humidity at the interface of wound and dressing, provide gaseous permeability (both water vapor and air), allow thermal insulation, protect the damaged site from penetration of microorganisms, should be nontoxic and nontraumatic in nature, should be tear-resistant and should not disintegrate in both wet and dry conditions, should be comfortable and sterilizable, cost effectiveness and requiring low frequency of dressing change (Joshi et al., 2014). Current efforts using polymer nano fibrous membranes as medical dressing are still in its infancy but electrospun materials meet most of the requirements outlined for wound-healing polymer devices.

Recent ongoing research activities aim to develop electrospun ultra-fine fibers containing drug formulations, which can be used not only as potential drug carriers but also to act smart to release drugs on demand (Joshi et al., 2014).

8.6.1.1 Bioactive Dressings

Infection and bacterial colonization are the most important factors in delayed wound healing. Hence antimicrobial dressings containing certain antibiotics or antiseptics are key for infection control as well as in promoting topical wound recovery (Joshi et al., 2014). Duan et al. (2007)

produced antimicrobial nanofibers of poly ε-caprolactone (PCL) by electrospinning PCL solution with small amounts of silver-loaded zirconium phosphate nanoparticles (nano AgZr) for potential use in wound-dressing applications. The result showed that the fibers possessed strong killing ability against the tested bacteria strains (gram-positive *Staphylococcus aureus* (ATCC 6538) and Gram-negative *E. coli* (ATCC 25922) and no discoloration was observed. The nanofibers biocompatibility was tested by culturing primary human dermal fibroblasts (HDFs) on the fibrous mats, the results indicated that the cells attached and proliferated as continuous layers on the nano AgZr-containing nanofibers and maintained the healthy morphology. Ignatova et al. (2007) showed that electrospinning of PVP – iodine complex and PEO/PVP – iodine complex as prospective route to antimicrobial wound dressing materials.

Wound-dressing material was prepared by electrospinning PVA/AgNO$_3$ aqueous solution into nonwoven fibers and then treating by heat or UV radiation. Heat treatment as well as UV radiation reduced the Ag ions in the electrospun PVA/AgNO$_3$ fibers into the Ag nanoparticles. Also the heat treatment improved the crystallinity of the electrospun PVA fiber web and made the web insoluble in moisture environment. Silver has long been recognized as a highly effective antimicrobial agent for treating wounds and burns (Agarwal et al., 2008).

Rho et al. (2006) have investigated the wound-healing properties of mats of electrospun type I collagen fibers on wounds in mice and they found that healing of the wounds was better with the nanofiber mats than with conventional wound care, especially in the early stages of the healing process. Powell et al. (2008) have compared freeze-dried and electrospun skin substitutes-based on natural polymer, collagen in a thymic mice. Cell distribution, proliferation, organization, maturation engraftment and healing of full thickness wounds were compared between the skin substitutes. Although no significant difference was observed in cell proliferation, surface hydration or cellular organization between freeze drying and electrospun scaffolds, wound contracted faster with electrospun scaffold.

Zong et al. (2002) fabricated bio-absorbable amorphous poly (D, L-lactic acid) and semicrystalline poly (L-lactic acid) nonwoven membranes containing mefoxin antibiotic. Initial burst drug release was observed due to the concentration gradient, which was useful for prevention of postoperative infection since most infections occur within the first few hours after surgery (Joshi et al., 2014).

Taepaiboon et al. (2006) developed mats of poly(vinyl alcohol) nanofibers with four different types of nonsteroidal antiinflammatory drug with variable water solubility; sodium salicylate, diclofenac sodium, naproxen, and indomethacin for transdermal drug delivery systems. Results indicated that with an increasing molecular weight of a drug, the rate and amount of drug release decreases. Also the drug-loaded mats exhibited better release kinetics as compared to solvent cast films due to a high porosity of the electrospun nonwoven mats. Supaphol et al., studied release characteristics of *Centella asiatica* herbal extract from gelatin nanofibers. *Centella asiatica* is well known for its wound healing ability. Depending on the unit weight of herbal constituents present in the fibers, total amount releasing from the fiber was higher and vice versa (Sikareepaisan et al., 2008).

Kenawy et al. (2009) studied the release profile of drug ketoprofen (nonsteroidal antiinflammatory drug) from nanofibers derived from either biodegradable PCL or nonbiodegradable polyurethane or from the blend of both. Results indicated that release rates from all the systems were almost similar. Kanawung et al. (2007) investigated the delivery of diclofenac sodium and tetracycline hydrochloride from poly ε-caprolactone and poly(vinyl alcohol) nano fibrous mats. With longer submerging time, the drug release from the mats were increased monotonically and the release became constant at long immersion times.

Many other synthetic and natural polymers like carboxyethyl chitosan/PVA/silk fibroin (Zhou et al., 2013), chitosan (Charernsriwilaiwat et al., 2012; Jayakumar et al., 2010, 2011), collagen/PLGA (Liu et al., 2010), silk fibroin (Calamak et al., 2014; Uttayarat et al., 2012), chitosan-poly (vinyl alcohol) blend (Kanani et al., 2010) with a nano fibrous membrane, biocompatible carboxyethyl chitosan/poly(vinyl alcohol)/silk fibroin (Zhou et al., 2013) composite membrane, blended chitosan and silk fibroin nano fibrous membrane (Cai et al., 2010), chitosan coated poly(vinyl alcohol) nano fibrous matrix (Kang et al., 2010) have been electrospun for wound-dressing applications. Other than wound-dressing applications, electrospun nano fibrous systems are also efficient in facilitating targeted drug delivery for treatment of cancer cells (Joshi et al., 2014).

8.6.2 DRUG DELIVERY SYSTEM FOR TREATING CANCER

Recently the electrospun fiber mats for local chemotherapy applications became popular. Compared with other dosage forms like liposomes, micelles, hydrogels, and nanoparticles, electrospun mats can reduce the system toxicity and increase the local drug concentration. Combination of surgical operation for removing the tumor with subsequent chemotherapy or radiation therapy is the normal procedure to reduce the probability of tumor recurrence (Hu et al., 2014).

8.6.2.1 Lung Cancer

Shao et al. (2011) reported the anticancer property of PCL/MWCNTs/green tea polyphenols (GTP) electrospun nanofibers. GTP is well known for its potential antioxidant, antiinflammatory and anticancer properties. The main problem with GTP are instability and lack of site specificity which hinders its utilization in treatment. To overcome this drawback, the GTP was loaded in MWCNTs to gain sustained release. The release behavior can be tailored by adjusting MWCNTs content-based on the requirement. The Alamar blue assay was carried out using osteoblasts and human lung epithelial cancer cells (A549) to present the toxicity of hypothesized delivery system against normal and cancer cells. Surprisingly, the cytotoxicity of PCL/GTP and PCL/MWCNTs/GTP remained at similar range for first two days, but on the fifth day the cytotoxicity of MWCNTs loaded GTP nanofiber was significantly higher. Hence, they concluded that GTP incorporated PCL/MWCNTs has commendable properties to destroy lung cancer cells in a controlled manner with only traceable cytotoxicity to normal cells.

Nadia et al. (2015) have reported an effective drug delivery system to safely deliver doxorubicin (DOX) at the site of cancerous tissue. Here, the drug was loaded into PLA/PEG/multiwalled carbon nanotube (MWCNT) composite. The incorporated MWCNTs improved drug loading and retains the efficiency of prepared nanofibers. The in vitro cytotoxicity study was carried out using A549 lung cancer cells, the results portrays that the prepared delivery system possesses significant antitumor activity and killed 65–92% of cultured cancer cells. DOX-loaded PLA/PEG/MWCNT nanofibers maintained the cytotoxicity effect for an extended period when compared to PLA/PEG/MWCNT nanofibers and far better than free DOX which has a half-life of only 16–18 h.

Ardeshirzadeh et al. (2015) proposed DOX integrated polyethylene oxide (PEO), chitosan (CS) and graphene oxide (GO) based electrospun scaffolds to avoid the side-effects while using free DOX for lung cancer treatment.

Apart from synthetic drugs, various natural agents incorporated nanofibers also significantly fought against cancer cells. Sridhar and his co-workers developed PCL based nano fibrous drug eluting device to deliver extracts of curcumin (CU), aloe vera (AV) and neem (NE). PCL/AV, PCL/CU, PCL/CU/NE and PCL/CU/AV meshes expressed 70%, 65%, 23%, 18% cytotoxicity against A459 cells respectively. In particular, the 1% AV and 5% CU coupled PCL nanofibers observed to have 15% more cytotoxicity when compared with 1% commercial drug (cis-platin) loaded PCL nanofiber (Sridhar et al., 2014).

In addition to the targeted and site-specific distribution, presently the drug delivery system also expected to act-based on bodily environmental conditions. The stimuli-responsive nanofibers was first reported by Salehi et al. (2013). They used poly(N-isopropylacrylamide) (PNIPAAm) to offer predetermined delivery of DOX. PNIPAAm has the ability to behave differently in response to temperature change, but the difficulty is its higher water solubility which hinders its application especially in the drug delivery system. To overcome this, PNIPAAm was cross-linked with hydrophilic polymers so as to yield water-insoluble fibers. DOX loaded PNIPAAm nanofibers exhibited a linear relationship between the drug release percentage and time (approximately) at different concentration of the drug loaded. In vitro cytotoxicity was studied against A549 cells (lung cancer cell lines). The DOX integrated PNIPAAm nanofibers retained the same level of toxicity for 72 h due to sustained release and eradicated more number of cancer cells when compared to that of control.

Yong et al. (2014) fabricated poly(propylene carbonate) nanofibers integrated with multiple drugs paclitaxel and cisplatin. Paclitaxel elicits a unique mechanism of action by targeting tubulin of the cells (Bharadwaj and Yu, 2004; Priyadarshini and Aparajitha, 2012; Stanton et al., 2011). Whereas cisplatin acts against cancer cells with its ability to crosslink with the purine base on the DNA, interfering DNA repair mechanisms, causing DNA damage, and later inducing apoptosis (Dasari and Tchounwou, 2014; Martins et al., 2011). The simultaneous administration of these two effective drugs produced promising results. Through in vitro studies performed on A549 lung cancer cells, it was depicted that Poly(propylene carbonate)/paclitaxel/cisplatin microfibers showed synergism and excellent cytotoxicity than the free drugs and fibers incorporated with single drug, respectively (Balaji et al., 2015).

8.6.2.2   Breast Cancer

Achille et al. (2012) tailored PCL electrospun fibers for delivering DNA plasmid to suppress targeted oncogenic expression and eradicate breast cancer cells. Three different scaffolds were framed namely, pristine PCL, PCL/pKD-Cdk2-v5 plasmid (Cdk2i) and PCL/pKDEGFP-v1 plasmid (EGFPi). To confirm delivering of DNA encodes for stimulating shRNA to disrupt the cell cycle of breast cancer cells by silencing Cdk2 genes, several test were carried out. Firstly, the Quantitative Real Time PCR (QRTPCR) showed that the suppression of Cdk2 genes was higher in the cells while exploiting pKD-Cdk2-v5 plasmid than others. Similarly, the visual monitoring through the microscope exposed decreased cell viability in the presence of Cdk2 shRNA encoded plasmid. Interestingly, they were able to produce the same when culturing human breast cancer cells (MCF-7) on synthesized scaffolds, it insinuates the fact that expression of Cdk2 genes were lower in cancer cells present on PCL/pKD-Cdk2-v5 plasmid scaffolds. Through LIVE/DEAD assay they also confirmed that the cell death observed was due to silencing of aspired oncogenic expression.

Laiva et al. (2015) developed a carrier system for the safe delivery of titanocene dichloride (TDC) by rectifying the complications reported during clinical trials. Pristine poly ε-caprolactone (PCL), PCL/silk fibroin (SF) and PCL/SF/TDC nanofibers were fabricated. The in vitro drug release studies delineates that proposed nanofibers maintained sustain release for a maximum of six days and PCL/SF system with higher concentration of TDC described to release more drug after incubation of 144 h. The results insinuate excellent stabilization of drug within nanofibers which offers organized release with minimum or negligible initial burst. On the other hand, the MTT assay exposed excellent inhibition of growth of MCF-7 cells cultured on high concentrated TDC coupled PCL/SF nanofiber system compared to that of both pristine PCL and free drug at the end of 3rd day.

Sundar and Sangeetha (2012) fabricated stimuli-responsive drug releasing scaffolds from collagen (CG)/ PNIPAA/CS loaded with widely used anticancer drug 5- Fluorouracil (5-FU) against breast cancer cell lines. The variable was the concentration of CS to yield different nanofiber mats. More percentage of the drug was deposited on nanofiber surface in the form of aggregates and only a few were assimilated inside resulting in initial burst release in all mats. However, by increasing the concentration of CS, slow degradation and diffusion of 5-FU molecules was observed. Further, the release kinetics of the drug also differed in response to pH of buffer solution. As expected, the in vitro cell viability of cultured MCF-7 cells decreased with increasing the drug concentration. Whereas normal L929 cells cultured on CG/PNIPAA/CS/5 FU nanofibers expressed typical cell morphology which proves oncogenic cell dependent activity of incorporated chemotherapeutic agent.

The administration of single chemotherapeutic agent is not sufficient to cure solid or advanced tumors. So, a blend of anticancer drugs is preferred to treat this clinical condition. Milena et al. (Ignatova et al., 2011) designed a multiple drug delivery system consisting of poly(L-lactide-*co*-D, L-lactide) (coPLA) as a carrier to deliver natural quaternized chitosan (QCh) and DOX to provide combination therapy for breast tumors. The carrier system was tested against MCF-7 breast cancer cells. MTT assay results delineates excellent cytotoxicity of synthesized nanofiber mats QCh/coPLA, coPLA/DOX and QCh/coPLA/DOX against the cultured MCF-7 cancer cells. The highest apoptotic percentage was reported on QCh/coPLA/DOX mats as expected due to the combined antitumor activity of biomolecules used. So, by using this integrated drug delivery system, the side-effects of using free DOX was nullified and also intense remedy was achieved.

8.6.2.3   Other Types of Cancer

Liu et al. (2013b) prepared DOX encapsulated PLLA nanofibers and examined its efficacy as a local chemotherapy system against secondary hepatic carcinoma (SHCC). The results indicated that the majority of the loaded DOX in the fibers were released and diffused into the tumor site underneath the fiber mat, leading to a great inhibitory effect on tumor growth and little damage to the surrounding organs. These results provide an encouraging prospect of using drug-loaded electrospun nanofibers in local chemotherapy, especially for those patients receiving complete tumor resection or cyto-reductive surgery.

Ranganath and Wang (2008) prepared paclitaxel-loaded nanofibers and evaluated their postsurgical chemotherapy effect against malignant glioma in vitro and in vivo. Poly(D, L-lactide-*co*-glycolide) (PLGA) 85:15 copolymer was used to fabricate microfiber disc (MFD) and microfiber sheet (MFS) and PLGA 50:50 copolymer was used to fabricate submicrofiber disc (SFD) and submicrofiber sheet (SFS). All the dosage forms showed sustained paclitaxel release over 80 days in vitro with a small initial burst. Sheets exhibited a relatively higher initial burst compared to discs probably due to the lower compactness. Also, submicrofibers showed higher release against microfiber due to higher surface area to volume ratio and higher degradation rate. Animal study confirmed inhibited tumor growth of 75, 78, 69 and 71% for MFD, SFD, MFS and SFS treated groups over placebo control groups after 24 days of tumor growth. Liu et al. (Liu et al., 2012) incorporated dichloroacetate (DCA) into the polylactide (PLA) nonwoven fabrics via electrospinning. These DCA-loaded electrospun mats were directly implanted to cover the solid tumor. Results indicated that total tumor suppression degree of 96% was achieved in less than 19 days.

Liu et al. (Liu et al., 2015) developed PCL nanofibers loaded with sodium dichloroacetate (DCA) and diisopropylamine dichloroacetate (DADA) to fight against colon cancer and it was administrated to C26 tumor-bearing mice separated into appropriate groups. The histopathology performed after 12th day of conduct revealed that animals applied with PLA/DCA and PLA/DADA accomplished the suppression of tumor growth by 75% and 84%. Approximately, more than 95% of cancer reduction was achieved at the end of 15th day.

Lin et al. (Lin et al., 2013) successfully demonstrated hyperthermic eradication of colon cancer cells by using magnetic nanoparticles or thermoseeds loaded chitosan nanofibers. The thermoseeds or ferrite magnetites are low toxic nanoparticles which possess super-paramagnetism. These nanoparticles are approved by FDA and already finds several applications in drug delivery, cancer tissue imaging, biomolecule separation, etc. The required heat to kill the cancer cells produced by applying strong electromagnetic field using the coils placed outside, followed by intravenous administration of magnetic nanoparticles.

Xu et al. (2006) encapsulated 1, 3-bis-2-chloroethyl-1-nitrosourea (BCNU) in PEG–PLLA diblock copolymeric fibers to treat malignant glioma. Drug release from these fibers were dependent on the initial drug-loading and primarily followed a diffusion route. In vitro cytotoxicity results showed the BCNU-loaded fibers retain its antitumor activity against the rat glioma C6 cells throughout the whole experiment, while pristine BCNU disappeared within 48 hours. Xu et al. (2009) incorporated paclitaxel (PTX) and DOX as model drugs into PEG–PLA mats by emulsion electrospinning. Hydrophilic DOX diffuses out at a faster rate from the fibers than hydrophobic PTX. In vitro cytotoxicity against rat C6 cells suggested that a dual drug combination imposed a higher rate of inhibition against C6 cells rather than a single drug-loaded system.

8.6.3 ELECTROSPUN NANOFIBERS INCORPORATED WITH SMALL MOLECULES OR BIOLOGICS

8.6.3.1 Protein

The incorporation of bioactive proteins into polymeric nanofibers with the purpose of controlled release is strongly preferred because of the unique structure of nano fibrous meshes which resembles that of the ECM. The polymeric nature of proteins makes them spinnable under certain preparation conditions, resulting in the generation of nanofibers incorporated with proteinic matrices (Sebe et al., 2015).

Sullivan et al. (2014) demonstrated the fabrication of electrospun nanofibers from whey proteins (WP) and its component beta lactoglobulin (BLG). Aqueous whey protein solutions either in native or denatured form yielded interesting micro structures; while the addition of poly(ethylene oxide) (PEO) to the system led to bead-free nanofiber formation. Further, electrospun WP/PEO blends at a pH of 2 showed reduced bacterial growth, increased product shelf life and served as a successful scaffold for stem cell proliferation.

Choi et al. *(*2008) reported that bovine serum albumin (BSA) immobilized nanofibers showed no obvious burst release, although the release was observed within 1 week. Using the same strategy, an electrospun scaffolds with epidermal growth factor (EGF) was prepared and succeeded in effective delivery of these bioactive compounds in vivo.

Human epidermal growth factor (hEGF) was immobilized by means of chemical conjugation on nanofibers of block polymers (PCL and PEG). The prepared fibers were tested in vivo in diabetic mice with dorsal wounds as a model of diabetic ulcers. The growth factor loaded fibers demonstrated significantly higher wound healing activity compared to that of growth factor solution alone (Sebe et al., 2015).

Silk fibroin (SF) fibers loaded with bone morphogenetic protein 2 (BMP2) and/or hydroxyapatite (Hap) exhibited a significant supportive effect on mineralization on human-bone-marrow-derived mesenchymal stem cells, indicating that fibers of this proteinic matrix were appropriate for the delivery of proteins (Sebe et al., 2015).

The problem in retaining proteins bioactivity depends on macromolecule distribution in the matrix. Changes in the secondary and tertiary structures induced by the applied solvent system may result in loss of its bioactivity. Fiber formulation often exploits the benefits of a crystalline-amorphous transition in the course of processing bioactive macromolecules; the exploitation of the crystalline dispersion of the active compounds offers a promising means of avoiding unfavorable alterations (Puhl et al., 2014).

8.6.3.2 Enzymes

Enzymes have a shortened shelf life when stored at ambient conditions and require lyophilization. Electrospinning, which is able to dehydrate samples in a timescale of milliseconds have emerged as a promising ideal alternative preservation method for biological samples (Dai et al., 2012a).

Dai et al. (2012a) have spun poly(vinylpyrrolidone) (PVP) with incorporation of the enzyme horse radish peroxidase (HRP). Fibers were tested for their change in activity post electrospinning and during storage. Colorimetric assay was used to characterize the activity of HRP incorporated in the nanofiber mats in a microtiter plate and monitoring the change in absorption over time. Post spinning, the HRP activity was decreased by approximately 20%. During a storage study over 280 days, 40% of the activity retained.

Wang et al. (2007) immobilized catalase and peroxidase on electrospun poly(acrylonitrile-*co*-acrylic acid) (PANCA) nanofibers. Since the enzyme catalase requires a supply of electron for its enzymatic activity, the fibrous mesh was filled with multiwalled carbon nanotubes (MWCNTs). A significant improvement in enzymatic activity was found due to the utilization of MWCNTs when compared with enzymes immobilized on the nanofiber mesh without MWCNT.

Charernsriwilaiwat et al. (2012) manufactured lysozyme-loaded electrospun nanofibers by applying a mixture of chitosan–ethylenediamine-tetraacetic acid (CS-EDTA). The fibers were intended for wound-healing purposes and were subjected to in vivo tests, in which a significant acceleration of wound healing was found compared to that of control. Further enzymes, such as α-chymotrypsin, lipase and lysozyme, have also been successfully immobilized on polymeric nanofibers (Sebe et al., 2015).

8.7 CONCLUSION

The nanotechnology empowered delivery systems will play a pivotal role in full filling therapeutic goals such as increased drug absorption, distribution and minimum excretion. The current clinical modalities are badly hammered by the high affinity of systemic side effects and recurrence of the disease. The current trends in drug delivery research and development clearly show that nanomaterials, especially nanofibers are increasingly being explored for possible applications in targeted drug delivery. Electrospinning seems to be a promising candidate for drug delivery applications. Solution and processing parameters such as viscosity, molecular weight, concentration of the polymer, applied voltage, tip to collector distance, conductivity, etc. significantly affect the fiber morphology and by manipulation of these parameters one can get desired properties for specific application.

Despite the increasing interest in electrospinning over the past decade, the widespread success of electrospun nanofiber meshes for tissue engineering and drug delivery systems remains somewhat limited. This is due to difficulties in creating identical scaffolds between research groups, primarily due to the extensive number of parameters that contribute to fiber properties and scaling up to large scale. An optimum range of all parameters is necessary to obtain the nanofibers with predetermined characteristics.

About 40% of newly discovered drugs and approximately 50% of the drugs available in the market belong to group of poorly soluble drugs. The solubility and bioavailability of the drugs can be enhanced through electrospinning by making a nanofiber delivery system. Future research in the area of plant-based nanomaterials will be beneficial in relation to cost, availability, and other commercial issues. Biomimetic nanofibers as drug delivery devices that are responsive to different stimuli, such as temperature, pH, light, and electric/magnetic field for controlled release of therapeutic substances are the new thrust area of research.

KEYWORDS

- nanofibers
- electrospinning
- transdermal drug delivery
- targeted drug delivery for cancer
- bioactive dressing, rotary jet spinning.

REFERENCES

Abdelkader, H.; Alany, R. G. Controlled and Continuous Release Ocular Drug Delivery Systems: Pros and Cons. *Current Drug Delivery*. **2012**, *9*, 421–430.

Aboutalebi Anaraki, N.; Roshanfekr Rad, L.; Irani, M.; Haririan, I. Fabrication of PLA/ PEG/MWCNT Electrospun Nanofibrous Scaffolds for Anticancer Drug Delivery. *J. Appl. Polym.* Sci. 2015, 132(3), 41286 (1–9).

Achille, C.; Sundaresh, S.; Chu, B.; Hadjiargyrou, M. Cdk2 Silencing via a DNA/PCL Electrospun Scaffold Suppresses Proliferation and Increases Death of Breast Cancer Cells. *PLoS One* **2012**, *7*(12), e52356 (1–9).

Agarwal, S.; Wendorff, J. H.; Greiner, A. Use of Electrospinning Technique for Biomedical Applications. *Polymer (Guildf)*. **2008**, *49*(26), 5603–5621.

Angammana, C. J.; Jayaram, S. H. Analysis of the Effects of Solution Conductivity on Electrospinning Process and Fiber Morphology. *IEEE Trans. Ind. Appl.* **2011**, *47*(3), 1109–1117.

Ardeshirzadeh, B.; Anaraki, N. A.; Irani, M.; Rad, L. R.; Shamshiri, S. Controlled Release of Doxorubicin from Electrospun PEO/chitosan/graphene Oxide Nanocomposite Nanofibrous Scaffolds. *Mater. Sci. Eng. C. Mater. Biol. Appl.* **2015**, *48*, 384–390.

Aulton, M. E.; Taylor, K. *Aulton's Pharmaceutics: The Design and Manufacture of Medicines*; Elsevier Health Sciences, London, **2007**, 3rd Edition, p. 5.

Badrossamay, M. R.; McIlwee, H. A.; Goss, J. A.; Parker, K. K. Nanofiber Assembly by Rotary Jet-Spinning. *Nano Lett.* **2010**, *10*(6), 2257–2261.

Badylak, S. F.; Freytes, D. O.; Gilbert, T. W. Extracellular Matrix as a Biological Scaffold Material: Structure and Function. *Acta Biomater.* **2009**, *5*(1), 1–13.

Balaji, A.; Vellayappan, M. V.; John, A. A.; Subramanian, A. P.; Jaganathan, S. K.; Supriyanto, Eko; Razak, S. I. A. An Insight on Electrospun-Nanofibers-Inspired Modern Drug Delivery System in the Treatment of Deadly Cancers. *RSC Adv.* **2015**, *5*, 57984–58004.

Banyal, S.; Malik, P.; Tuli, H. S.; Mukherjee, T. K. Advances in Nanotechnology for Diagnosis and Treatment of Tuberculosis. *Curr. Opin. Pulm. Med.* **2013**, *19*(3), 289–297.

Basile, L.; Pignatello, R.; Passirani, C. Active Targeting Strategies for Anticancer Drug Nanocarriers. *Current Drug Delivery.* **2012**, *9*(3), 255–268.

Beachley, V.; Wen, X. Polymer Nanofibrous Structures: Fabrication, Biofunctionalization, and Cell Interactions. *Prog. Polym. Sci.* **2010**, *35*(7), 868–892.

Bharadwaj, R.; Yu, H. The Spindle Checkpoint, Aneuploidy, and Cancer. *Oncogene* **2004**, *23*(11), 2016–2027.

Bhardwaj, N.; Kundu, S. C. Electrospinning: A Fascinating Fiber Fabrication Technique. *Biotechnol. Adv.* **2010**, *28*(3), 325–347.

Bhattarai, N.; Edmondson, D.; Veiseh, O.; Matsen, F. A.; Zhang, M. Electrospun Chitosan-Based Nanofibers and Their Cellular Compatibility. *Biomaterials* **2005**, *26*(31), 6176–6184.

Cai, Z. X.; Mo, X. M.; Zhang, K. H.; Fan, L. P.; Yin, A. L.; He, C. L.; Wang, H. S. Fabrication of Chitosan/silk Fibroin Composite Nanofibers for Wound-Dressing Applications. *Int. J. Mol. Sci.* **2010**, *11*(9), 3529–3539.

Calamak, S.; Erdogdu, C.; Ozalp, M.; Ulubayram, K. Silk Fibroin Based Antibacterial Bionanotextiles as Wound Dressing Materials. *Mater. Sci. Eng. C. Mater. Biol. Appl.* **2014**, *43*, 11–20.

Caracciolo, P. C.; Tornello, P. C. R.; Ballarin, F. M.; Abraham, G. A. Development of Electrospun Nanofibers for Biomedical Applications: State of the Art in Latin America. *J. Biomater. Tissue Eng.* **2013**, *3*(1), 39–60.

Carbone, E. J.; Jiang, T.; Nelson, C.; Henry, N.; Lo, K. W. H. Small Molecule Delivery through Nanofibrous Scaffolds for Musculoskeletal Regenerative Engineering. *Nanomedicine* **2014**, *10*(8), 1691–1699.

Carrier, R. L.; Waterman, K. C. *Handbook of Biodegradable Polymeric Materials and Their Applications*, Volume 2. Mallapragada, S. K.; Narasimhan, B.; Eds.; American scientific publishers: Los Angeles, USA, **2005**.

Casasola, R.; Thomas, N. L.; Trybala, A.; Georgiadou, S. Electrospun Poly Lactic Acid (PLA) Fibers: Effect of Different Solvent Systems on Fiber Morphology and Diameter. *Polymer (Guildf).* **2014**, *55*(18), 4728–4737.

Casper, C. L.; Stephens, J. S.; Tassi, N. G.; Chase, D. B.; Rabolt, J. F. Controlling Surface Morphology of Electrospun Polystyrene Fibers: Effect of Humidity and Molecular Weight in the Electrospinning Process. *Macromolecules* **2004**, *37*(2), 573–578.

Charernsriwilaiwat, N.; Opanasopit, P.; Rojanarata, T.; Ngawhirunpat, T. Lysozyme-Loaded, Electrospun Chitosan-Based Nanofiber Mats for Wound Healing. *Int. J. Pharm.* **2012**, *427*(2), 379–384.

Chaudhary, S.; Garg, T.; Murthy, R. S. R.; Rath, G.; Goyal, A. K. Recent Approaches of Lipid-Based Delivery System for Lymphatic Targeting via Oral Route. *J. Drug Target.* **2014**, *22*(10), 871–882.

Chen, M.; Li, Y. F.; Besenbacher, F. Electrospun Nanofibers-Mediated on-Demand Drug Release. *Adv. Healthc. Mater.* **2014**, *3*(11), 1721–1732.

Choi, J. S.; Leong, K. W.; Yoo, H. S. In Vivo Wound Healing of Diabetic Ulcers Using Electrospun Nanofibers Immobilized with Human Epidermal Growth Factor (EGF). *Biomaterials* **2008**, *29*(5), 587–596.

Chong, E. J.; Phan, T. T.; Lim, I. J.; Zhang, Y. Z.; Bay, B. H.; Ramakrishna, S.; Lim, C. T. Evaluation of Electrospun PCL/gelatin Nanofibrous Scaffold for Wound Healing and Layered Dermal Reconstitution. *Acta Biomater.* **2007**, *3*(3), 321–330.

Chuangcote, S.; Supaphol, P. Fabrication of Aligned Poly (Vinyl Alcohol) Nanofibers by Electrospinning. *J. Nanosci. Nanotechnol.* **2006**, *6*(1), 125–129.

Cramariuc, B.; Cramariuc, R.; Scarlet, R.; Manea, L. R.; Lupu, I. G.; Cramariuc, O. Fiber Diameter in Electrospinning Process. *J. Electrostat.* **2013**, *71*(3), 189–198.

Cui Yong; Liu Mingliang; Wu Bingqun; Duan Xinchun G. M. Antitumor activity of pacli-taxel or/and cisplatin drug delivery system against lung cancer cells A549 in vitro. *J. Cap. Med. Univ.* **2014**, *35*(6), 694–697.

Dai, M.; Jin, S.; Nugen, S. R. Water-Soluble Electrospun Nanofibers as a Method for On-Chip Reagent Storage. *Biosensors* **2012a**, *2*(4), 388–395.

Dai, X. Y.; Nie, W.; Wang, Y. C.; Shen, Y.; Li, Y.; Gan, S. J. Electrospun Emodin Polyvinylpyrrolidone Blended Nanofibrous Membrane: A Novel Medicated Biomaterial for Drug Delivery and Accelerated Wound Healing. *J. Mater. Sci. Mater. Med.* **2012b**, *23*(11), 2709–2716.

Dasari, S.; Bernard Tchounwou, P. Cisplatin in Cancer Therapy: Molecular Mechanisms of Action. *Eur. J. Pharmacol.* **2014**, *740*, 364–378.

De Vrieze, S.; Van Camp, T.; Nelvig, A.; Hagström, B.; Westbroek, P.; De Clerck, K. The Effect of Temperature and Humidity on Electrospinning. *J. Mater. Sci.* **2008**, *44*(5), 1357–1362.

Deitzel, J. M.; Kleinmeyer, J.; Harris, D.; Beck Tan, N. C. The Effect of Processing Variables on the Morphology of Electrospun Nanofibers and Textiles. *Polymer (Guildf)*. **2001**, *42*(1), 261–272.

Demirci, S.; Celebioglu, A.; Aytac, Z.; Uyar, T. pH-Responsive Nanofibers with Controlled Drug Release Properties. *Polym. Chem.* **2014**, *5*(6), 2050–2056.

Dhinakaran, M. K.; Soundarajan, K.; Mohan Das, T. Self-Assembly of Novel Benzimidazole N-Glycosylamines into Nanofibers and Nanospheres. *New J. Chem.* **2014**, *38*(7), 2874.

Doshi, J.; Reneker, D. H. Electrospinning Process and Applications of Electrospun Fibers. *J. Electrostat.* **1995**, *35*(2–3), 151–160.

Duan, Y.; Jia, J.; Wang, S.; Yan, W.; Jin, L.; Wang, Z. Preparation of Antimicrobial Poly(ε-Caprolactone) Electrospun Nanofibers Containing Silver-Loaded Zirconium Phosphate Nanoparticles. *J. Appl. Polym. Sci.* **2007**, *106*(2), 1208–1214.

Ellison, C. J.; Phatak, A.; Giles, D. W.; Macosko, C. W.; Bates, F. S. Melt Blown Nanofibers: Fiber Diameter Distributions and Onset of Fiber Breakup. *Polymer (Guildf)*. **2007**, *48*(11), 3306–3316.

Endres, T.; Zheng, M.; Beck-Broichsitter, M.; Samsonova, O.; Debus, H.; Kissel, T. Optimizing the Self-Assembly of siRNA Loaded PEG-PCL-lPEI Nano-Carriers Employing Different Preparation Techniques. *J. Control. Release* **2012**, *160*(3), 583–591.

Farhana, S. A.; Shantakumar, S. M.; Shyale, S.; Shalam, M.; Narasu, L. Sustained Release of Verapamil Hydrochloride from Sodium Alginate Microcapsules. *Curr. Drug Deliv.* **2010**, *7*(2), 98–108.

Florence, A. T.; Attwood, D. *Physicochemical Principles of Pharmacy*; Pharmaceutical Press, London, **2006**, 4th Edition, 341–355.

Fortunato, G.; Guex, A. G.; Popa, A. M.; Rossi, R. M.; Hufenus, R. Molecular Weight Driven Structure Formation of PEG Based E-Spun Polymer Blend Fibres. *Polymer (Guildf)*. **2014**, *55*(14), 3139–3148.

Ganesh, N.; Jayakumar, R.; Koyakutty, M.; Mony, U.; Nair, S. V. Embedded Silica Nanoparticles in Poly(caprolactone) Nanofibrous Scaffolds Enhanced Osteogenic Potential for Bone Tissue Engineering. *Tissue Eng. Part A* **2012**, *18*(17–18), 1867–1881.

Garg, T.; Rath, G.; Goyal, A. K. Biomaterials-Based Nanofiber Scaffold: Targeted and Controlled Carrier for Cell and Drug Delivery. *J. Drug Target.* **2014**, *23*(3), 202–221.

Garg, T.; Singh, O.; Arora, S.; Murthy, R. S. R. Scaffold: A Novel Carrier for Cell and Drug Delivery. *Critical Reviews in Therapeutic Drug Carrier Systems.* **2012**, 29(1), 1–63.

Gates, B. D.; Xu, Q.; Stewart, M.; Ryan, D.; Willson, C. G.; Whitesides, G. M. New Approaches to Nanofabrication: Molding, Printing, and Other Techniques. *Chem. Rev.* **2005**, *105*(4), 1171–1196.

Geng, X.; Kwon, O. H.; Jang, J. Electrospinning of Chitosan Dissolved in Concentrated Acetic Acid Solution. *Biomaterials* **2005**, *26*(27), 5427–5432.

Girlich, C.; Scholmerich, J. Topical Delivery of Steroids in Inflammatory Bowel Disease. *Current Drug Delivery.* **2012**, *9*(4), 345–349.

Grimm, S.; Giesa, R.; Sklarek, K.; Langner, A.; Gösele, U.; Schmidt, H.-W.; Steinhart, M. Nondestructive Replication of Self-Ordered Nanoporous Alumina Membranes via Cross-Linked Polyacrylate Nanofiber Arrays. *Nano Lett.* **2008**, *8*(7), 1954–1959.

Gupta, P.; Elkins, C.; Long, T. E.; Wilkes, G. L. Electrospinning of Linear Homopolymers of Poly(methyl Methacrylate): Exploring Relationships between Fiber Formation, Viscosity, Molecular Weight and Concentration in a Good Solvent. *Polymer (Guildf)*. **2005**, *46*(13), 4799–4810.

Haghi, A. K.; Akbari, M. Trends in Electrospinning of Natural Nanofibers. *Phys. Status Solidi* **2007**, *204*(6), 1830–1834.

Hartgerink, J. D.; Beniash, E.; Stupp, S. I. Self-Assembly and Mineralization of Peptide-Amphiphile Nanofibers. *Science* **2001**, *294*(5547), 1684–1688.

Hiremath, N.; Bhat, G. Melt Blown Polymeric Nanofibers for Medical Applications- An Overview. *Nanosci. Technoloogy* **2015**, *2*(1), 1–9.

Hohman, M. M.; Shin, M.; Rutledge, G.; Brenner, M. P. Electrospinning and Electrically Forced Jets. I. Stability Theory. *Phys. Fluids* **2001a**, *13*(8), 2201.

Hohman, M. M.; Shin, M.; Rutledge, G.; Brenner, M. P. Electrospinning and Electrically Forced Jets. II. Applications. *Phys. Fluids* **2001b**, *13*(8), 2221.

Hu, X.; Liu, S.; Zhou, G.; Huang, Y.; Xie, Z.; Jing, X. Electrospinning of Polymeric Nanofibers for Drug Delivery Applications. *J. Control. Release* **2014**, *185*(1), 12–21.

Huang, Z. M.; Zhang, Y. Z.; Kotaki, M.; Ramakrishna, S. A Review on Polymer Nanofibers by Electrospinning and Their Applications in Nanocomposites. *Compos. Sci. Technol.* **2003**, *63*(15), 2223–2253.

Ignatova, M.; Manolova, N.; Rashkov, I. Electrospinning of Poly(vinyl Pyrrolidone)–iodine Complex and Poly(ethylene Oxide)/poly(vinyl Pyrrolidone)–iodine Complex – a Prospective Route to Antimicrobial Wound Dressing Materials. *Eur. Polym. J.* **2007**, *43*(5), 1609–1623.

Ignatova, M.; Yossifova, L.; Gardeva, E.; Manolova, N.; Toshkova, R.; Rashkov, I.; Alexandrov, M. Antiproliferative Activity of Nanofibers Containing Quaternized Chitosan And/or Doxorubicin against MCF-7 Human Breast Carcinoma Cell Line by Apoptosis. *J. Bioact. Compat. Polym.* **2011**, *26*(6), 539–551.

Jalili, R.; Morshed, M.; Ravandi, S. A. H. Fundamental Parameters Affecting Electrospinning of PAN Nanofibers as Uniaxially Aligned Fibers. *J. Appl. Polym. Sci.* **2006**, *101*(6), 4350–4357.

Jang, J. H.; Castano, O.; Kim, H. W. Electrospun Materials as Potential Platforms for Bone Tissue Engineering. *Adv. Drug Deliv. Rev.* **2009**, *61*(12), 1065–1083.

Jankovic, B.; Pelipenko, J.; Škarabot, M.; Musevic, I.; Kristl, J. The Design Trend in Tissue-Engineering Scaffolds Based on Nanomechanical Properties of Individual Electrospun Nanofibers. *Int. J. Pharm.* **2013**, *455*(1–2), 338–347.

Jarusuwannapoom, T.; Hongrojjanawiwat, W.; Jitjaicham, S.; Wannatong, L.; Nithitanakul, M.; Pattamaprom, C.; Koombhongse, P.; Rangkupan, R.; Supaphol, P. Effect of Solvents on Electro-Spinnability of Polystyrene Solutions and Morphological Appearance of Resulting Electrospun Polystyrene Fibers. *Eur. Polym. J.* **2005**, *41*(3), 409–421.

Jayakumar, R.; Prabaharan, M.; Nair, S. V; Tamura, H. Novel Chitin and Chitosan Nanofibers in Biomedical Applications. *Biotechnol. Adv.* **2010**, *28*(1), 142–150.

Jayakumar, R.; Prabaharan, M.; Sudheesh Kumar, P. T.; Nair, S. V; Tamura, H. Biomaterials Based on Chitin and Chitosan in Wound Dressing Applications. *Biotechnol. Adv.* **2011**, *29*(3), 322–337.

Jayaraman, K.; Kotaki, M.; Zhang, Y.; Mo, X.; Ramakrishna, S. Recent Advances in Polymer Nanofibers. *Journal of Nanoscience and Nanotechnology*. **2004**, *4*(1), 52–65.

Jiang, H.; Wang, L.; Zhu, K. Coaxial Electrospinning for Encapsulation and Controlled Release of Fragile Water-Soluble Bioactive Agents. *J. Control. Release* **2014a**, *193*, 296–303.

Jiang, T.; Deng, M.; James, R.; Nair, L. S.; Laurencin, C. T. Micro- and Nanofabrication of Chitosan Structures for Regenerative Engineering. *Acta Biomater.* **2014b**, *10*(4), 1632–1645.

Joshi, M.; Butola, B. S.; Saha, K. Advances in Topical Drug Delivery System: Micro to Nanofibrous Structures. *J. Nanosci. Nanotechnol.* **2014**, *14*(1), 853–867.

Kanani, A. G.; Bahrami, S. H.; Taftei, H. A.; Rabbani, S.; Sotoudeh, M. Effect of Chitosan-Polyvinyl Alcohol Blend Nanofibrous Web on the Healing of Excision and Incision Full Thickness Wounds. *IET nanobiotechnology* **2010**, *4*(4), 109–117.

Kanawung, K.; Panitchanapan, K.; Puangmalee, S.; Utok, W.; Kreua-ongarjnukool, N.; Rangkupan, R.; Meechaisue, C.; Supaphol, P. Preparation and Characterization of Polycaprolactone/Diclofenac Sodium and Poly(vinyl alcohol)/Tetracycline Hydrochloride Fiber Mats and Their Release of the Model Drugs. *Polym. J.* **2007**, *39*(4), 369–378.

Kang, Y. O.; Yoon, I. S.; Lee, S. Y.; Kim, D. D.; Lee, S. J.; Park, W. H.; Hudson, S. M. Chitosan-Coated Poly(vinyl Alcohol) Nanofibers for Wound Dressings. *J. Biomed. Mater. Res. B. Appl. Biomater.* **2010**, *92*(2), 568–576.

Kenawy, E. R.; Abdel-Hay, F. I.; El-Newehy, M. H.; Wnek, G. E. Processing of Polymer Nanofibers through Electrospinning as Drug Delivery Systems. *Mater. Chem. Phys.* **2009**, *113*(1), 296–302.

Khadka, D. B.; Haynie, D. T. Protein- and Peptide-Based Electrospun Nanofibers in Medical Biomaterials. *Nanomedicine* **2012**, *8*(8), 1242–1262.

Khamforoush, M.; Mahjob, M. Modification of the Rotating Jet Method to Generate Highly Aligned Electrospun Nanofibers. *Mater. Lett.* **2011**, *65*(3), 453–455.

Kriegel, C.; Kit, K. M.; McClements, D. J.; Weiss, J. Electrospinning of Chitosan–poly(ethylene Oxide) Blend Nanofibers in the Presence of Micellar Surfactant Solutions. *Polymer (Guildf).* **2009**, *50*(1), 189–200.

Laiva, A. L.; Venugopal, J. R.; Karuppuswamy, P.; Navaneethan, B.; Gora, A.; Ramakrishna, S. Controlled Release of Titanocene into the Hybrid Nanofibrous Scaffolds to Prevent the Proliferation of Breast Cancer Cells. *Int. J. Pharm.* **2015**, *483*(1–2), 115–123.

Li, J.; He, A.; Zheng, J.; Han, C. C. Gelatin and Gelatin-Hyaluronic Acid Nanofibrous Membranes Produced by Electrospinning of Their Aqueous Solutions. *Biomacromolecules* **2006**, *7*(7), 2243–2247.

Li, L.; Qian, Y.; Jiang, C.; Lv, Y.; Liu, W.; Zhong, L.; Cai, K.; Li, S.; Yang, L. The Use of Hyaluronan to Regulate Protein Adsorption and Cell Infiltration in Nanofibrous Scaffolds. *Biomaterials* **2012**, *33*(12), 3428–3445.

Liang, D.; Hsiao, B. S.; Chu, B. Functional Electrospun Nanofibrous Scaffolds for Biomedical Applications. *Adv. Drug Deliv. Rev.* **2007**, *59*(14), 1392–1412.

Lin, T. C.; Lin, F. H.; Lin, J. C. In Vitro Characterization of Magnetic Electrospun IDA-Grafted Chitosan Nanofiber Composite for Hyperthermic Tumor Cell Treatment. *J. Biomater. Sci. Polym. Ed.* **2013**, *24*(9), 1152–1163.

Liu, D.; Liu, S.; Jing, X.; Li, X.; Li, W.; Huang, Y. Necrosis of Cervical Carcinoma by Dichloroacetate Released from Electrospun Polylactide Mats. *Biomaterials* **2012**, *33*(17), 4362–4369.

Liu, D.; Wang, F.; Yue, J.; Jing, X.; Huang, Y. Metabolism Targeting Therapy of Dichloroacetate-Loaded Electrospun Mats on Colorectal Cancer. *Drug Deliv.* **2015**, *22*(1), 136–143.

Liu, H.; Ding, X.; Zhou, G.; Li, P.; Wei, X.; Fan, Y. Electrospinning of Nanofibers for Tissue Engineering Applications. *J. Nanomater.* **2013a**, 2013, 1–11.

Liu, J.; Liu, J.; Xu, H.; Zhang, Y.; Chu, L.; Liu, Q.; Song, N.; Yang, C. Novel Tumor-Targeting, Self-Assembling Peptide Nanofiber as a Carrier for Effective Curcumin Delivery. *Int. J. Nanomedicine* **2014**, *9*(1), 197–207.

Liu, S. J.; Kau, Y. C.; Chou, C. Y.; Chen, J. K.; Wu, R. C.; Yeh, W. L. Electrospun PLGA/collagen Nanofibrous Membrane as Early Stage Wound Dressing. *J. Memb. Sci.* **2010**, *355*(1–2), 53–59.

Liu, S.; Zhou, G.; Liu, D.; Xie, Z.; Huang, Y.; Wang, X.; Wu, W.; Jing, X. Inhibition of Orthotopic Secondary Hepatic Carcinoma in Mice by Doxorubicin-Loaded Electrospun Polylactide Nanofibers. *J. Mater. Chem. B* **2013b**, *1*(1), 101–109.

Luu, Y. K.; Kim, K.; Hsiao, B. S.; Chu, B.; Hadjiargyrou, M. Development of a Nanostructured DNA Delivery Scaffold via Electrospinning of PLGA and PLA–PEG Block Copolymers. *J. Control. Release* **2003**, *89*(2), 341–353.

Ma, Z.; Kotaki, M.; Inai, R.; Ramakrishna, S. Potential of Nanofiber Matrix as Tissue-Engineering Scaffolds. *Tissue Eng.* **2005**, *11*(1–2), 101–109.

Maira, F.; Catania, A.; Candido, S.; Russo, A. E.; McCubrey, J. A.; Libra, M.; Malaponte, G.; Fenga, C. Molecular Targeted Therapy in Melanoma: A Way to Reverse Resistance to Conventional Drugs. *Current Drug Delivery.* **2012**, *9*(1), 17–29.

Maleki, M.; Latifi, M.; Amani-Tehran, M.; Mathur, S. Electrospun Core-Shell Nanofibers for Drug Encapsulation and Sustained Release. *Polym. Eng. Sci.* **2013**, *53*(8), 1770–1779.

Maleknia, L.; Majdi, Z. Electrospinning of Gelatin Nanofiber for Biomedical Application. *Orient. J. Chem.* **2014**, *30*(4), 2043–2048.

Malik, R.; Garg, T.; Goyal, A. K.; Rath, G. Polymeric Nanofibers: Targeted Gastro-Retentive Drug Delivery Systems. *J. Drug Target.* **2015**, *23*(2), 109–124.

Martins, I.; Kepp, O.; Schlemmer, F.; Adjemian, S.; Tailler, M.; Shen, S.; Michaud, M.; Menger, L.; Gdoura, A.; Tajeddine, N.; et al., Restoration of the Immunogenicity of Cisplatin-Induced Cancer Cell Death by Endoplasmic Reticulum Stress. *Oncogene* **2011**, *30*(10), 1147–1158.

McKee, M. G.; Layman, J. M.; Cashion, M. P.; Long, T. E. Phospholipid Nonwoven Electrospun Membranes. *Science* **2006**, *311*(5759), 353–355.

McKee, M. G.; Wilkes, G. L.; Colby, R. H.; Long, T. E. Correlations of Solution Rheology with Electrospun Fiber Formation of Linear and Branched Polyesters. *Macromolecules* **2004**, *37*(5), 1760–1767.

McManus, M. C.; Boland, E. D.; Simpson, D. G.; Barnes, C. P.; Bowlin, G. L. Electrospun Fibrinogen: Feasibility as a Tissue Engineering Scaffold in a Rat Cell Culture Model. *J. Biomed. Mater. Res. A* **2007**, *81*(2), 299–309.

Medeiros, E. S.; Mattoso, L. H. C.; Offeman, R. D.; Wood, D. F.; Orts, W. J. Effect of Relative Humidity on the Morphology of Electrospun Polymer Fibers. *Can. J. Chem.* **2008**, *86*(6), 590–599.

Megelski, S.; Stephens, J. S.; Chase, D. B.; Rabolt, J. F. Micro- and Nanostructured Surface Morphology on Electrospun Polymer Fibers. *Macromolecules* **2002**, *35*(22), 8456–8466.

Meng, Z. X.; Xu, X. X.; Zheng, W.; Zhou, H. M.; Li, L.; Zheng, Y. F.; Lou, X. Preparation and Characterization of Electrospun PLGA/gelatin Nanofibers as a Potential Drug Delivery System. *Colloids Surf. B. Biointerfaces* **2011**, *84*(1), 97–102.

Mit-uppatham, C.; Nithitanakul, M.; Supaphol, P. Ultrafine Electrospun Polyamide-6 Fibers: Effect of Solution Conditions on Morphology and Average Fiber Diameter. *Macromol. Chem. Phys.* **2004**, *205*(17), 2327–2338.

Murugan, R.; Ramakrishna, S. Design Strategies of Tissue Engineering Scaffolds with Controlled Fiber Orientation. *Tissue Eng.* **2007**, *13*(8), 1845–1866.

Nalluri, S. K. M.; Shivarova, N.; Kanibolotsky, A. L.; Zelzer, M.; Gupta, S.; Frederix, P. W. J. M.; Skabara, P. J.; Gleskova, H.; Ulijn, R. V. Conducting Nanofibers and Organogels Derived from the Self-Assembly of Tetrathiafulvalene-Appended Dipeptides. *Langmuir* **2014**, *30*(41), 12429–12437.

Nayak, R.; Padhye, R.; Kyratzis, I. L.; Truong, Y. B.; Arnold, L. Recent Advances in Nanofiber Fabrication Techniques. *Text. Res. J.* **2011**, *82*(2), 129–147.

Ning Zhu; Xiongbiao Chen. Biofabrication of Tissue Scaffolds, Advances in Biomaterials Science and Biomedical Applications, Prof. Rosario Pignatello (Ed.), InTech, **2013**. doi: 10.5772/54125. Available from: http://www.intechopen.com/books/advances-in-biomaterials-science-and-biomedical-applications/biofabrication-of tissue scaffolds.

Ohkawa, K.; Cha, D.; Kim, H.; Nishida, A.; Yamamoto, H. Electrospinning of Chitosan. *Macromol. Rapid Commun.* **2004**, *25*(18), 1600–1605.

Paaver, U.; Heinämäki, J.; Laidmäe, I.; Lust, A.; Kozlova, J.; Sillaste, E.; Kirsimäe, K.; Veski, P.; Kogermann, K. Electrospun Nanofibers as a Potential Controlled-Release Solid Dispersion System for Poorly Water-Soluble Drugs. *Int. J. Pharm.* **2015**, *479*(1), 252–260.

Parhi, R.; Suresh, P.; Mondal, S.; Kumar, P. M. Novel Penetration Enhancers for Skin Applications: A Review. *Current Drug Delivery.* **2012**, *9*(2), 219–230.

Pelipenko, J.; Kocbek, P.; Kristl, J. Critical Attributes of Nanofibers: Preparation, Drug Loading, and Tissue Regeneration. *Int. J. Pharm.* **2015**, *484*(1–2), 57–74.

Pelipenko, J.; Kristl, J.; Rosic, R.; Baumgartner, S.; Kocbek, P. Interfacial Rheology: An Overview of Measuring Techniques and Its Role in Dispersions and Electrospinning. *Acta Pharm.* **2012**, *62*(2), 123–140.

Pham, Q. P.; Sharma, U.; Mikos, A. G. Electrospinning of Polymeric Nanofibers for Tissue Engineering Applications: A Review. *Tissue Eng.* **2006**, *12*(5), 1197–1211.

Powell, H. M.; Supp, D. M.; Boyce, S. T. Influence of Electrospun Collagen on Wound Contraction of Engineered Skin Substitutes. *Biomaterials* **2008**, *29*(7), 834–843.

Priyadarshini K1 and Keerthi Aparajitha. Paclitaxel Against Cancer: A Short Review. *Med. Chem. (Los. Angeles).* **2012**, *2*(7), 142–146.

Puhl, S.; Li, L.; Meinel, L.; Germershaus, O. Controlled Protein Delivery from Electrospun Non-Wovens: Novel Combination of Protein Crystals and a Biodegradable Release Matrix. *Mol. Pharm.* **2014**, *11*(7), 2372–2380.

Ramakrishna, S.; Fujihara, K.; Teo, W.-E.; Lim, T. C.; Ma, Z. *An Introduction to Electrospinning Process*; World Scientific Publication Co. Pvt. Ltd, Singapore, **2005**.

Ramakrishna, S.; Fujihara, K.; Teo, W.-E.; Yong, T.; Ma, Z.; Ramaseshan, R. Electrospun Nanofibers: Solving Global Issues. *Mater. Today* **2006**, *9*(3), 40–50.

Ranganath, S. H.; Wang, C. H. Biodegradable Microfiber Implants Delivering Paclitaxel for Post-Surgical Chemotherapy against Malignant Glioma. *Biomaterials* **2008**, *29*(20), 2996–3003.

Ravichandran, R.; Seitz, V.; Reddy Venugopal, J.; Sridhar, R.; Sundarrajan, S.; Mukherjee, S.; Wintermantel, E.; Ramakrishna, S. Mimicking Native Extracellular Matrix with Phytic Acid-Cross-linked Protein Nanofibers for Cardiac Tissue Engineering. *Macromol. Biosci.* **2013**, *13*(3), 366–375.

Ravikumar M. N. V. *Handbook of Particulate Drug Delivery*, Vol. 1; American Scientific Publishers, Los Angeles, USA, **2008**.

Reneker, D. H.; Yarin, A. L.; Fong, H.; Koombhongse, S. Bending Instability of Electrically Charged Liquid Jets of Polymer Solutions in Electrospinning. *J. Appl. Phys.* **2000**, *87*(9I), 4531–4547.

Rosic, R.; Pelipenko, J.; Kocbek, P.; Baumgartner, S.; Bešter-Rogač, M.; Kristl, J. The Role of Rheology of Polymer Solutions in Predicting Nanofiber Formation by Electrospinning. *Eur. Polym. J.* **2012b**, *48*(8), 1374–1384.

Rosic, R.; Pelipenko, J.; Kristl, J.; Kocbek, P.; Baumgartner, S. Properties, Engineering and Applications of Polymeric Nanofibers: Current Research and Future Advances. *Chem. Biochem. Eng. Q.* **2012a**, *26*(4), 417–425.

Rosic, R.; Pelipenko, J.; Kristl, J.; Kocbek, P.; Bešter-Rogač, M.; Baumgartner, S. Physical Characteristics of Poly(Vinyl Alcohol) Solutions in Relation to Electrospun Nanofiber Formation. *Eur. Polym. J.* **2013**, *49*(2), 290–298.

Salehi, R.; Irani, M.; Rashidi, M.-R.; Aroujalian, A.; Raisi, A.; Eskandani, M.; Haririan, I.; Davaran, S. Stimuli-Responsive Nanofibers Prepared from Poly(N-Isopropylacrylamide-Acrylamide-Vinylpyrrolidone) by Electrospinning as an Anticancer Drug Delivery. *Des. Monomers Polym.* **2013**, *16*(6), 515–527.

Sebe, I.; Szabo, P.; Kállai-Szabo, B.; Zelko, R. Incorporating Small Molecules or Biologics into Nanofibers for Optimized Drug Release: A Review. *Int. J. Pharm.* **2015**, *494*(1), 516–530.

Shalumon, K. T.; Sowmya, S.; Sathish, D.; Chennazhi, K. P.; Nair, S. V.; Jayakumar, R. Effect of Incorporation of Nanoscale Bioactive Glass and Hydroxyapatite in PCL/Chitosan Nanofibers for Bone and Periodontal Tissue Engineering. *J. Biomed. Nanotechnol.* **2013**, *9*(3), 430–440.

Shanmuga Sundar, S.; Sangeetha, D. Fabrication and Evaluation of Electrospun collagen/poly(N-Isopropyl Acrylamide)/chitosan Mat as Blood-Contacting Biomaterials for Drug Delivery. *J. Mater. Sci., Mater. Med.* **2012**, *23*(6), 1421–1430.

Shao, S.; Li, L.; Yang, G.; Li, J.; Luo, C.; Gong, T.; Zhou, S. Controlled Green Tea Polyphenols Release from Electrospun PCL/MWCNTs Composite Nanofibers. *Int. J. Pharm.* **2011**, *421*(2), 310–320.

Shin, Y. M.; Hohman, M. M.; Brenner, M. P.; Rutledge, G. C. Electrospinning: A Whipping Fluid Jet Generates Submicron Polymer Fibers. *Appl. Phys. Lett.* **2001**, *78*(8), 1149.

Sikareepaisan, P.; Suksamrarn, A.; Supaphol, P. Electrospun Gelatin Fiber Mats Containing a Herbal-Centella Asiatica-Extract and Release Characteristic of Asiaticoside. *Nanotechnology* **2008**, *19*(1), 15102.

Sill, T. J.; von Recum, H. A. Electrospinning: Applications in Drug Delivery and Tissue Engineering. *Biomaterials* **2008**, *29*(13), 1989–2006.

Smith, L. A.; Ma, P. X. Nano-Fibrous Scaffolds for Tissue Engineering. *Colloids Surf. B. Biointerfaces* **2004**, *39*(3), 125–131.

Sokolsky-Papkov, M.; Agashi, K.; Olaye, A.; Shakesheff, K.; Domb, A. J. Polymer Carriers for Drug Delivery in Tissue Engineering. *Adv. Drug Deliv. Rev.* **2007**, *59*(4–5), 187–206.

Son, W. K.; Youk, J. H.; Lee, T. S.; Park, W. H. The Effects of Solution Properties and Polyelectrolyte on Electrospinning of Ultrafine Poly(ethylene Oxide) Fibers. *Polymer (Guildf).* **2004**, *45*(9), 2959–2966.

Sridhar, R.; Lakshminarayanan, R.; Madhaiyan, K.; Amutha Barathi, V.; Lim, K. H. C.; Ramakrishna, S. Electrosprayed Nanoparticles and Electrospun Nanofibers Based on Natural Materials: Applications in Tissue Regeneration, Drug Delivery and Pharmaceuticals. *Chem. Soc. Rev.* **2015**, *44*(3), 790–814.

Sridhar, R.; Ravanan, S.; Venugopal, J. R.; Sundarrajan, S.; Pliszka, D.; Sivasubramanian, S.; Gunasekaran, P.; Prabhakaran, M.; Madhaiyan, K.; Sahayaraj, A.; et al., Curcumin- and Natural Extract-Loaded Nanofibers for Potential Treatment of Lung and Breast Cancer: In Vitro Efficacy Evaluation. *J. Biomater. Sci. Polym. Ed.* **2014**, *25*(10), 985–998.

Stanton, R. A.; Gernert, K. M.; Nettles, J. H.; Aneja, R. Drugs That Target Dynamic Microtubules: A New Molecular Perspective. *Med. Res. Rev.* **2011**, *31*(3), 443–481.

Su, Y.; Lu, B.; Xie, Y.; Ma, Z.; Liu, L.; Zhao, H.; Zhang, J.; Duan, H.; Zhang, H.; Li, J.; et al., Temperature Effect on Electrospinning of Nanobelts: The Case of Hafnium Oxide. *Nanotechnology* **2011**, *22*(28), 285609.

Subbiah, T.; Bhat, G. S.; Tock, R. W.; Parameswaran, S.; Ramkumar, S. S. Electrospinning of Nanofibers. *J. Appl. Polym. Sci.* **2005**, *96*(2), 557–569.

Subramanian, A.; Krishnan, U. M.; Sethuraman, S. Fabrication, Characterization and in Vitro Evaluation of Aligned PLGA-PCL Nanofibers for Neural Regeneration. *Ann. Biomed. Eng.* **2012**, *40*(10), 2098–2110.

Sukigara, S.; Gandhi, M.; Ayutsede, J.; Micklus, M.; Ko, F. Regeneration of Bombyx Mori Silk by Electrospinning—part 1: Processing Parameters and Geometric Properties. *Polymer (Guildf)*. **2003**, *44*(19), 5721–5727.

Sullivan, S. T.; Tang, C.; Kennedy, A.; Talwar, S.; Khan, S. A. Electrospinning and Heat Treatment of Whey Protein Nanofibers. *Food Hydrocoll.* **2014**, *35*, 36–50.

Sultana, S.; Khan, M. R.; Kumar, M.; Kumar, S.; Ali, M. Nanoparticles-Mediated Drug Delivery Approaches for Cancer Targeting: A Review. *J. Drug Target.* **2013**, *21*(2), 107–125.

Taepaiboon, P.; Rungsardthong, U.; Supaphol, P. Drug-Loaded Electrospun Mats of Poly(vinyl Alcohol) Fibres and Their Release Characteristics of Four Model Drugs. *Nanotechnology* **2006**, *17*(9), 2317–2329.

Theron, S. A.; Yarin, A. L.; Zussman, E.; Kroll, E. Multiple Jets in Electrospinning: Experiment and Modeling. *Polymer (Guildf)*. **2005**, *46*(9), 2889–2899.

Torchilin, V. P. Recent Advances with Liposomes as Pharmaceutical Carriers. *Nat. Rev. Drug Discov.* **2005**, *4*(2), 145–160.

Tripatanasuwan, S.; Zhong, Z.; Reneker, D. H. Effect of Evaporation and Solidification of the Charged Jet in Electrospinning of Poly(ethylene Oxide) Aqueous Solution. *Polymer (Guildf)*. **2007**, *48*(19), 5742–5746.

Tsai, P. P.; Schreuder-Gibson, H.; Gibson, P. Different Electrostatic Methods for Making Electret Filters. *J. Electrostat.* **2002**, *54*(3–4), 333–341.

Uppal, R.; Ramaswamy, G. N.; Arnold, C.; Goodband, R.; Wang, Y. Hyaluronic Acid Nanofiber Wound Dressing–Production, Characterization, and in Vivo Behavior. *J. Biomed. Mater. Res. B. Appl. Biomater.* **2011**, *97*(1), 20–29.

Utreja, P.; Jain, S.; Tiwary, A. K. Novel Drug Delivery Systems for Sustained and Targeted Delivery of Anti-Cancer Drugs: Current Status and Future Prospects. *Current Drug Delivery.* **2010**, *7*(2), 152–161.

Uttayarat, P.; Jetawattana, S.; Suwanmala, P.; Eamsiri, J.; Tangthong, T.; Pongpat, S. Antimicrobial Electrospun Silk Fibroin Mats with Silver Nanoparticles for Wound Dressing Application. *Fibers Polym.* **2012**, *13*(8), 999–1006.

Vasita, R.; Katti, D. S. Nanofibers and Their Applications in Tissue Engineering. *Int. J. Nanomedicine* **2006**, *1*(1), 15–30.

Verreck, G.; Chun, I.; Peeters, J.; Rosenblatt, J.; Brewster, M. E. Preparation and Characterization of Nanofibers Containing Amorphous Drug Dispersions Generated by Electrostatic Spinning. *Pharm. Res.* **2003**, *20*(5), 810–817.

Wang, B.; Qiao, W.; Wang, Y.; Yang, L.; Zhang, Y.; Shao, P. Cancer Therapy Based on Nanomaterials and Nanocarrier Systems. *J. Nanomater.* **2010**, 2010, 1–9.

Wannatong, L.; Sirivat, A.; Supaphol, P. Effects of Solvents on Electrospun Polymeric Fibers: Preliminary Study on Polystyrene. *Polym. Int.* **2004**, *53*(11), 1851–1859.

Welle, A.; Kröger, M.; Döring, M.; Niederer, K.; Pindel, E.; Chronakis, I. S. Electrospun Aliphatic Polycarbonates as Tailored Tissue Scaffold Materials. *Biomaterials* **2007**, *28*(13), 2211–2219.

Wnek, G. E.; Carr, M. E.; Simpson, D. G.; Bowlin, G. L. Electrospinning of Nanofiber Fibrinogen Structures. *Nano Lett.* **2003**, *3*(2), 213–216.

Xing, X.; Wang, Y.; Li, B. Nanofibers Drawing and Nanodevices Assembly in Poly(trimethylene Terephthalate). *Opt. Express* **2008**, *16*(14), 10815.

Xing, X.; Yu, H.; Zhu, D.; Zheng, J.; Chen, H.; Chen, W.; Cai, J. Sub-wavelength and Nanometer Diameter Optical Polymer Fibers as Building Blocks for Miniaturized Photonics Integration, Optical Communication; Dr. Narottam Das (Ed.), InTech, 2012. DOI: 10.5772/47822. Available from: http://www.intechopen.com/books/optical-communication/subwavelength-and-nanometer-diameter-optical-polymer-fibers-as-building-blocks-for-miniaturized-phot.

Xu, X.; Chen, X.; Wang, Z.; Jing, X. Ultrafine PEG-PLA Fibers Loaded with Both Paclitaxel and Doxorubicin Hydrochloride and Their in Vitro Cytotoxicity. *Eur. J. Pharm. Biopharm.* **2009**, *72*(1), 18–25.

Xu, X.; Chen, X.; Xu, X.; Lu, T.; Wang, X.; Yang, L.; Jing, X. BCNU-Loaded PEG-PLLA Ultrafine Fibers and Their in Vitro Antitumor Activity against Glioma C6 Cells. *J. Control. Release* **2006**, *114*(3), 307–316.

Yang, D.; Li, Y.; Nie, J. Preparation of gelatin/PVA Nanofibers and Their Potential Application in Controlled Release of Drugs. *Carbohydr. Polym.* **2007**, *69*(3), 538–543.

Yang, G.; Wang, J.; Li, L.; Ding, S.; Zhou, S. Electrospun Micelles/drug-Loaded Nanofibers for Time-Programmed Multi-Agent Release. *Macromol. Biosci.* **2014**, *14*(7), 965–976.

Yao, J.; Bastiaansen, C.; Peijs, T. High Strength and High Modulus Electrospun Nanofibers. *Fibers* **2014**, *2*(2), 158–186.

Yarin, A. L.; Koombhongse, S.; Reneker, D. H. Bending Instability in Electrospinning of Nanofibers. *J. Appl. Phys.* **2001**, *89*(5), 3018.

Yördem, O. S.; Papila, M.; Menceloğlu, Y. Z. Effects of Electrospinning Parameters on Polyacrylonitrile Nanofiber Diameter: An Investigation by Response Surface Methodology. *Mater. Des.* **2008**, *29*(1), 34–44.

Yu, D. G.; Chian, W.; Wang, X.; Li, X.-Y.; Li, Y.; Liao, Y.-Z. Linear Drug Release Membrane Prepared by a Modified Coaxial Electrospinning Process. *J. Memb. Sci.* **2013**, *428*, 150–156.

Yu, D. G.; White, K.; Yang, J. H.; Wang, X.; Qian, W.; Li, Y. PVP Nanofibers Prepared Using Co-Axial Electrospinning with Salt Solution as Sheath Fluid. *Mater. Lett.* **2012**, *67*(1), 78–80.

Yuan, X.; Zhang, Y.; Dong, C.; Sheng, J. Morphology of Ultrafine Polysulfone Fibers Prepared by Electrospinning. *Polym. Int.* **2004**, *53*(11), 1704–1710.

Zamani, M.; Morshed, M.; Varshosaz, J.; Jannesari, M. Controlled Release of Metronidazole Benzoate from Poly ε-Caprolactone Electrospun Nanofibers for Periodontal Diseases. *Eur. J. Pharm. Biopharm.* **2010**, *75*(2), 179–185.

Zeng, J.; Aigner, A.; Czubayko, F.; Kissel, T.; Wendorff, J. H.; Greiner, A. Poly(vinyl Alcohol) Nanofibers by Electrospinning as a Protein Delivery System and the Retardation of Enzyme Release by Additional Polymer Coatings. *Biomacromolecules* **2005**, *6*(3), 1484–1488.

Zhang, S. Fabrication of Novel Biomaterials through Molecular Self-Assembly. *Nat. Biotechnol.* **2003**, *21*(10), 1171–1178.

Zhang, X.; Reagan, M. R.; Kaplan, D. L. Electrospun Silk Biomaterial Scaffolds for Regenerative Medicine. *Adv. Drug Deliv. Rev.* **2009**, *61*(12), 988–1006.

Zhen Gang Wang; Bei Ke; Zhi Kang Xu. Covalent Immobilization of Redox Enzyme on Electrospun Nonwoven Poly(Acrylonitrile-Co-Acrylic Acid) Nanofiber Mesh Filled With Carbon Nanotubes: A Comprehensive Study. *Biotechnol. Bioeng.* **2007**, *97*, 708–720.

Zong X. H., Kim K, Fang D. F, Ran S. F, Hsiao B. S and Chu B, "Structure and Process Relationship of Electrospun Bioabsorbable Nanofiber Membranes," Polymer, **2002**, 43(16), 4403-4412.

Zhou, Y.; Yang, H.; Liu, X.; Mao, J.; Gu, S.; Xu, W. Electrospinning of Carboxyethyl Chitosan/poly(vinyl Alcohol)/silk Fibroin Nanoparticles for Wound Dressings. *Int. J. Biol. Macromol.* **2013**, *53*, 88–92.

Zuo, W.; Zhu, M.; Yang, W.; Yu, H.; Chen, Y.; Zhang, Y. Experimental Study on Relationship between Jet Instability and Formation of Beaded Fibers during Electrospinning. *Polym. Eng. Sci.* **2005**, *45*(5), 704–709.

Zussman, E.; Rittel, D.; Yarin, A. L. Failure Modes of Electrospun Nanofibers. *Appl. Phys. Lett.* **2003**, *82*(22), 3958.

CHAPTER 9

NANOCOMPOSITE MICROPARTICLES (nCmP) FOR PULMONARY DRUG DELIVERY APPLICATIONS

ZIMENG WANG,[1] ELISA A. TORRICO-GUZMÁN,[1]
SWETA K. GUPTA,[1] and SAMANTHA A. MEENACH[1,2]*

[1]Department of Chemical Engineering, University of Rhode Island, Kingston, RI 02881, USA

[2]Department of Biomedical and Pharmaceutical Sciences, University of Rhode Island, Kingston, RI 02881, USA, Tel.: +1 401-874-4303; *E-mail: smeenach@uri.edu

CONTENTS

ABSTRACT

Despite the many advances in drug delivery and pulmonary therapeutics, the lungs present a challenge where physiological barriers need to be overcome in order to facilitate effective delivery of active pharmaceutical ingredients. Nanocomposite microparticles (nCmP) have the potential to overcome the physiological limitations of the lungs. Comprised of nanoparticles (NP) entrapped or encapsulated into a microparticle system, nCmP are dry powder aerosols that can deposit in various areas of the lungs depending on their size. Once the nCmP reach the lungs, they decompose into their native NP, allowing for delivery of a therapeutic encapsulated in the NP. The microparticle size allows for effective pulmonary delivery while the NP allow for the effective penetration into physiological barriers such as the mucosa and/or pulmonary epithelium. This book chapter outlines the advantages, methods of preparation, optimization, drug release, and applications of nCmP.

9.1   INTRODUCTION

9.1.1   PULMONARY DELIVERY OF DRY POWDER PARTICLE-BASED THERAPEUTICS

Aerosols in their dry and wet forms, including steam, gas, and smoke, have been used for medical purposes for decades. In 1956, the first metered-dose inhaler became available and in 1970 the first dry powder inhaler reached the market (Bell et al., 1971). Such aerosol systems have been used for many therapeutic applications including local administration of chemopreventive agents (Lubet et al., 2004), activators of the local immune system (Skubitz and Anderson, 2000; Wylam et al., 2006), and antibiotics, among others. Aerosol delivery of therapeutics to the lungs offers an attractive way to deliver high drug concentrations directly to the site of disease, reducing toxicity while improving the therapeutic potential. Pulmonary delivery is also an attractive route for systemic administration with a more rapid onset of action than is traditionally seen when given orally. The rapid onset is due to fast absorption of the therapeutic by the massive surface area of the alveolar region, abundant pulmonary vasculature, thin air–blood barrier, high solute permeability, and the avoidance of first pass metabolism, which degrades many proteins that have shown promise for use as biotherapeutics (Hess, 2008; Sung et al., 2007).

There is great promise for advances in pulmonary drug delivery using inhaled aerosols for the targeted treatment of respiratory diseases such as chronic obstructive pulmonary disease (COPD), asthma, cystic fibrosis, lung cancer, and infectious diseases (e.g., tuberculosis) (Mansour et al., 2011). To date, several types of particle-based aerosol therapeutics have been developed including nanoparticulates suspended in aqueous form from inhalers and nebulizers as well as microparticle-based liquid and dry powder formulations. Both nanoparticle and microparticle-only formulations exhibit limitations that will be discussed in detail later, leading to a need in the development of advanced formulations using particle-engineering techniques. In particular, it has been shown that solid drug-loaded nanoparticles can be effectively delivered to the lungs when they are encapsulated in a carrier system that dissolves after coming in contact with the aqueous environment of the lung epithelium. These so-called nanocomposite microparticles (nCmP) can enhance the delivery of therapeutics via the aerosol route as described in the following sections. In this chapter, a general overview of pulmonary delivery and the current state-of-the-art related to nCmP will be followed by an overview of current nCmP systems and their design considerations.

9.1.2 DRY POWDER AEROSOL DELIVERY DEVICES

The most common inhaler devices are nebulizers, pressurized metered-dose inhalers (MDI), and dry powder inhalers (DPI). Nebulizers deliver liquid medication in a steady steam of tiny droplets, do not require patient coordination, and can deliver larger doses compared to the other devices (Hess, 2008). MDIs are the most popular devices used to treat local respiratory diseases and their mechanism uses a valve designed to deliver a precise aerosol amount each time the device is actuated (Cummings, 1999; Vaswani and Creticos, 1998). DPIs are portable devices that deliver medication in the form of a dry powder directly to airways (Sanchis et al., 2013). Traditional dry powder blends are typically comprised of micronized drug particles with a mass median aerodynamic diameter (MMAD) less than 5 μm blended with inactive excipients of larger sizes (Islam and Gladki, 2008; Lippmann et al., 1980; Newman and Busse, 2002). The drug is delivered when a patient inhales, pulling air through a punctured capsule, blister, or reservoir.

Advantages of DPIs are that they are portable, compact, and have multidose functionality. DPIs are breath-activated and unlike MDIs do not require any outside energy source or propellant, eliminating the need to coordinate actuation and inhalation (Laube et al., 2011). Another practical advantage of using a dry powder system is the far slower degradation of drugs in the dry state as compared to their suspension in a liquid form, resulting in higher long-term stability and sterility (Sung et al., 2007). Dry formulations of nanoparticles also reduce interparticle attractive forces as does blending with larger carrier molecules, which improves aerodynamic performance (Willis et al., 2012). The use of biodegradable particles in DPIs may increase the bioavailability of therapeutic agents and change the pharmacokinetic plasma profile, making the agent more suitable for pulmonary delivery (Edwards et al., 1998).

DPIs have many advantages with respect to medication delivery; however, they are not without their disadvantages. These disadvantages include poor deposition in the lower bronchioles and alveoli as well as the limited availability of drugs from the device itself. In many cases, DPIs are a relatively new technology still in the development stage, in comparison to the more established nebulizer and MDI technology. A more detailed overview of aerosol delivery devices can be found elsewhere (Al-Hallak et al., 2011; Vehring, 2008; Xu et al., 2011; Ziffels et al., 2015). Overall, the ease of use and powder storage advantages discussed further in this chapter provides strong reasoning for the application of DPIs to delivery advanced dry powder formulations.

9.1.3 OVERVIEW OF CURRENT DRY POWDER AEROSOL THERAPEUTICS

As seen in Table 9.1, diseases that most commonly incorporate DPIs in treatment regimens are asthma and chronic obstructive pulmonary disease (COPD). COPD affects over 5% of the U.S. population and kills 120,000 individuals each year, ranking it as the third-leading cause of death.

TABLE 9.1 Current Dry Powder Inhaler (DPI) Formulations Approved By the FDA Including the Approval Date, Brand Date, Active Pharmaceutical Ingredient (API), Disease, and Company

Date	Brand Name	API	Disease	Company
2015	ProAir Respiclick	Albuterol sulfate	Asthma	Teva
2014	Arnuity Ellipta	Fluticasone furoate	Asthma	GlaxoSmithKline
2014	Incruse Ellipta	Umeclidinium	COPD	GlaxoSmithKline
2013	Tobi Podhaler	Tobramycin	Cystic fibrosis	Novartis
2013	Breo Ellipta	Fluticasone furoate, vilanterol	COPD	GlaxoSmithKline
2013	Anoro Ellipta	Umeclidinium, vilanterol	COPD	GlaxoSmithKline
2012	Tudorza Pressair	Aclidinium bromide	COPD	Forest Labs
2011	Arcapta	Indacaterol	COPD	Novartis
2010	Aridol	Mannitol	Asthma (testing)	Pharmaxis
2006	Pulmicort Flexhaler	Budesonide	Asthma	Astrazeneca
2005	Asmanex Twisthaler	Mometasone furoate	Asthma	Merck
2004	Spiriva	Tiotropium	COPD	Boehringer Ingelheim
2001	Foradil Aerolizer	Formoterol fumarate	Asthma, COPD	Novartis
2000	Advair Diskus	Fluticasone propionate, salmeterol xinafoate	Asthma, COPD	Glaxo
2000	Flovent Diskus	Fluticasone	Asthma	GlaxoSmithKline
1999	Relenza	Zanamivir	Influenza	GlaxoSmithKline
1986	Provocholine	Methacholine Chloride	Diagnosis of bronchial airway hyper reactivity	Methapharm

Management for asthma and COPD consists of short-acting bronchodilators for acute exacerbations and in more severe disease, daily maintenance therapy with bronchodilators and anti-inflammatory medications. Tiotropium bromide is an anticholinergic bronchodilator given to COPD patients via DPI in the U.S.A. meta-analysis of various clinical trials found tiotropium to significantly improve mean quality of life and significantly reduce the number of participants suffering from exacerbations (Karner et al., 2014).

Combination therapy with an umeclidinium-vilanterol DPI is approved for once-daily use for COPD patients in the U.S. Compared to individual agents, this formulation results in greater increases in forced expiratory volume over 1-second (FEV1) at both trough and mean peak levels (Donohue et al., 2013). In addition, a once-daily fluticasone-salmeterol DPI is FDA approved for COPD treatment (Martinez et al., 2013). While not yet FDA-approved, a once-daily DPI containing glycopyrronium and indacaterol is approved for use in Europe and Japan (Frampton, 2014). Also, a twice-daily combination DPI comprised of aclidinium-formoterol is approved for use in Europe, the United Kingdom, and Canada (Bateman et al., 2015).

Asthma maintenance therapy is often similar to that seen in COPD with the goals of bronchodilation and reduction in inflammation. Some examples of these DPIs can be seen in Table 9.1. Recently, in April 2015, the FDA approved a new DPI device for albuterol sulfate administration. Albuterol, traditionally delivered via MDI, is a short-acting beta-agonist (SABA) that is commonly used as a rapid and effective treatment for acute asthma exacerbations, in which the throat and airways can eventually close off if left untreated. In a news release, David I. Bernstein was quoted, "ProAir RespiClick" is the first and only breath-actuated dry-powder rescue inhaler to be approved by the FDA for the treatment of acute asthma symptoms (Mannix and Meir, 2015). In addition to the treatment of reactive airway diseases, DPIs can also be used in diagnostic applications. Aridol is a DPI that delivers mannitol, a sugar alcohol, to the lungs to assess for hyperresponsiveness, which indicates whether or not a patient has asthma. Research has been conducted to develop particles for DPIs including theophyllyne with sodium stearate and salmeterol xinafoate with lactose, both for the treatment of asthma.

Cystic Fibrosis (CF) is a genetic disorder caused by mutations in the cystic fibrosis transmembrane conductance regulator (CFTR) protein.

Current U.S. guidelines suggest treating all CF patients older than 6 years with chronic pulmonary infections with inhaled antibiotics, first tobramycin, then aztreonam, and finally colistin. Aerosolized antibiotic therapy is traditionally given through a nebulizer; however, in March 2013 the TOBI Podhaler was approved as a DPI alternative for inhaled tobramycin. A trial comparing the tobramycin-inhaled powder (TIP) versus nebulized solution found that TIP was generally well tolerated, provided a significantly more convenient treatment option, and resulted in a higher discontinuation rate (Konstan et al., 2011a, 2011b; Parkins and Elborn, 2011). Colistin, while considered third-line by U.S. guidelines, is frequently used as first choice in Europe for CF patients (Smyth et al., 2014). The DPI form of colistin (Colobreathe) is approved for use in Europe.

Antibiotic DPI formulations including vancomycin and clarithromycin with dipalmitoylphosphatidylcholine have been developed for the treatment of pulmonary infections (Park et al., 2013). In addition, azithromycin microparticles have been developed successfully in an inhalable aerodynamic range (<10 μm) (Li et al., 2014).

9.1.4 OBSTACLES IN PULMONARY DELIVERY OF PARTICLE-BASED THERAPEUTICS

The human lungs have the innate ability to remove aerosolized particulates, which can decrease the delivered aerosol drug load. The natural removal mechanisms include mucociliary clearance, phagocytosis, and enzymatic degradation. Mucociliary clearance is a critical host defense mechanism of the airways to clear locally produced debris, excessive secretions, or unwanted inhaled particles. It consists of ciliated epithelial cells present from the naso/oropharynx and the upper tracheobronchial regions down to the most peripheral terminal bronchioles (Suarez and Hickey, 2000). Effective mucociliary clearance requires appropriate mucus production and coordinated ciliary activity (Antunes and Cohen, 2007), including the "mucociliary escalator," where ciliated epithelia sweep particles in the upper airways away from the lungs towards the mouth (Pavia, 1984).

Phagocytosis occurs primarily in the deep lungs where alveolar sacs rich in macrophages play a key role in innate immune defense. Particles

1–5 µm in size are typical of bacteria and are readily engulfed via phagocytosis, thereby limiting the bioavailability of therapeutics (Mahmud and Discher, 2011). Phagocytic activity is maximized for particles 1–2 µm in diameter and less so for smaller or larger particles outside of this range (Li et al., 2010). The contribution of pulmonary endocytosis to the overall lung clearance is determined by the particle size and shape, solubility, particle burden, and chemical nature of the inhaled aerosols.

Macromolecular drugs are also susceptible to enzymatic degradation. The contribution of drug breakdown along with other pulmonary clearance mechanisms is minimal in comparison with the gastrointestinal tract; however, it requires consideration in enzyme-sensitive compounds. Therefore, to protect from rapid clearance or degradation and achieve sustained release within the lung, encapsulation into nanoparticles or microparticles has been pursued.

Systemic diseases can be treated via pulmonary delivery as formulations and their released drugs can translocate from lung tissue to the cardiovascular endothelium. The mechanisms by which therapeutics are translocated from the lung occurs by either transcellular or paracellular transport. In transcellular transport, absorption of the therapeutic typically occurs through receptor/carrier-mediated endocytosis. This is a slow process (hours to days) and occurs for larger particles (>5–6 nm) having molecular weights of more than 40 kDa (Mahmud and Discher, 2011). Transcellular transport involves the internalization of caveolin-1, transcellular channels, and vesicular trafficking. Paracellular transport involves diffusion of drug through alveoli, which is a fast process (5 to 90 min) and happens in the case of smaller molecules (<5–6 nm). Several studies have evaluated the mechanisms of nanoparticle interaction with the lung and translocation of nanoparticles from nanocomposite microparticles through the epithelial barrier (McIntosh et al., 2002; Yacobi et al., 2010).

Nanocomposite microparticles offer the ability to overcome many of the barriers associated with the aerosol delivery of therapeutics. First, nCmP are typically an appropriate size for effective pulmonary delivery (e.g., micrometer) and thereby can be delivered to all locations in the lungs. While micro-sized particles are usually removed from the lungs via mucociliary clearance and/or phagocytosis, nCmP can be designed to rapidly release nanoparticles from the nCmP bulk in a very quick fashion to avoid these clearance mechanisms. In addition to avoidance of pulmonary clearance, the nanoparticles released from nCmP offer their own advantages including the ability to translocate through mucus, penetrate into cells, and translocate across the pulmonary epithelium/cardiovascular endothelium barrier (Chen et al., 2006; Tang et al., 2009). At this point, the use of nCmP is the only effective way to delivery dry nanoparticle-based therapeutics.

9.1.5 DESIGN FOR PULMONARY DEPOSITION OF PARTICLES

With respect to aerosol formulation and design, drug distribution and deposition along the respiratory tract depends on many characteristics of the inhaled formulation including the diameter, size distribution, shape, charge, density, and hygroscopicity of the particulate system. The mass median aerodynamic diameter (MMAD) and geometric standard deviation are what determine the site of deposition in the respiratory tract (Vanbever et al., 1999).

Dry powder particle formulations used in pulmonary applications must exhibit specific physical properties for successful implementation. Respirable particles with an MMAD between 0.5 and 5 µm undergo deposition in the alveolar region, which can facilitate systemic bioavailability (Bosquillon et al., 2001; Coates and O'Callaghan, 2006). Particles larger than this may potentially be deposited in the inhaler, oropharynx, and/or larynx, or they may not reach the desired site within the lung due to size constraints. For example, it was shown that decreasing the particle size from 5.4 µm to 2.7 µm reduced the total throat deposition by half and increased the total lung deposition by over two-fold for mannitol particles (Glover et al., 2008).

The process of particle deposition in human airways includes: a) inertial impaction, which is dominant in the upper airways where velocities are at a maximum and where many particles impact and stick to the pulmonary surface, b) sedimentation, which is predominant throughout the central and distal tract, where the particles settle on the surface of the lung due to gravitational forces and air resistance, and c) diffusion, which is the most important for submicrometer-sized particles (<0.5 µm), which are in random motion and deposit on the lung walls via Brownian motion (Hinds, 1999; Suarez and Hickey, 2000; Vanbever et al., 1999). Previous studies have shown that the size and distribution site of particles inhaled in the lungs is as follows: particles smaller than 1 μm tend to diffuse, remaining suspended in the airways, and are typically exhaled; particles 1–3 μm in diameter deposit primarily in the alveolar region, 8–10 μm-sized particles undergo tracheobronchial deposition; and particle larger than 10 μm often exhibit deposition in the mouth (Mahmud and Discher, 2011).

9.1.6 LONG-TERM STORAGE CONSIDERATIONS

Another major concern in the development of powder aerosols is in the storage of particles, which tend to agglomerate (Edwards et al., 1998). In general, the more uniformly the particles are distributed, the more stable their shelf life is (Muller et al., 2000). Spray-drying, freeze-drying, and spray-freeze-drying often improve long-term particle stability during storage for dry formulations in comparison to aqueous formulations. For example, freeze-dried poly(methylidene malonate) nanoparticles were stored for 12 months at different temperatures and showed no significant variations in pH, size, turbidity and cytotoxicity (Roy et al., 1997). Suspensions of freeze-dried PLGA microspheres, either alone or loaded with cyclosporine, were stored at 8°C and room temperature (RT) for six months. The suspensions were substantially instable while the freeze-dried particles showed an alternative for long-term stability at low temperatures (Chacón et al., 1996). In another study, a poorly soluble drug, celecoxib, was formulated into solid phospholipid nanoparticles and showed that the amorphous matrix generated via spray-drying and freeze-drying significantly enhanced the dissolution rates and apparent solubility (Fong et al., 2015).

Another technique applied to enhance stability and solubility of particles is spray-drying. TIP used for the treatment of cystic fibrosis was designed and formulated considering the process parameters related to critical temperature transitions. A room temperature stable product was obtained that requires no refrigeration (Miller et al., 2015). Gentamicin particles for cystic fibrosis developed with L-leucine showed no significant degradation for up to 6 months of storage (Aquino et al., 2012). Spray-freeze-drying is a relatively new method of producing biopharmaceutical powder preparations. It combines the advantages of freeze- and spray-drying techniques in order to get a stable product and increase the solubility of poorly water-soluble drugs. This technique was used to obtain a fine stable probiotic powder of *Lactobacillus casei* in mannose and $CaCO_3$ (Her, Kim and Lee, 2015). Finally, reaggregation of palygorskite nanofibers containing ofloxacin was successfully overcome by freeze-drying. Comparing with the traditional oven-dried sample, the freeze-dried sample showed enhanced dispersion stability of palygorskite in deionized water (Wang et al., 2014).

9.2 NANOCOMPOSITE MICROPARTICLES: A SOLUTION FOR OVERCOMING THE OBSTACLES OF PARTICLE-BASED PULMONARY DELIVERY

Nanocomposites are an emerging class of materials used in pulmonary drug delivery. Such systems are typically multiphase structures, comprised of a single continuous phase or matrix (e.g., polymers) with one or more discontinuous or dispersed phases (e.g., drugs). A formulation of nanocomposite microparticles (nCmP) uses a combination of phase or matrix components, active pharmaceutical ingredients (APIs), and inert components (excipients) as seen in Figure 9.1. Although the overall size of the nanocomposite microparticles can vary between 0.1–100 μm, a fraction of these particles having aerodynamic diameters 1–2 μm have been reported to be more effective in depositing into alveolar spaces (Ungaro

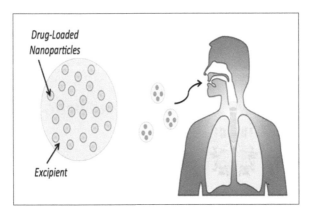

FIGURE 9.1 Schematic of nanocomposite microparticle (nCmP) system used for pulmonary delivery applications.

et al., 2012). The properties of nanocomposite materials typically differ from conventional composite materials owing to the 'critical size' or 'nanometer range' of the dispersed drugs in the matrix, which enhances the properties of the nanocomposite particle.

A list of nCmP developed for the treatment of pulmonary diseases is outlined in Table 9.2. The examples in this table not only cover a wide variety of pulmonary applications but also include different encapsulated nanoparticle (NP) systems, APIs, and carriers. The loaded NP are primarily polymer-based, however, systems with metals (iron oxide) and pure drug are also present. The table encompasses a comprehensive overview of the current pulmonary nCmP systems that have been designed and evaluated and one significant conclusion is that the majority of the systems were fabricated via spray drying, which is a fairly straightforward, accessible, and high throughput synthesis method.

As seen in Figure 9.2, nCmP formulations exhibit a variety of morphologies ranging from nanocomposite microcrystals, solid nanoparticle spheres, hollow nanoparticles spheroids, nanoparticle agglomerates, and nanoparticles dispersed in a carrier. The morphology can be tuned via synthesis parameters in order to ensure effective API delivery, stability, and and/or targeted pulmonary deposition.

9.3 PREPARATION OF NANOCOMPOSITE MICROPARTICLES (NCMP)

Particle engineering is a key factor in the development of nCmP systems with desirable properties (Muralidharan et al., 2015). In this section, we review techniques used in the preparation of nanoparticles that are used in nCmP formulations as well as techniques used to prepare the nCmP systems themselves.

9.3.1 TECHNIQUES USED TO PREPARE DRUG-LOADED NANOPARTICLES

9.3.1.1 Solvent Evaporation

In this method, two main strategies are used to form emulsions including single emulsions with oil-in-water (o/w) and double emulsions with

TABLE 9.2 Overview of Nanocomposite Microparticles (nCmP) Systems Used to Deliver Therapeutics for Pulmonary Applications with Their Active Pharmaceutical Ingredient (API), Treatment Application, and Key Findings

Encapsulated NP	API/Payload	Carrier/Excipient	Synthesis Method	Application/Disease Treated
Haloperidol-bovine serum albumin NP (Varshosaz et al., 2015)	Doxorubicin	Trehalose, L-leucin, mannitol	Spray drying of NP suspension	Cancer
Iron oxide magnetic NP (Stocke et al., 2015)	-	D-mannitol	Spray drying of NP suspension	Thermal treatment of the lung diseases
Poly(glycerol adipate-cov-pentadecalacton) NP (Kunda et al., 2015a, 2015b)	Antigen PspA or BSA	L-leucine	Spray drying of NP suspension	Pneumonia, delivery of protein
Poly(e-caprolactone) NP (Lee et al., 2013)	Hydrocortisone acetate	Poly(lactic-coglycolic) acid (PLGA)	Double emulsion	Cancer
PLGA NP (Yang et al., 2012)	Salmon calcitonin	Chitosan	Spray drying fluidized bed granulation and dry powder coating techniques	Enhanced pulmonary absorption of drug
PLGA NP and alginate/PLGA NP (Ungaro et al., 2012)	Tobramycin	Lactose	Spray drying of NP suspension	Delivery of antibiotic
Chitosan/tripolyphosphate (TPP) NP (Jafarinejad et al., 2012)	Itraconazole	Lactose, mannitol, L-leucine	Spray drying of NP suspension	Delivery of antifungal drug
Itraconazole nanosuspension (Duret et al., 2012)	Itraconazole	Mannitol, sodium taurocholate	Spray drying of NP suspension	Invasive pulmonary aspergillosis
PLGA NP (Beck-Broichsitter, et al., 2012)	Coumarin 6	-	Spray drying of NP suspension	Alveolar tissue targeting

TABLE 9.2 (Continued)

Encapsulated NP	API/Payload	Carrier/Excipient	Synthesis Method	Application/Disease Treated
Chitosan/TPP NP (Pourshahab et al., 2011)	Isoniazid	Lactose, mannitol, maltodextrin, leucine	Spray drying of NP suspension	Tuberculosis
PLGA NP (Cheow et al., 2011)	Levofloxacin	poly(vinyl alcohol) (PVA), leucine	Spray drying of NP suspension	Delivery of antibiotic
Glyceryl monostearate/soybean phosphatidylcholine solid lipid NP (SLNs) (Li et al., 2010)	Thymopentin	Mannitol, leucine	Spray drying of SLN suspension	Delivery of peptide
PLGA NP (Jensen et al., 2010)	siRNA	Trehalose, lactose, mannitol	Spray drying of NP suspension	Delivery of siRNA
PLGA NP (Sung et al., 2009)	Rifampicin	L-leucine	Spray drying of NP suspension	Tuberculosis
Tobramycin nanosuspension (Pilcer et al., 2009)	Tobramycin	-	Spray drying of NP suspension	Delivery of antibiotic
PLGA NP formed during spray drying (Ohashi et al., 2009)	Rifampicin	Mannitol	Four-fluid nozzle spray drying	Tuberculosis
PLGA NP (Tomoda et al., 2008)	Rifampicin	Trehalose, lactose	Spray drying of NP suspension	Tuberculosis
PLGA/Polyethylenimine (PEI) nanospheres suspension (Takashima et al., 2007)	Plasmid DNA (pCMV-Luc)	Mannitol	Spray drying of nanosphere suspension directly	Gene delivery, vaccination
Pranlukast hemihydrate particles formed during spray drying (Mizoe et al., 2007)	Pranlukast hemihydrate	Mannitol	Four-fluid nozzle spray drying	Asthma

TABLE 9.2 (Continued)

Encapsulated NP	API/Payload	Carrier/Excipient	Synthesis Method	Application/Disease Treated
Polyacrylate NP and silica NP (Hadinoto et al., 2006, 2007, 2007)	-	Phospholipids	Spray drying of NP suspension	Pulmonary delivery
Chitosan/TPP NP (Grenha et al., 2007)	FITC-BSA	Mannitol	Spray drying of NP suspension	Delivery of protein
Poly(butylcyanoacrylate) (PBCA) NP (Ely et al., 2007)	FITC-dextran	Lactose	Spray drying of NP suspension	Pulmonary delivery
PBCA NP (Azarmi et al., 2006)	Doxorubicin	Lactose	Spray freeze-drying of NP suspension	Cancer
Chitosan/TPP NP (Grenha et al., 2005)	Insulin	Mannitol, lactose	Spray drying of NP suspension	Delivery of protein
Terbutaline sulfate nanosuspension (Cook et al., 2005)	Terbutaline sulfate	Glyceryl behenate/ Tripalmitin or Hydrogenated palm oil	Spray drying of NP suspension	Delivery of drug with sustained release
Gelatin or PBCA NP (Sham et al., 2004)	Dextran	Lactose	Spray drying of NP suspension	Delivery of drugs and diagnostics
Polystyrene NP (Tsapis et al., 2002)	-	Phospholipids	Spray drying of NP suspension	Pulmonary drug delivery

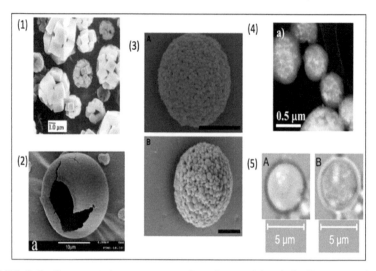

FIGURE 9.2 Representative nanocomposite microparticles (nCmP) systems imaged via scanning electron microscopy (SEM), transmission electron microscopy (TEM), and fluorescence imaging. The nCmP systems and imaging techniques are: (1) spray-dried NaCl particles (SEM), (2) hollow spherical LPNP (SEM), (3) poly(styrene) (top) and poly(lactide-coglycolide) (bottom) where scale bar is 1 μm (SEM), (4) magnetic nCmP (TEM), and (5) hollow fluorescently labeled polycyanoacrylate NP (fluorescence microscopy) (Beck-Broichsitter, Merkel and Kissel, 2012; Schafroth, Arpagaus, Jadhav, Makne and Douroumis, 2012; Sham et al., 2004; Stocke et al., 2015)

water-in-oil-in-water (w/o/w) phases. For single emulsion, drugs and polymers are dissolved in oil phase, while in double emulsion, drugs are usually in inner water phase. Volatile solvents such as dichloromethane, chloroform, and ethyl acetate are mostly used as oil phase. The emulsions are formulated using high-speed homogenization or ultrasonication and converted into nanoparticle suspension by solvent evaporation in a spinning solution with surfactant (Nagavarma, 2012).

9.3.1.2 Nanoprecipitation

Nanoprecipitation, also called solvent displacement, is used to prepare nanoparticles by precipitating polymer at the interface between the water and the organic solvent. The precipitation is driven by the diffusion of the organic solvent into the aqueous phase with or without surfactant. Therefore, water-miscible solvents like acetone, methanol, or ethanol are recommended. The nanoprecipitation method often results in high drug loading and encapsulation efficacy for hydrophobic drugs, however, it is not an efficient means to encapsulate water-soluble drugs because of the high diffusion rate of drug from organic solvent into water (Nagavarma, 2012).

9.3.1.3   Emulsification/Solvent Diffusion

Emulsification/solvent diffusion is a modified version of the solvent evaporation method. In this method, an oil phase is formed by dissolving drug and polymer in a partially water soluble solvent such as propylene carbonate. This mixture is saturated with water to ensure the initial thermodynamic equilibrium of both liquids and then emulsified in an aqueous solution of surfactant. The solvent will diffuse into the aqueous phase leading to nanoparticle or nanocapsule formation and will then be eliminated by evaporation or filtration, according to its boiling point. This method also allows a high drug loading and encapsulation efficacy for hydrophobic drugs as well as for hydrophilic ones (Nagavarma, 2012).

9.3.1.4   Salting Out

Salting out, regarded as the modified version of emulsification/solvent diffusion, is-based on the separation of a water-miscible solvent from aqueous solution caused by a salting out effect. In this method, drug and polymer are dissolved in a water-miscible solvent such as acetone. The mixture is emulsified in an aqueous gel containing a salting-out agent such as magnesium chloride, calcium chloride, magnesium acetate, or sucrose as well as surfactant and is then diluted with a sufficient volume of aqueous solution to enhance the diffusion of solvent into the aqueous phase. Nanoparticles are formed as the result of solvent diffusion. The salting out agent is the key factor in the method influencing the efficacy of encapsulation but should be eliminated with solvent in the end. This method exhibits an advantage over others by avoiding high stress and increased temperature, which is suitable for protein encapsulation. Salting out is ineffective for hydrophobic drug and requires extensive washing steps (Jung et al., 2000; Lambert et al., 2001; Nagavarma, 2012).

9.3.1.5 Dialysis

Dialysis can be used as a simple and effective method for the preparation of nanoparticles of small size and narrow distribution (Nagavarma, 2012). However, the mechanism of this method is not completely understood right now and is regarded as similar to that of nanoprecipitation (Fessi et al., 1989). In dialysis, polymer and drug are dissolved in an organic solvent inside a dialysis tube with a proper molecular weight cut off. As the solvent displacement progresses, aggregation of polymer happens due to loss of solubility, thus forming a nanoparticle suspension. The properties of resulting nanoparticles are influenced by the type of solvent used in salting out.

9.3.1.6 Supercritical Fluid (SCF) Technology

A supercritical fluid is defined as any substance at a temperature and pressure above its critical point. Supercritical fluids are defined as compressed gases or liquids above their critical pressures and temperatures, which exhibit several fundamental advantages as solvent or nonsolvent. For pharmaceutical manufacturing, supercritical CO_2 is widely used due to its low critical temperature (31.1°C) and moderate pressure (73.8 bar), nontoxic inert nature, and low cost. Supercritical fluid technology offers an environmentally friendly way to make nanoparticles due to efficient extraction and separation of organic solvents in this process. The resulting nanoparticles are in pure dry form or as pure aqueous suspensions (York, 1999).

9.3.1.7 Preparation of Nanoparticles From Polymerization of Monomers

Nanoparticles with desired properties can be prepared by polymerization of monomers. This method can be subdivided into emulsion polymerization, mini-emulsion polymerization, interfacial polymerization, and controlled/living radical polymerization. For this method, toxic organic solvents, surfactants, monomers and initiators must be removed from formed nanoparticles, which limit the use of the method (Nagavarma, 2012).

9.3.2 TECHNIQUES USED TO PREPARE NCMP

The aerosol performance of nCmP is directly related to their preparation techniques. In this section, both traditional and more advanced methods used in the preparation of nCmP are discussed (Bailey and Berkland, 2009). These methods are classified into two categories, including top-down as milling larger particles to reduce size and bottom-up precipitation of nanoparticles out of solution. Techniques such as the modification of spray drying also show promise to prepare nCmP (Pilcer and Amighi, 2010).

9.3.2.1 Milling

Milling is the traditional method of drug powder micronization, which has been intensively studied. The milling method can involve the use of ball mills, colloid mills, hammer mills, and jet or fluid energy mills, in which the jet mill is applied to prepare most inhalation powders. Jet milling is defined as a process of micronization by interparticle collision and attrition. In this method, coarse particles carried by a high velocity gas as milling medium pass through a nozzle into the jet mill. The particles are then suspended in the turbulent gas stream where they break into smaller particles due to interparticle collisions. The fine particles are then taken up by the gas stream out through the exit, while the larger ones remain in the mill for further micronization by more interparticle collision. The application of jet milling is limited by lack of control over resulting particle properties and the high energy in the process increasing the risk of degradation of therapeutics (Pilcer and Amighi, 2010; Shoyele and Cawthorne, 2006).

9.3.2.2 Supercritical Fluid Technology (SCF)

Supercritical fluid technology can also be used for preparation of micro-size inhalation as discussed previously. More specific discussion of SCF could be found in several review articles (Muralidharan et al., 2015; Pilcer and Amighi, 2010; Shoyele and Cawthorne, 2006).

9.3.2.3   Spray Drying

Spray drying is the most commonly used method to prepare dry powders for inhalation (Wu et al., 2013). In spray drying, nanoparticle suspensions with or without excipients are atomized at the nozzle. The droplets of the suspension go through the drying chamber with heated gas, resulting in evaporation of solvent and generation of nCmP. The resulting nCmP are separated in a cyclone separator and collected in the collector (Irngartinger et al., 2004). The spray drying method provides easily tunable process conditions and high reproducible properties of resulting particles. However, challenges exist to determine optimal processing conditions to achieve desirable properties, due to the fact that the drying process depends on complex parameters including the solvent used, solute concentration, inlet temperature, outlet temperature, atomizing pressure, feed properties, pump rate, drying gas type, and gas flow rate (Calvo et al., 1997; Corkery, 2000; Grenha et al., 2005; Wu, X. et al., 2013).

9.3.2.4   Spray-Freeze Drying (SFD)

Spray freeze-drying is a modified version of spray-drying, in which the suspension or solution is atomized into a cryogenic liquid such as liquid nitrogen generating droplets (Shoyele and Cawthorne, 2006). The droplets are then lyophilized, resulting in porous spherical particles suitable for inhalation. Application of SFD is limited by several disadvantages such as causing irreversible damage to the proteins by drying and freezing stress, time consumption, and high expense (Codrons et al., 2003).

9.3.3   MATERIALS AND EXCIPIENTS USED IN THE PREPARATION OF NCMP

9.3.3.1   Drug-Loaded Nanoparticle Materials

The materials used in the polymer matrix of nanoparticles should exhibit biodegradable and biocompatible properties in themselves as well as their degradation products. These polymers could be natural polymers such as chitosan gelatin, sodium alginate, and albumin, or synthetic polymers including polylactides (PLA), polyglycolides (PGA), poly(lactide-co-glycolides) (PLGA), poly(methylmethacrylate) (PMM), poly(cyanoacrylate) (PCA), poly(caprolactone) (PCL), and poly(ethylene glycol) (PEG)-conjugated- PLGA. To select a polymer as nanoparticle matrix, the nanoparticle formation techniques, the encapsulated drugs, the properties of polymers, and further nCmP preparation methods should be taken into consideration. In addition, the desired properties of nanoparticles are also an important concern when choosing polymers. Detailed discussion on the materials used for nanoparticle formation can be found in former reviews (Azarmi et al., 2008; De Jong and Borm, 2008; Jung et al., 2000; Mudshinge et al., 2011; Nagavarma, 2012; Parveen et al., 2012; Pulliam et al., 2007; Shim et al., 2007; Sun et al., 2014; Sung et al., 2007; Tsapis et al., 2002; Wilczewska et al., 2012; Zhang et al., 2011).

9.3.3.2 Excipients Used in nCmP Synthesis

The nCmP carriers and excipients are the inactive ingredients used to improve the delivery efficacy and stability of nCmP formulations for pulmonary delivery. The excipients have no therapeutic effects, but still have a significant role to form a desired delivery system of therapeutics by enhancing the mechanical properties of dry powder, physical or chemical stabilization of drug, and improving redispersion and dissolution properties of encapsulated nanoparticles (Pilcer and Amighi, 2010). The Food and Drug Administration (FDA) recommends the use of excipients that are either commercially established or "generally recognized as safe" (GRAS). Hence, the choice of excipients used for nCmP formation is limited since the current excipients approved for pulmonary drug delivery are very limited in number (Baldrick, 2000). Detailed discussion on the excipients used for inhalation application can be found in other review and research articles (Bosquillon et al., 2001; Minne et al., 2008; Pilcer and Amighi, 2010; Rohani et al., 2014).

9.4 OPTIMIZATION OF NCMP MANUFACTURING VIA SPRAY DRYING

While there are obviously many methods available for the synthesis of nCmP, spray drying has been the primary method to date. Spray drying is a straightforward and high throughput method allowing for easy reproduction of nCmP and thus the remaining sections of this book chapter will focus on the optimization of nCmP via spray drying. In a spray drying process, several parameters can be optimized to achieve desirable properties for nCmP formulations. In this section, these parameters are classified into two major categories: 1) spray drying conditions and 2) feed solution composition. The spray drying conditions include the inlet temperature, aspirator, spray gas flow, feed speed (pump rate), which can be easily tuned through the control panel of spray dryer, and humidity of drying gas, which can be modified by incorporating a dehumidifier or using dry nitrogen or compressed air. The feed solution composition covers the ratio of nanoparticles to carriers (excipients), solid concentration (concentration of total dry materials), properties of carriers, properties of nanoparticles, and solvent composition (Ameri and Maa, 2006; Kaur et al., 2015; Littringer et al., 2011; Vehring, 2008).

The relevant parameters for the spray process correlate with and depend on each other. Therefore, to design an optimal nCmP system with parameters suitable for desirable resulting particles, a comprehensive evaluation of all parameters should be taken. However, difficulties exist in the selection of the process conditions, lying in the facts that: (a) a single change in parameter may influence several other properties of spray drying products; (b) several parameters may influence one property of spray drying products and these parameters depend on each other; and (c) a parameter may have positive effects on products in one range and negative effects or no influence in another range. In summary, there is no best spray drying condition for any given system, but only an optimal condition that is most suited for a product intended for a certain therapeutic application. This section aims at providing a strategy in the design of optimal process conditions and to aid in the understanding of the internal relationship between the process parameters and corresponding properties of resulting particles.

9.4.1   HOW SPRAY DRYER SETTINGS INFLUENCE DRY POWDER FORMULATIONS

There are many spray dryer settings that influence the characteristics of dry powder formulations like nCmP (Das et al., 2014; Maa et al., 1997).

These settings include the inlet temperature, aspiration rate, feed (pump) rate, spray-drying gas flow rate, composition of feed solution, and other external parameters discussed below.

The inlet temperature is directly proportional to outlet temperature, which is the actual temperature at which the spray-dried materials are exposed. A higher inlet temperature is able to reduce the relative humidity in the drying gas, resulting in dryer and less sticky powders, thus increasing the powder yield. Aspiration, which provides a gas flow depending on the pressure drop of the overall system, reflects the drying energy that the system exerts on the drying process. The aspiration rate leads to a positive effect on increasing outlet temperature, reducing the amount of residual moisture in the product, and thus improving the degree of product separation in the cyclone. The feed rate affects the amount of time that materials undergo the drying process. A higher feed rate results in larger droplet dispersion in the drying chamber, which decreases the outlet temperature. An increase in the feed rate also increases the moisture content in the gas, resulting in more humid or moist products that could adhere to the glassware (cyclone or collector), thereby decreasing the yield. The spray-drying gas flow rate is another parameter that affects the products. Higher gas leads to a reduction in outlet temperature and reduces the droplet size, thus decreasing the size of resulting particles.

While the previous parameters can be tuned through the control panel of the spray dryer, the feed composition of the spray-dried solution can also affect the final powder product. A direct result of increasing the feed concentration is a decrease in the solvent amount, which results in less liquid evaporation, leading to higher outlet temperature. An increase in the feed concentration also reduces the partial pressure of solvent in the gas, leading to less water in the final product. The feed concentration exerts a positive influence on the particle size due to an increase of solid in droplets and a higher yield contributes to the ease in the collection of larger particles. The use of an organic solvent for the feed solution will result in smaller particles due to lower surface tension of these solvents. The organic solvent also requires less energy to vaporize, leading to higher outlet temperatures, thus resulting in dry products. Finally, the humidity of the drying gas can be reduced through the use of a dehumidifier or using ultra dry nitrogen, and this has a negative influence on the vapor uptake capability in the gas stream, resulting in positive effects on outlet temperature. As a result, higher drying gas humidity leads to higher humidity of the final products with lower yield.

9.4.2 A THEORETICAL FRAMEWORK OF NANOCOMPOSITE MICROPARTICLE (NCMP) FORMATION

After a greater understanding of how spray dryer parameter settings influence the spray-drying process and resulting materials, we can move on to the more complicated spray-drying process with nanoparticles. Nanoparticles with various properties add complexity into the drying process in the droplet of feed solution dispersed in the drying chamber, while the formation process of droplet is affected by parameters such as the feed speed, aspirator speed, and drying gas flow rate. The engineering of nCmP formation requires an overall understanding of microbiology, chemistry, formulation science, colloid and interface science, heat and mass transfer, solid state physics, aerosol and powder science, and nanotechnology (Vehring, 2008).

Figure 9.3 shows nCmP formation from a droplet of feed solution in the spray-drying process. Unlike traditional spray-dried formulations, where particle formation is due to the micronization and drying process of dry materials in feed solution, novel nCmP systems exhibit sophisticated inner attributes and substructures. During nCmP formation, the feed solution is atomized through the nozzle, forming droplets of the nanoparticle suspension in the carrier solution. In ideal conditions, the nanoparticle dispersion remains stable, where the carrier molecules and nanoparticles are dispersed uniformly in the solvent. Once the droplets come into contact with hot drying air, evaporation of the solvent at the droplet surface will occur. Accompanied by the shrinkage of the droplets, two driving forces are generated, which in combination are responsible for the separation of the components in the droplets. The first driving force is the local temperature gradient created at the droplet surface contributed by the flux of heat caused by water evaporation from the surface of the droplet into the gas stream. As a result, thermophoresis of the nanoparticles occurs, leading to their movement towards the surface of the droplets (Iskandar et al., 2003; Kim, 2008; Vehring, 2008). A detailed mathematical model of this phenomenon can be found in Ferry Iskandar's research (Iskandar et al., 2003b, 2003b).

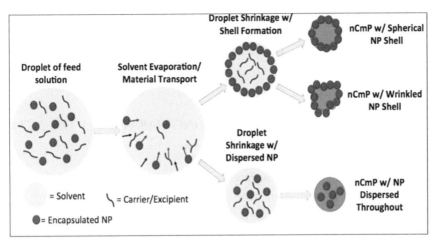

FIGURE 9.3   Schematic of nanocomposite microparticle (nCmP) formation during spray drying from a droplet of feed solution of nanoparticles (NP) and excipients in a solvent.

The other driving force is caused by the concentration gradients in the droplet. The evaporation of solvent at the droplet surface increases the concentration of components in the same plane, causing diffusion of the solute and nanoparticles towards the droplet center (shown in Figure 9.3) (Vehring, 2008; Vehring et al., 2007; Verkman et al., 2003). For the solute or nanoparticles that diffuse fast enough compared to the shrinkage of droplet caused by evaporation, their concentration gradient along the droplet radius will be insignificant, resulting in relatively uniform distribution of the dry materials in the nanoparticles. However, if the diffusion is slower than the surface shrinkage, there will be a higher concentration of solute or nanoparticles at the droplet surface compared to the droplet center, resulting in a shell of particles. The ratio of evaporation rate to diffusion rate can be described by the dimensionless Péclet number, which is an important factor influencing the size, shape, surface morphology, and component distribution of resulting particles. The Péclet number is well discussed in other review and research articles (Littringer et al., 2013; Vehring, 2008; Vehring et al., 2007) and can be used to explain how the inlet temperature influences the properties nCmP. In the final step during nCmP formation, solvent evaporation finalizes, and the particles undergo changes in morphology and thermal states. Crystallization or recrystallization may occur, resulting in crystals with different sizes located inside or at the surface of resulting particles. Whether nCmP exhibit spherical or wrinkled morphology is also determined in this final step (Littringer et al., 2013; Tewa-Tagne et al., 2007; Vehring, 2008).

9.4.3 INFLUENCE OF SPRAY DRYING PARAMETERS ON NCMP FORMATION

After covering the basics on how spray-drying parameters influence the properties of nCmP and an overall theoretical framework of nCmP formation, we can continue with a detailed discussion of these parameters. As previously mentioned, the parameters relevant to the spray process correlate with and depend on each other. In this section we discuss the influences of these parameters in terms of primary parameters, secondary parameters, physicochemical properties, and application properties (desirable properties). It is assumed that while discussing a given parameter, that the others remain constant.

9.4.3.1 Influence of Inlet Temperature

The inlet temperature proportionally determines the outlet temperature, which is close to the actual temperatures that spray-dried materials are exposed to. At high spray drying temperatures evaporation is rapid, resulting in a higher Péclet number thus indicating a fast increase of the solute or nanoparticle concentration at the droplet surface. Once a critical concentration is reached at the surface, precipitation will take place. Early precipitation results in smaller differences between the solid particle shell and the initial droplet size, producing larger shells. In the preparation of nCmP, suspended nanoparticles will form a composite shell in early processes and insufficient solids remain to fill the internal space of the shell, thus forming large and hollow nCmP. On the contrary, lower temperatures lead to slower evaporation that may lag behind the solid diffusion rate, causing delayed shell formation and smaller nCmP. In general, higher inlet temperature leads to higher processing temperature, improving the formation of larger and hollow particles, which tend to exhibit favorable aerosol performance. However, exceptions have been reported indicating that temperature may not influence the particle size (Maas et al., 2011). In that case, a wider temperature range could be applied for study on influence of temperature (Littringer et al., 2013). In another study, temperature firstly has a negative effect on size of nCmP, then presents a positive effect (Tomoda et al., 2008). In that case, an optimal inlet temperature should be chosen-based on comprehensive consideration on other parameters.

The inlet temperature also influences the crystallization of carrier molecules during solvent evaporation at the droplet surface, thereby changing the surface roughness of the resulting particles. At very high temperatures, carriers may not have enough time to crystallize during the precipitation process, thus forming into an amorphous state. Given enough time for crystallization, lower temperatures ensure larger crystals due to low nucleation rates from the solution, resulting in more roughness at particle surfaces. On the contrary, spray drying at higher temperatures leads to higher nucleation rates with more nuclei, resulting in smaller crystals and smooth surfaces. However, converse results have also been reported, in which higher temperature leads to larger crystals, while at lower temperature smaller crystals are produced. This latter case usually happens in the spray drying of small droplets at lab scale, where high temperature results in higher possibility of supersaturation at the surface than in large droplets. Therefore, delayed crystallization from a highly supersaturated viscous liquid or even water-free melt may occur, leading to larger crystals and rougher surfaces. In summary, temperature is able to influence the crystalline states of carrier molecules, which may further affect the performance of pulmonary delivery (Littringer et al., 2012; Maas et al., 2010). More detailed discussions on the influence of temperature on crystalline states of spray-dried particles can be found elsewhere (Littringer et al., 2011; Maas et al., 2011; Raula et al., 2004; Ståhl et al., 2002).

Another consideration with process temperature is its influence on the bioactivity of the powder products. Although therapeutics are encapsulated and protected by nanoparticles and microcarriers, the process temperature should be run at a safe range to keep bioactivity of the payloads (Ameri and Maa, 2006; Jensen et al., 2010; Kaur et al., 2015; Takashima et al., 2007). In addition, temperature can impact the redispersion of the nanoparticles from the dissolved nCmP. At high temperatures, the melting of dry materials can occur, which will lead to the fusion of nanoparticles when they diffuse towards the droplet center and form a nanocomposite shell. Poor redispersion can cause nanoparticles to lose their favorable properties as drug delivery systems and impair the efficacy of nCmP.

9.4.3.2 Influence of Ratio and Properties of Nanoparticles

In the absence of other driving forces, the freely soluble carrier molecules driven by diffusion exhibit even distribution in the droplet during the evaporation. The initial saturation of carriers is small and the characteristic time for precipitation at the droplet surface is close to the droplet lifetime. Given that no nanoparticles are included, solid particles with a density close to the true density of the dry materials tend to form (Burger et al., 2000; Das et al., 2014; Ho et al., 2015; Littringer et al., 2011 et al., 2012, 2013; Maas et al., 2011; Maury et al., 2005; Shi and Zhong, 2015; Wu et al., 2014). In the case of when a system with lower ratio of nanoparticles to carrier molecules mixture is spray-dried, nanoparticles in the droplet may not be able to form a shell (Kim et al., 2003). When the nanoparticles account for a larger part in the dry materials, a nanocomposite shell is able to form and this determines the shape and size of resulting nCmP. Although the formation of nCmP is initiated from the same initial droplet solidification at surface, the morphology varies significantly. For nanoparticles capable of building rigid shells quickly, solid hollow spherical nCmP are formed. Otherwise, dimpled or wrinkled particles are formed. This difference may be caused by the properties of the suspended nanoparticles and thus it varies significantly in different spray drying conditions and feed solutions. No universal conclusion can be elucidated from this conclusion, yet optimization should be performed-based on specific conditions by referring to former studies. Extensive studies have been done reporting or discussing the variety in the morphology of spray-dried nCmP (Baras et al., 2000; Fu et al., 2001; Hadinoto et al., 2007; Jensen et al., 2010; Li and Birchall, 2006; Mu and Feng, 2001; Ting et al., 1992; Tsapis et al., 2002) or particles-based on other colloids (Ameri and Maa, 2006; Chew and Chan, 2001; Kim et al., 2003; Maa et al., 1997; Maa et al., 1998; Maury et al., 2005; Zijlstra et al., 2004). The ratio of nanoparticles to carriers also influences the redispersion of the nanoparticles from nCmP, as well as the particle size (Tomoda et al., 2008). Properties of nanoparticles including particle size and surface properties can influence the properties of resulting nCmP. No overall review of this field has been performed and thus further studies may be warranted (Al-Hallak et al., 2011; Azarmi et al., 2008; Jensen et al., 2010, 2012; Menon et al., 2014; Pulliam et al., 2007; Shim et al., 2007; Stocke et al., 2015; Sung et al., 2007; Tomoda et al., 2008; Ungaro et al., 2012; Zhang et al., 2011).

9.4.3.3   Influence of Excipients

Carriers (excipients) are inactive materials that are applied in spray drying to enhance the physical or chemical stability of the active pharmaceutical ingredient. Sugars are widely used as excipients of nCmP, as they provide advantages such as rapid dissolution in aqueous environment, leading to the immediate release of encapsulated nanoparticles. These excipients have been well studied in aerosol manufacturing processes with highly reproducible products (Pilcer and Amighi, 2010; Smyth and Hickey, 2005; Telko and Hickey, 2005). Safety should be a primary consideration in choosing the carrier excipient for nCmP. Many excipients that can be used in drug formulations for delivery in nonpulmonary routes may not suitable for pulmonary delivery due to their potential to injure the lungs (Telko and Hickey, 2005). Excipients approved for pulmonary delivery or presented as interesting alternatives for DPI formulations should be considered in high priority. Carriers also have significant impact on properties of nCmP, including the size, surface morphology, water content, and redispersion potential of nanoparticles. Comparisons of the influence of different excipients on dry powder formulation that can provide insight into how the excipient properties will affect nCmP have been studied (Bosquillon et al., 2001; Minne et al., 2008; Pilcer and Amighi, 2010).

9.4.3.4   Summary

Several parameters in the spray drying process can be tuned to achieve desirable properties of nCmP. These parameters, which correlate with and depend on each other, should be considered comprehensively to design optimized particles. Some parameters may have variable effects on the nCmP systems, and not all will be regarded as favorable. In this case, some of the properties may be abandoned in an overall consideration.

9.4.4 OPTIMIZATION OF NCMP FORMATION PROCESS

Pulmonary delivery of therapeutics using nCmP is increasingly recommended for the treatment of lung disease due to their high efficacy of delivery, more convenient administration, and more flexible storage conditions. These advantages are a result of effective particle engineering of the delivery system, which includes optimal aerodynamic size, stability, and recovery of primary nanoparticles. In this section, strategies to achieve these favorable properties are discussed.

9.4.4.1 Optimization of Aerodynamic Size and Morphology

A schematic showing the parameters involved in optimization of aerodynamic size of nCmP is shown in Figure 9.4. The proper aerodynamic diameter will be dependent on the delivery strategy and final particle location for a given application. Since higher temperatures may impair the redispersion of nanoparticles and bioactivity of encapsulated therapeutics, relatively lower temperature should be given priority if possible. A way to prepare nCmP with small aerodynamic sizing is to prepare particles with small geometric sizing, since these sizes are positively proportional to one another. This goal can be achieved by reducing the feed speed and feed solution concentration during spray drying. The former way results in smaller droplet size with lower dry material content, while the latter one decreases the dry materials in droplet directly, thus leading to smaller particles. nCmP can be designed with rough surface morphology and low water content to allow for high aerosol dispersion and low cohesion of the resulting nCmP. Overall, the aerodynamic diameter can be influenced by the geometric diameter, structure, processing conditions, and materials used in the preparation of nCmP.

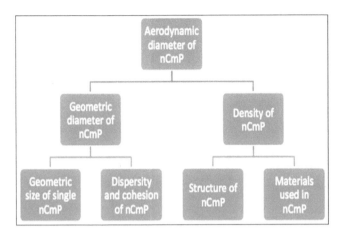

FIGURE 9.4   Schematic of processing parameters affecting the aerodynamic diameter of nanocomposite microparticles (nCmP).

9.4.4.2   Optimization of Stability

Figure 9.5 is a schematic outlining the factors important in the stability of nCmP. The stability of pulmonary delivery systems-based on nCmP includes not only activity of encapsulated therapeutics, but also the integrated structure of the delivery system. The drying temperature should be set to maintain the activity of therapeutic(s). Minimum water content should be a target to enhance stability of the structure of nCmP so that the favorable amorphous state of the nCmP is maintained.

9.4.4.3   Optimization of Primary Nanoparticle Recovery

A schematic outlining the optimization of nanoparticle recovery from nCmP is shown in Figure 9.6. To combine the advantages of both nanoparticle and microparticle systems, recovery of primary nanoparticles plays a significant role. An increase in the amount of nanoparticles released from nCmP can be achieved by increasing the ratio of nanoparticles to carrier (excipient). Meanwhile, the high redispersion of nanoparticles can be achieved by applying a suitable coating on the nanoparticles and by using a lower drying temperature. At last, improvement of overall nCmP yield may be regarded as an indirect way to obtain higher primary nanoparticle recovery.

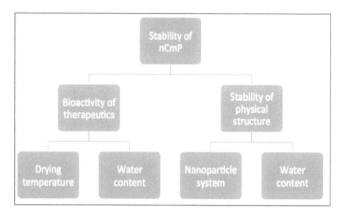

FIGURE 9.5   Schematic outlining the parameters that can influence the stability of nCmP.

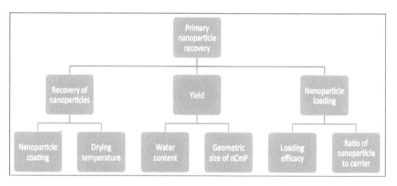

FIGURE 9.6   Schematic of the parameters that affect the recovery of redispersed nanoparticles from nCmP.

9.5   DRUG RELEASE BEHAVIOR FROM NANOCOMPOSITE MICROPARTICLE (NCMP) SYSTEMS

As with many other drug delivery systems, nCmP is comprised of mainly carriers and API. The carrier often serves as the most important component for successful and targeted drug delivery. Significant amounts of research has been underway to explore the potential of nanoparticles as drug delivery carriers as they offer several advantages including enhanced cellular uptake, increased drug stability, decreased side-effects and prolonged drug release (Mansour et al., 2009). Drugs encapsulated in a carrier are released from the nanoparticles by various physicochemical processes such as diffusion, dissolution, osmosis, magnetically controlled and targeted delivery, and the mechanism of drug release depends on the selection of carrier system and the solubility of the drug. The general physicochemical process for drug release from NP in nCmP are discussed as follows:

1. Diffusion: As a result of hydration, polymeric NP swell and release drugs through diffusion. The rate of drug release from the inner core of the water insoluble polymeric material follows Fick's law of diffusion (Gaede and Gawrisch, 2003; Guenneau and Puvirajesinghe, 2013; Wolfe et al., 1998). The diffusion of drug from the nanoparticle depends on the type of polymer, thickness of polymer coating, solution-diffusion membrane, and rate of permeation.

2. Dissolution: NP dissolution is a process that involves three steps: (1) release of drugs from the surface, (2) slight degradation of the polymer entrapping the remaining drugs, and (3) complete disintegration of polymer membrane, resulting in the rapid release of the entrapped drug. The degradation of the polymer membrane can be achieved using chemical or enzymatic reaction (Hillery, 2002).

3. Solvent controlled (osmosis): When semipermeable polymers are used to encapsulate drug molecules, water crosses the polymer membrane and dissolves the encapsulated drug since the osmotic pressure builds up in the NP interior and drugs are forced to dissociate from the polymer and release outside of the particles (Hillery, 2002).

4. Magnetically controlled: If a component of the nCmP is comprised of superparamagnetic nanoparticles (Fe_3O_4), a high frequency alternating magnetic field can be used to release the encapsulated drugs. Stocke et al., demonstrated the incorporation of iron oxide magnetic nanoparticles in magnetic nCmP (MnMs), which could enhance drug release with the presence of an alternating magnetic field (Stocke et al., 2015).

9.6 THERAPEUTIC APPLICATIONS OF NANOCOMPOSITE MICROPARTICLES: IMPACT IN PULMONARY AND NONPULMONARY DISEASES

Over the past few years, research in the field of pulmonary drug delivery has gained momentum for treatment of pulmonary and nonpulmonary diseases. Pulmonary route for drug delivery is an attractive target, yet it poses serious challenges for investigation. Particles are being engineered in such a way to overcome the lung clearance mechanism, targeting specific regions of the lungs and being retained in the lung for longer time period. The nCmP formulations form an attractive interdisciplinary area that brings together polymer/material science, nanotechnology and biology for a multitude of biomedical applications. These are 'intelligent' systems widely used to deliver small molecules, genes, protein/peptides or drug nanoparticles to various targeted locations in the body. nCmP help in protecting them from degradation during delivery, targeting them to specific sites and regulating their release properties with respect to time. There are various nCmP that have been used to deliver therapeutic agents to the lungs (pulmonary drug delivery) and other body parts (nonpulmonary drug delivery) with their key findings discussed in Tables 9.1 and 9.2.

KEYWORDS

- aerosols
- drug delivery
- nanocomposite microparticles
- pharmaceutical formulations
- pulmonary
- spray drying

REFERENCES

Al-Hallak, M. H.; Sarfraz, M. K.; Azarmi, S.; Roa, W. H.; Finlay, W. H.; Lobenberg, R. Pulmonary delivery of inhalable nanoparticles: dry powder inhalers. *Ther. Deliv.* **2011**, *2*(10), 1313–1324.

Ameri, M.; Maa, Y. F. Spray Drying of Biopharmaceuticals: Stability and Process Considerations. *Drying Technol.* **2006**, *24*(6), 763–768.

Antunes, M. B.; Cohen, N. A. Mucociliary clearance–a critical upper airway host defense mechanism and methods of assessment. *Curr Opin Allergy Clin Immunol.* **2007**, *7*(1), 5–10.

Aquino, R. P.; Prota, L.; Auriemma, G.; Santoro, A.; Mencherini, T.; Colombo, G.; Russo, P. Dry powder inhalers of gentamicin and leucine: formulation parameters, aerosol performance and in vitro toxicity on CuFi1 cells. *Int J Pharm.* **2012**, *426*(1–2), 100–107.

Azarmi, S.; Roa, W. H.; Löbenberg, R. Targeted delivery of nanoparticles for the treatment of lung diseases. *Adv Drug Deliv Rev.* **2008**, *60*(8), 863–875.

Azarmi, S.; Tao, X.; Chen, H.; Wang, Z.; Finlay, W. H.; Löbenberg, R.; Roa, W. H. Formulation and cytotoxicity of doxorubicin nanoparticles carried by dry powder aerosol particles. *Int J Pharm.* **2006**, *319*(1), 155–161.

Bailey, M. M.; Berkland, C. J. Nanoparticle formulations in pulmonary drug delivery. *Med Res Rev.* **2009**, *29*(1), 196–212.

Baldrick, P. Pharmaceutical excipient development: the need for preclinical guidance. *Regul Toxicol Pharmacol.* **2000**, *32*(2), 210–218.

Baras, B.; Benoit, M. A.; Gillard, J. Parameters influencing the antigen release from spray-dried poly(dl-lactide) microparticles. *Int J Pharm.* **2000**, *200*(1), 133–145.

Bateman, E. D.; Chapman, K. R.; Singh, D.; D'Urzo, A. D.; Molins, E.; Leselbaum, A.; Gil, E. G. Aclidinium bromide and formoterol fumarate as a fixed-dose combination in COPD: pooled analysis of symptoms and exacerbations from two six-month, multicenter, randomized studies (ACLIFORM and AUGMENT). *Respir Res.* **2015**, *16*, 92.

Beck-Broichsitter, M.; Merkel, O. M.; Kissel, T. Controlled pulmonary drug and gene delivery using polymeric nano-carriers. *J Control Release.* **2012**, *161*(2), 214–224.

Beck-Broichsitter, M.; Schweiger, C.; Schmehl, T.; Gessler, T.; Seeger, W.; Kissel, T. Characterization of novel spray-dried polymeric particles for controlled pulmonary drug delivery. *J Control Release.* **2012**, *158*(2), 329–335.

Bell, J. H.; Hartley, P. S.; Cox, J. S. Dry powder aerosols. I. A new powder inhalation device. *J Pharm Sci.* **1971**, *60*(10), 1559–1564.

Bosquillon, C.; Lombry, C.; Preat, V.; Vanbever, R. Influence of formulation excipients and physical characteristics of inhalation dry powders on their aerosolization performance. *J Control Release.* **2001**, *70*(3), 329–339.

Burger, A.; Henck, J. O.; Hetz, S.; Rollinger, J. M.; Weissnicht, A. A.; Stöttner, H. Energy/temperature diagram and compression behavior of the polymorphs of D-mannitol. *J Pharm Sci.* **2000**, *89*(4), 457–468.

Calvo, P.; Remuñán-López, C.; Vila-Jato, J. L.; Alonso, M. J. Novel hydrophilic chitosan-polyethylene oxide nanoparticles as protein carriers. *Journal of Applied Polymer Science.* **1997**, *63*(1), 125–132.

Chacón, M.; Berges, L.; Molpeceres, J.; Aberturas, M. R.; Guzman, M. Optimized preparation of poly d, l (lactic-glycolic) microspheres and nanoparticles for oral administration. *Int J Pharm.* **1996**, *141*(1–2), 81–91.

Chen, J. M.; Tan, M. G.; Nemmar, A.; Song, W. M.; Dong, M.; Zhang, G. L.; Li, Y. Quantification of extrapulmonary translocation of intratracheal-instilled particles in vivo in rats: Effect of lipopolysaccharide. *Toxicology.* **2006**, *222*(3), 195–201.

Cheow, W. S.; Chang, M. W.; Hadinoto, K. The roles of lipid in antibiofilm efficacy of lipid–polymer hybrid nanoparticles encapsulating antibiotics. *Colloids Surf A Physicochem Eng Asp*. **2011**, *389*(1–3), 158–165.

Chew, N. Y.; Chan, H. K. Use of solid corrugated particles to enhance powder aerosol performance. *Pharm Res*. **2001**, *18*(11), 1570–1577.

Coates, A. L.; O'Callaghan, C. *Drug administration by aerosol in children*. Saunders-Elsevier: Philadelphia, PA, 2006.

Codrons, V.; Vanderbist, F.; Verbeeck, R. K.; Arras, M.; Lison, D.; Preat, V.; Vanbever, R. Systemic delivery of parathyroid hormone (1–34) using inhalation dry powders in rats. *J Pharm Sci*. **2003**, *92*(5), 938–950.

Cook, R. O.; Pannu, R. K.; Kellaway, I. W. Novel sustained release microspheres for pulmonary drug delivery. *J Control Release*. **2005**, *104*(1), 79–90.

Corkery, K. Inhalable drugs for systemic therapy. *Respir Care*. **2000**, *45*(7), 831–835.

Cummings, R. H. Pressurized metered dose inhalers: chlorofluorocarbon to hydrofluoroalkane transition-valve performance. *J Allergy Clin Immunol*. **1999**, *104*(6), S230–236.

Das, D.; Wang, E.; Langrish, T. A. G. Solid-phase crystallization of spray-dried glucose powders: A perspective and comparison with lactose and sucrose. *Adv Powder Technol*. **2014**, *25*(4), 1234–1239.

De Jong, W. H.; Borm, P. J. A. Drug delivery and nanoparticles: Applications and hazards. *Int J Nanomedicine* **2008**, *3*(2), 133–149.

Donohue, J. F.; Maleki-Yazdi, M. R.; Kilbride, S.; Mehta, R.; Kalberg, C.; Church, A. Efficacy and safety of once-daily umeclidinium/vilanterol 62.5/25 mcg in COPD. *Respr Med*. **2013**, *107*(10), 1538–1546.

Duret, C.; Wauthoz, N.; Sebti, T.; Vanderbist, F.; Amighi, K. New inhalation-optimized itraconazole nanoparticle-based dry powders for the treatment of invasive pulmonary aspergillosis. *Int J Nanomedicine*. **2012**, *7*, 5475–5489.

Edwards, D. A.; Ben-Jebria, A.; Langer, R. Recent advances in pulmonary drug delivery using large, porous inhaled particles. *J Appl Physiol*. **1998**, *85*(2), 379–385.

Ely, L.; Roa, W.; Finlay, W. H.; Löbenberg, R. Effervescent dry powder for respiratory drug delivery. *Eur J Pharm Biopharm*. **2007**, *65*(3), 346–353.

Fessi, H.; Puisieux, F.; Devissaguet, J. P.; Ammoury, N.; Benita, S. Nanocapsule formation by interfacial polymer deposition following solvent displacement. *Int J Pharm*. **1989**, *55*(1), R1–R4.

Fong, S. Y. K.; Ibisogly, A.; Bauer-Brandl, A. Solubility enhancement of BCS Class II drug by solid phospholipid dispersions: Spray drying versus freeze-drying. *Int J Pharm*. **2015**, *496*(2), 382-391.

Frampton, J. E. QVA149(indacaterol/glycopyrronium fixed-dose combination): a review of its use in patients with chronic obstructive pulmonary disease. *Drugs*. **2014**, *74*(4), 465–488.

Fu, F. L.; Mi, T. B.; Wong, S. S.; Shyu, Y. J. Characteristic and controlled release of anti-cancer drug loaded poly (D, L-lactide) microparticles prepared by spray drying technique. *J Microencapsul*. **2001**, *18*(6), 733–747.

Gaede, H. C.; Gawrisch, K. Lateral diffusion rates of lipid, water, and a hydrophobic drug in a multilamellar liposome. *Biophys J*. **2003**, *85*(3), 1734–1740.

Glover, W.; Chan, H. K.; Eberl, S.; Daviskas, E.; Verschuer, J. Effect of particle size of dry powder mannitol on the lung deposition in healthy volunteers. *Int J Pharm.* **2008**, *349*(1–2), 314–322.

Grenha, A.; Grainger, C. I.; Dailey, L. A.; Seijo, B.; Martin, G. P.; Remuñán-López, C.; Forbes, B. Chitosan nanoparticles are compatible with respiratory epithelial cells in vitro. *J Pharm SciEur J Pharm SciEur J Pharm Sci.* **2007**, *31*(2), 73–84.

Grenha, A.; Seijo, B.; Remuñán-López, C. Microencapsulated chitosan nanoparticles for lung protein delivery. *J Pharm SciEur J Pharm Sci.* **2005**, *25*(4–5), 427–437.

Guenneau, S.; Puvirajesinghe, T. M. Fick's second law transformed: one path to cloaking in mass diffusion. *J R Soc Interface.* **2013**, *10*(83), 20130106.

Hadinoto, K.; Phanapavudhikul, P.; Kewu, Z.; Tan, R. B. H. Dry powder aerosol delivery of large hollow nanoparticulate aggregates as prospective carriers of nanoparticulate drugs: Effects of phospholipids. *Int J Pharm.* **2007**, *333*(1–2), 187–198.

Hadinoto, K.; Phanapavudhikul, P.; Kewu, Z.; Tan, R. B. H. Novel Formulation of Large Hollow Nanoparticles Aggregates as Potential Carriers in Inhaled Delivery of Nanoparticulate Drugs. *Ind Eng Chem Res.* **2006**, *45*(10), 3697–3706.

Hadinoto, K.; Zhu, K.; Tan, R. B. H. Drug release study of large hollow nanoparticulate aggregates carrier particles for pulmonary delivery. *Int J Pharm.* **2007**, *341*(1–2), 195–206.

Her, J. Y.; Kim, M. S.; Lee, K. G. Preparation of probiotic powder by the spray freeze-drying method. *J Food Eng.* **2015**, *150*, 70–74.

Hess, D. R. Aerosol delivery devices in the treatment of asthma. *Respir Care.* **2008**, *53*(6), 699–723; discussion 723–695.

Hillery, A. M. *Advanced Drug Delivery and targeting: An Introduction.* CRC Press, 2002.

Hinds, W. C. *Aerosol Technology: Properties, Behavior, and Measurement of Airborne Particles* (2nd ed.). Wiley-InterScience, 1999.

Ho, T. M.; Howes, T.; Bhandari, B. R. Characterization of crystalline and spray-dried amorphous α-cyclodextrin powders. *Powder Technol.* **2015**, *284*, 585–594.

Irngartinger, M.; Camuglia, V.; Damm, M.; Goede, J.; Frijlink, H. W. Pulmonary delivery of therapeutic peptides via dry powder inhalation: effects of micronization and manufacturing. *Eur J Pharm Biopharm.* **2004**, *58*(1), 7–14.

Iskandar, F.; Chang, H. W.; Okuyama, K. Preparation of microencapsulated powders by an aerosol spray method and their optical properties. *Adv Powder Technol.* **2003**, *14*(3), 349–367.

Iskandar, F.; Gradon, L.; Okuyama, K. Control of the morphology of nanostructured particles prepared by the spray drying of a nanoparticle sol. *J Colloid Interface Sci.* **2003**, *265*(2), 296–303.

Islam, N.; Gladki, E. Dry powder inhalers (DPIs)–a review of device reliability and innovation. *Int J Pharm.* **2008**, *360*(1–2), 1–11.

Jafarinejad, S.; Gilani, K.; Moazeni, E.; Ghazi-Khansari, M.; Najafabadi, A. R.; Mohajel, N. Development of chitosan-based nanoparticles for pulmonary delivery of itraconazole as dry powder formulation. *Powder Technol.* **2012**, *222*, 65–70.

Jensen, D. K.; Jensen, L. B.; Koocheki, S.; Bengtson, L.; Cun, D.; Nielsen, H. M.; Foged, C. Design of an inhalable dry powder formulation of DOTAP-modified PLGA nanoparticles loaded with siRNA. *J Control Release.* **2012**, *157*(1), 141–148.

Jensen, D. M. K.; Cun, D.; Maltesen, M. J.; Frokjaer, S.; Nielsen, H. M.; Foged, C. Spray drying of siRNA-containing PLGA nanoparticles intended for inhalation. *J Control Release*. **2010**, *142*(1), 138–145.

Jung, T.; Kamm, W.; Breitenbach, A.; Kaiserling, E.; Xiao, J. X.; Kissel, T. Biodegradable nanoparticles for oral delivery of peptides: is there a role for polymers to affect mucosal uptake? *Eur J Pharm Biopharm*. **2000**, *50*(1), 147–160.

Karner, C.; Chong, J.; Poole, P. Tiotropium versus placebo for chronic obstructive pulmonary disease. *Cochrane Database Syst Rev*. **2014**, *7*, CD009285.

Kaur, P.; Singh, S. K.; Garg, V.; Gulati, M.; Vaidya, Y. Optimization of spray drying process for formulation of solid dispersion containing polypeptide-k powder through quality by design approach. *Powder Technol*. **2015**, *284*, 1–11.

Kim, E. H. J. (2008). *Surface Composition of Industrial Spray-Dried Dairy Powders and Its Formation Mechanisms*. (PhD), University of Auckland.

Kim, E. H. J.; Chen, X. D.; Pearce, D. On the Mechanisms of Surface Formation and the Surface Compositions of Industrial Milk Powders. *Drying Technol*. **2003**, *21*(2), 265–278.

Konstan, M. W.; Flume, P. A.; Kappler, M.; Chiron, R.; Higgins, M.; Brockhaus, F.; Zhang, J.; Angyalosi, G.; He, E.; Geller, D. E. Safety, efficacy and convenience of tobramycin inhalation powder in cystic fibrosis patients: The EAGER trial. *J Cyst Fibros*. **2011**, *10*(1), 54–61.

Konstan, M. W.; Geller, D. E.; Minic, P.; Brockhaus, F.; Zhang, J.; Angyalosi, G. Tobramycin Inhalation Powder for P. aeruginosa Infection in Cystic Fibrosis: The EVOLVE Trial. *Pediatr Pulmonol*. **2011**, *46*(3), 230–238.

Kunda, N. K.; Alfagih, I. M.; Dennison, S. R.; Somavarapu, S.; Merchant, Z.; Hutcheon, G. A.; Saleem, I. Y. Dry powder pulmonary delivery of cationic PGA-coPDL nanoparticles with surface adsorbed model protein. *Int J Pharm*. **2015a**, *492*(1–2), 213–222.

Kunda, N. K.; Alfagih, I. M.; Miyaji, E. N.; Figueiredo, D. B.; Goncalves, V. M.; Ferreira, D. M.; Dennison, S. R.; Somavarapu, S.; Hutcheon, G. A.; Saleem, I. Y. Pulmonary Dry Powder Vaccine of Pneumococcal Antigen Loaded Nanoparticles. *Int J Pharm*. **2015b**.

Lambert, G.; Fattal, E.; Couvreur, P. Nanoparticulate systems for the delivery of antisense oligonucleotides. *Adv. Drug Deliv. Rev*. **2001**, *47*(1), 99–112.

Laube, B. L.; Janssens, H. M.; de Jongh, F. H. C.; Devadason, S. G.; Dhand, R.; Diot, P.; Everard, M. L.; Horvath, I.; Navalesi, P.; Voshaar, T.; Chrystyn, H. What the pulmonary specialist should know about inhalation therapies. *Eur Respir J*. **2011**, *37*, 1308–1331.

Lee, Y. S.; Johnson, P. J.; Robbins, P. T.; Bridson, R. H. Production of nanoparticles-in-microparticles by a double emulsion method: A comprehensive study. *Eur J Pharm Biopharm*. **2013**, *83*(2), 168–173.

Li, H.-Y.; Birchall, J. Chitosan-Modified Dry Powder Formulations for Pulmonary Gene Delivery. *Pharm Res*. **2006**, *23*(5), 941–950.

Li, X.; Vogt, F. G.; Hayes, Jr., D.; Mansour, H. M. Physicochemical characterization and aerosol dispersion performance of organic solution advanced spray-dried microparticulate/nanoparticulate antibiotic dry powders of tobramycin and azithromycin for pulmonary inhalation aerosol delivery. *Eur J Pharm Sci*. **2014**, *52*, 191–205.

Li, Y.-Z.; Sun, X.; Gong, T.; Liu, J.; Zuo, J.; Zhang, Z.-R. Inhalable microparticles as carriers for pulmonary delivery of thymopentin-loaded solid lipid nanoparticles. *Pharm Res.* **2010**, *27*, 1977–1986.

Lippmann, M.; Yeates, D. B.; Albert, R. E. Deposition, retention, and clearance of inhaled particles. *Br J Ind Med.* **1980**, *37*(4), 337–362.

Littringer, E. M.; Mescher, A.; Eckhard, S.; Schröttner, H.; Langes, C.; Fries, M.; Griesser, U.; Walzel, P.; Urbanetz, N. A. Spray Drying of Mannitol as a Drug Carrier—The Impact of Process Parameters on Product Properties. *Drying Technol.* **2011**, *30*(1), 114–124.

Littringer, E. M.; Mescher, A.; Schroettner, H.; Achelis, L.; Walzel, P.; Urbanetz, N. A. Spray dried mannitol carrier particles with tailored surface properties – The influence of carrier surface roughness and shape. *Eur J Pharm Biopharm.* **2012**, *82*(1), 194–204.

Littringer, E. M.; Paus, R.; Mescher, A.; Schroettner, H.; Walzel, P.; Urbanetz, N. A. The morphology of spray dried mannitol particles — The vital importance of droplet size. *Powder Technol.* **2013**, *239*, 162–174.

Lubet, R. A.; Zhang, Z.; Wang, Y.; You, M. Chemoprevention of lung cancer in transgenic mice. *Chest.* **2004**, *125*(5 Suppl), 144S–147S.

Maa, Y. F.; Nguyen, P. A.; Andya, J. D.; Dasovich, N.; Sweeney, T. D.; Shire, S. J.; Hsu, C. C. Effect of spray drying and subsequent processing conditions on residual moisture content and physical/biochemical stability of protein inhalation powders. *Pharm Res.* **1998**, *15*(5), 768–775.

Maa, Y.-F.; Costantino, H. R.; Nguyen, P.-A.; Hsu, C. C. The Effect of Operating and Formulation Variables on the Morphology of Spray-Dried Protein Particles. *Pharm Dev Technol.* **1997**, *2*(3), 213–223.

Maa, Y.-F.; Nguyen, P.-A. T.; Hsu, S. W. Spray-drying of air–liquid interface sensitive recombinant human growth hormone. *J Pharm Sci.* **1998**, *87*(2), 152–159.

Maas, S. G.; Schaldach, G.; Littringer, E. M.; Mescher, A.; Griesser, U. J.; Braun, D. E.; Walzel, P. E.; Urbanetz, N. A. The impact of spray drying outlet temperature on the particle morphology of mannitol. *Powder Technol.* **2011**, *213*(1–3), 27–35.

Maas, S. G.; Schaldach, G.; Walzel, P.; Urbanetz, N. A. Tailoring dry powder inhaler performance by modifying carrier surface topography by spray drying. *Atomization and Sprays.* **2010**, *20*(9), 763–774.

Mahmud, A.; Discher, D. E. Lung Vascular Targeting Through Inhalation Delivery: Insight from Filamentous Viruses and Other Shapes. *Iubmb Life.* **2011**, *63*(8), 607–612.

Mannix, K. C.; Meir, R. (2015). Teva Announces FDA Approval of ProAir® RespiClick. http://www.tevapharm.com/news/teva_announces_fda_approval_of_proair_respiclick_04_15.aspx

Mansour, H. M.; Rhee, Y. S.; Park, C. W.; DeLuca, P. P. *Lipid Nanoparticulate Drug Delivery and Nanomedicine.* In A. Moghis (Ed.), *Lipids in Nanotechnology* (pp. 221–268). American Oil Chemists Society (AOCS) Press: Urbana, Illinois, 2011.

Mansour, H. M.; Rhee, Y. S.; Wu, X. Nanomedicine in pulmonary delivery. *Int J Nanomedicine* **2009**, *4*, 299–319.

Martinez, F. J.; Boscia, J.; Feldman, G.; Scott-Wilson, C.; Kilbride, S.; Fabbri, L.; Crim, C.; Calverley, P. M. Fluticasone furoate/vilanterol (100/25; 200/25 mug) improves lung function in COPD: a randomized trial. *Respir Med.* **2013**, *107*(4), 550–559.

Maury, M.; Murphy, K.; Kumar, S.; Mauerer, A.; Lee, G. Spray-drying of proteins: effects of sorbitol and trehalose on aggregation and FTIR amide I spectrum of an immunoglobulin G. *Eur J Pharm Biopharm*. **2005**, *59*(2), 251–261.

Maury, M.; Murphy, K.; Kumar, S.; Shi, L.; Lee, G. Effects of process variables on the powder yield of spray-dried trehalose on a laboratory spray-dryer. *Eur J Pharm Biopharm*. **2005**, *59*(3), 565–573.

McIntosh, D. P.; Tan, X. Y.; Oh, P.; Schnitzer, J. E. Targeting endothelium and its dynamic caveolae for tissue-specific transcytosis in vivo: a pathway to overcome cell barriers to drug and gene delivery. *Proc Natl Acad Sci U S A*. **2002**, *99*(4), 1996–2001.

Menon, J. U.; Ravikumar, P.; Pise, A.; Gyawali, D.; Hsia, C. C. W.; Nguyen, K. T. Polymeric nanoparticles for pulmonary protein and DNA delivery. *Acta Biomaterialia*. **2014**, *10*(6), 2643–2652.

Miller, D. P.; Tan, T.; Tarara, T. E.; Nakamura, J.; Malcolmson, R. J.; Weers, J. G. Physical Characterization of Tobramycin Inhalation Powder: I. Rational Design of a Stable Engineered-Particle Formulation for Delivery to the Lungs. *Mol Pharm*. **2015**, *12*(8), 2582–2593.

Minne, A.; Boireau, H.; Horta, M. J.; Vanbever, R. Optimization of the aerosolization properties of an inhalation dry powder-based on selection of excipients. *Eur J Pharm Biopharm*. **2008**, *70*(3), 839–844.

Mizoe, T.; Ozeki, T.; Okada, H. Preparation of drug nanoparticle-containing microparticles using a 4-fluid nozzle spray drier for oral, pulmonary, and injection dosage forms. *J Control Release*. **2007**, *122*(1), 10–15.

Mu, L.; Feng, S. S. Fabrication, characterization and in vitro release of paclitaxel (Taxol®) loaded poly (lactic-coglycolic acid) microspheres prepared by spray drying technique with lipid/cholesterol emulsifiers. *J Control Release*. **2001**, *76*(3), 239–254.

Mudshinge, S. R.; Deore, A. B.; Patil, S.; Bhalgat, C. M. Nanoparticles: Emerging carriers for drug delivery. *Saudi Pharm J*. **2011**, *19*(3), 129–141.

Muller, R. H.; Mader, K.; Gohla, S. Solid lipid nanoparticles (SLN) for controlled drug delivery – a review of the state-of-the-art. *Eur J Pharm Biopharm*. **2000**, *50*(1), 161–177.

Muralidharan, P.; Malapit, M.; Mallory, E.; Hayes, Jr, D.; Mansour, H. M. Inhalable nanoparticulate powders for respiratory delivery. *Nanomedicine*. **2015**, *11*(5), 1189–1199.

Nagavarma, B. V. N. Y.; Hemant, K. S.; Ayaz, A.; Vasudha, L. S.; Shivakumar, H. G. Different Techniques for Preparation of Polymeric Nanoparticles: A Review. *Asian J Pharm Clin Res*. **2012**, 5.

Newman, S. P.; Busse, W. W. Evolution of dry powder inhaler design, formulation, and performance. *Respir Med*. **2002**, *96*(5), 293–304.

Ohashi, K.; Kabasawa, T.; Ozeki, T.; Okada, H. One-step preparation of rifampicin/ poly(lactic-coglycolic acid) nanoparticle-containing mannitol microspheres using a four-fluid nozzle spray drier for inhalation therapy of tuberculosis. *J Control Release*. **2009**, *135*(1), 19–24.

Park, C. W.; Li, X.; Vogt, F. G.; Hayes, D.; Jr.; Zwischenberger, J. B.; Park, E. S.; Mansour, H. M. Advanced spray-dried design, physicochemical characterization, and aerosol dispersion performance of vancomycin and clarithromycin multifunctional controlled release particles for targeted respiratory delivery as dry powder inhalation aerosols. *Int J Pharm.* **2013**, *455*(1–2), 374–392.

Parkins, M. D.; Elborn, J. S. Tobramycin Inhalation Powder: a novel drug delivery system for treating chronic Pseudomonas aeruginosa infection in cystic fibrosis. *Expert Rev Respir Med.* **2011**, *5*(5), 609–622.

Parveen, S.; Misra, R.; Sahoo, S. K. Nanoparticles: a boon to drug delivery, therapeutics, diagnostics and imaging. *Nanomedicine.* **2012**, *8*(2), 147–166.

Pavia, D. *Aerosols and the Lung: Clinical and Experimental Aspects.* In: S. W. Clarke and D. Pavia (Eds.), Butterworths, London, 1984, pp. 200–229.

Pilcer, G.; Amighi, K. Formulation strategy and use of excipients in pulmonary drug delivery. *Int J Pharm.* **2010**, *392*(1–2), 1–19.

Pilcer, G.; Vanderbist, F.; Amighi, K. Preparation and characterization of spray-dried tobramycin powders containing nanoparticles for pulmonary delivery. *Int J Pharm.* **2009**, *365*(1–2), 162–169.

Pourshahab, P. S.; Gilani, K.; Moazeni, E.; Eslahi, H.; Fazeli, M. R.; Jamalifar, H. Preparation and characterization of spray dried inhalable powders containing chitosan nanoparticles for pulmonary delivery of isoniazid. *J Microencapsul.* **2011**, *28*(7), 605–613.

Pulliam, B.; Sung, J. C.; Edwards, D. A. Design of nanoparticle-based dry powder pulmonary vaccines. *Expert Opin Drug Deliv.* **2007**, *4*(6), 651–663.

Raula, J.; Eerikäinen, H.; Kauppinen, E. I. Influence of the solvent composition on the aerosol synthesis of pharmaceutical polymer nanoparticles. *Int J Pharm.* **2004**, *284*(1–2), 13–21.

Razavi Rohani, S. S.; Abnous, K.; Tafaghodi, M. Preparation and characterization of spray-dried powders intended for pulmonary delivery of Insulin with regard to the selection of excipients. *Int J Pharm.* **2014**, *465*(1–2), 464–478.

Roy, D.; Guillon, X.; Lescure, F.; Couvreur, P.; Bru, N.; Breton, P. On shelf stability of freeze-dried poly(methylidene malonate 2.1.2) nanoparticles. *Int J Pharm.* **1997**, *148*(2), 165–175.

Sanchis, J.; Corrigan, C.; Levy, M. L.; Viejo, J. L.; Group, A. Inhaler devices – from theory to practice. *Respir Med.* **2013**, *107*(4), 495–502.

Schafroth, N.; Arpagaus, C.; Jadhav, U. Y.; Makne, S.; Douroumis, D. Nano and microparticle engineering of water insoluble drugs using a novel spray-drying process. *Colloids Surf., B.* **2012**, *90*, 8–15.

Sham, J. O. H.; Zhang, Y.; Finlay, W. H.; Roa, W. H.; Löbenberg, R. Formulation and characterization of spray-dried powders containing nanoparticles for aerosol delivery to the lung. *Int J Pharm.* **2004**, *269*(2), 457–467.

Shi, X.; Zhong, Q. Crystallinity and quality of spray-dried lactose powder improved by soluble soybean polysaccharide. LWT-Food Sci Technol. **2015**, *62*(1, Part 1), 89–96.

Shim, W. S.; Kim, J.-H.; Kim, K.; Kim, Y.-S.; Park, R.-W.; Kim, I.-S.; Kwon, I. C.; Lee, D. S. pH- and tempearture-sensitive, injectable, biodegradable block copolymer hydrogels as carriers for paclitaxel. *Int J Pharm.* **2007**, *331*, 11–18.

Shoyele, S. A.; Cawthorne, S. Particle engineering techniques for inhaled biopharmaceuticals. *Adv Drug Deliv Rev.* **2006**, *58*(9–10), 1009–1029.

Skubitz, K. M.; Anderson, P. M. Inhalational interleukin-2 liposomes for pulmonary metastases: a phase I clinical trial. *Anticancer Drugs.* **2000**, *11*(7), 555–563.

Smyth, A. R.; Bell, S. C.; Bojcin, S.; Bryon, M.; Duff, A.; Flume, P.; Kashirskaya, N.; Munck, A.; Ratjen, F.; Schwarzenberg, S. J.; Sermet-Gaudelus, I.; Southern, K. W.; Taccetti, G.; Ullrich, G.; Wolfe, S.; European Cystic Fibrosis, S. European Cystic Fibrosis Society Standards of Care: Best Practice guidelines. *J Cyst Fibros*. **2014**, *13* (Suppl 1), S23–42.

Smyth, H. C.; Hickey, A. J. Carriers in drug powder delivery. *AJADD*. **2005**, *3*(2), 117–132.

Ståhl, K.; Claesson, M.; Lilliehorn, P.; Lindén, H.; Bäckström, K. The effect of process variables on the degradation and physical properties of spray dried insulin intended for inhalation. *Int J Pharm*. **2002**, *233*(1–2), 227–237.

Stocke, N. A.; Meenach, S. A.; Arnold, S. M.; Mansour, H. M.; Hilt, J. Z. Formulation and characterization of inhalable magnetic nanocomposite microparticles (MnMs) for targeted pulmonary delivery via spray drying. *Int J Pharm*. **2015**, *479*(2), 320–328.

Suarez, S.; Hickey, A. J. Drug properties affecting aerosol behavior. *Respir Care*. **2000**, *45*(6), 652–666.

Sun, T.; Zhang, Y. S.; Pang, B.; Hyun, D. C.; Yang, M.; Xia, Y. Engineered Nanoparticles for Drug Delivery in Cancer Therapy. *Adv Mater*. **2014**, *53*(46), 12320–12364.

Sung, J. C.; Padilla, D. J.; Garcia-Contreras, L.; Verberkmoes, J. L.; Durbin, D.; Peloquin, C. A.; Elbert, K. J.; Hickey, A. J.; Edwards, D. A. Formulation and pharmacokinetics of self-assembled rifampicin nanoparticle systems for pulmonary delivery. *Pharm Res*. **2009**, *26*(8), 1847–1855.

Sung, J. C.; Pulliam, B. L.; Edwards, D. A. Nanoparticles for drug delivery to the lungs. *Trends Biotechnol*. **2007**, *25*(12), 563–570.

Takashima, Y.; Saito, R.; Nakajima, A.; Oda, M.; Kimura, A.; Kanazawa, T.; Okada, H. Spray-drying preparation of microparticles containing cationic PLGA nanospheres as gene carriers for avoiding aggregation of nanospheres. *Int J Pharm*. **2007**, *343*(1–2), 262–269.

Tang, B. C.; Dawson, M.; Lai, S. K.; Wang, Y. Y.; Suk, J. S.; Yang, M.; Zeitlin, P.; Boyle, M. P.; Fu, J.; Hanes, J. Biodegradable polymer nanoparticles that rapidly penetrate the human mucus barrier. *Proc Natl Acad Sci U S A*. **2009**, *106*(46), 19268–19273.

Telko, M. J.; Hickey, A. J. Dry Powder Inhaler Formulation. *Respir Care*. **2005**, *50*(9), 1209–1227.

Tewa-Tagne, P.; Briançon, S.; Fessi, H. Preparation of redispersible dry nanocapsules by means of spray-drying: Development and characterization. *J Pharm SciEur J Pharm Sci*. **2007**, *30*(2), 124–135.

Ting, T.-Y.; Gonda, I.; Gipps, E. Microparticles of Polyvinyl Alcohol for Nasal Delivery. I. Generation by Spray-Drying and Spray-Desolvation. *Pharm Res*. **1992**, *9*(10), 1330–1335.

Tomoda, K.; Ohkoshi, T.; Kawai, Y.; Nishiwaki, M.; Nakajima, T.; Makino, K. Preparation and properties of inhalable nanocomposite particles: Effects of the temperature at a spray-dryer inlet upon the properties of particles. *Colloids Surf., B*. **2008**, *61*(2), 138–144.

Tsapis, N.; Bennett, D.; Jackson, B.; Weitz, D. A.; Edwards, D. A. Trojan particles: Large porous carriers of nanoparticles for drug delivery. *Proc Natl Acad Sci U S A*. **2002**, *99*(19), 12001–12005.

Ungaro, F.; d'Angelo, I.; Coletta, C.; d'Emmanuele di Villa Bianca, R.; Sorrentino, R.; Perfetto, B.; Tufano, M. A.; Miro, A.; La Rotonda, M. I.; Quaglia, F. Dry powders-based on PLGA nanoparticles for pulmonary delivery of antibiotics:

Modulation of encapsulation efficiency, release rate and lung deposition pattern by hydrophilic polymers. *J Control Release*. **2012**, *157*(1), 149–159.

Ungaro, F.; d'Angelo, I.; Miro, A.; La Rotonda, M. I.; Quaglia, F. Engineered PLGA nano- and microcarriers for pulmonary delivery: challenges and promises. *J Pharm Pharmacol*. **2012**, *64*(9), 1217–1235.

Vanbever, R.; Mintzes, J. D.; Wang, J.; Nice, J.; Chen, D.; Batycky, R.; Langer, R.; Edwards, D. A. Formulation and physical characterization of large porous particles for inhalation. *Pharm Res*. **1999**, *16*(11), 1735–1742.

Varshosaz, J.; Hassanzadeh, F.; Mardani, A.; Rostami, M. Feasibility of haloperidol-anchored albumin nanoparticles loaded with doxorubicin as dry powder inhaler for pulmonary delivery. *Pharm Dev Technol*. **2015**, *20*(2), 183–196.

Vaswani, S. K.; Creticos, P. S. Metered dose inhaler: past, present, and future. *Ann Allergy Asthma Immunol*. **1998**, *80*(1), 11–19; quiz 19–20.

Vehring, R. Pharmaceutical particle engineering via spray drying. *Pharm Res*. **2008**, *25*(5), 999–1022.

Vehring, R.; Foss, W. R.; Lechuga-Ballesteros, D. Particle formation in spray drying. *J Aerosol Sci*. **2007**, *38*(7), 728–746.

Verkman, A. S.; Song, Y. L.; Thiagarajah, J. R. Role of airway surface liquid and submucosal glands in cystic fibrosis lung disease. *Am J Physiol Cell Physiol*. **2003**, *284*(1), C2-C15.

Wang, Q.; Zhang, J.; Wang, A. Freeze-drying: A versatile method to overcome reaggregation and improve dispersion stability of palygorskite for sustained release of ofloxacin. *Appl Clay Sci*. **2014**, *87*, 7–13.

Wilczewska, A. Z.; Niemirowicz, K.; Markiewicz, K. H.; Car, H. Nanoparticles as drug delivery systems. *Pharmacol Rep*. **2012**, *64*(5), 1020–1037.

Willis, L.; Hayes, D.; Mansour, H. M. Therapeutic Liposomal Dry Powder Inhalation Aerosols for Targeted Lung Delivery. *Lung*. **2012**, *190*(3), 251–262.

Wolfe, C. A.; James, P. S.; Mackie, A. R.; Ladha, S.; Jones, R. Regionalized lipid diffusion in the plasma membrane of mammalian spermatozoa. *Biol Reprod*. **1998**, *59*(6), 1506–1514.

Wu, L.; Miao, X.; Shan, Z.; Huang, Y.; Li, L.; Pan, X.; Yao, Q.; Li, G.; Wu, C. Studies on the spray dried lactose as carrier for dry powder inhalation. *Asian J Pharm Sci*. **2014**, *9*(6), 336–341.

Wu, X.; Hayes, D.; Jr.; Zwischenberger, J. B.; Kuhn, R. J.; Mansour, H. M. Design and physicochemical characterization of advanced spray-dried tacrolimus multifunctional particles for inhalation. *Drug Des Devel Ther*. **2013**, *7*, 59–72.

Wylam, M. E.; Ten, R.; Prakash, U. B.; Nadrous, H. F.; Clawson, M. L.; Anderson, P. M. Aerosol granulocyte-macrophage colony-stimulating factor for pulmonary alveolar proteinosis. *Eur Respir J*. **2006**, *27*(3), 585–593.

Xu, Z.; Mansour, H. M.; Hickey, A. J. Particle Interations in Dry Powder Inhaler Unit Processes. *J Adhes Sci Technol*. **2011**, *25*(4/5), 451–482.

Yacobi, N. R.; Malmstadt, N.; Fazllollahi, F.; DeMaio, L.; Marchelletta, R.; Hamm-Alvarez, S. F.; Borok, Z.; Kim, K.-J.; Crandall, E. D. Mechanisms of Alveolar Epithelial Translocation of a Defined Population of Nanoparticles. *American Journal of Respiratory Cell and Molecular Biology*. **2010**, *42*(5), 604–614.

Yang, M. S.; Yamamoto, H.; Kurashima, H.; Takeuchi, H.; Yokoyama, T.; Tsujimoto, H.; Kawashima, Y. Design and evaluation of inhalable chitosan-modified poly

(DL-lactic-coglycolic acid) nanocomposite particles. *Eur J Pharm Sci.* **2012**, *47*(1), 235–243.

York, P. Strategies for particle design using supercritical fluid technologies. *Pharmaceutical Science and Technology Today.* **1999**, *2*(11), 430–440.

Zhang, J.; Wu, L.; Chan, H.-K.; Watanabe, W. Formation, characterization, and fate of inhaled drug nanoparticles. *Adv Drug Deliv Rev.* **2011**, *63*(6), 441–455.

Ziffels, S.; Bemelmans, N. L.; Durham, P. G.; Hickey, A. J. In vitro dry powder inhaler formulation performance considerations. *J Control Release.* **2015**, *199*, 45–52.

Zijlstra, G. S.; Hinrichs, W. L. J.; Boer, A. H. D.; Frijlink, H. W. The role of particle engineering in relation to formulation and de-agglomeration principle in the development of a dry powder formulation for inhalation of cetrorelix. *Eur J Pharm Sci.* **2004**, *23*(2), 139–149.

SOLID LIPID NANOPARTICLES FOR TOPICAL DRUG DELIVERY

SONIA TROMBINO* and ROBERTA CASSANO

Department of Pharmacy, Health and Nutritional Sciences,
University of Calabria, 87036 Arcavacata di Rende, Cosenza, Italy,
*E-mail: sonia.trombino@unical.it

CONTENTS

ABSTRACT

One of the main purposes of pharmaceutical field is to obtain an effective
and target specific drug delivery. Usually the drugs are administered
orally and the route of administration is convenient, but it is not always
the best choice. For example, it is not suggested for sensitive drugs due
to gastrointestinal or liver metabolism. In some cases, topical drug

administration, could be a viable alternative to oral route. On the other hand, the protective function of human skin involves physicochemical limitations depending on the type of permeant that can traverse the barrier. A drug that needs to be delivered through the skin, must have certain requirements, such as adequate lipophilicity and also a molecular weight <500 Da. In the last years various strategies are emerging to optimize the delivery across the skin. In particular, formulations that enhance the penetration of drugs through the stratum corneum, are essential to improve their efficacy as topical agents. In this context the solid lipid nanoparticles (SLNs) are attracting great attention. The aim of this chapter is, precisely, highlight the role of SLNs as topical drug carriers of antimicotic, antinflammatory, antibiotic and antioxidant molecules.

10.1 INTRODUCTION

The choice of a route of drug administration is very important, because it can affect the entire pharmacokinetic process of the molecule introduced in the organism. Usually, the drugs are administered orally as this route is convenient (Fasinu, 2011), but it is not always the best choice, for example it is not suggested in the case of drugs sensitive to gastrointestinal or liver metabolism (Pang, 2003). Often the topical drug administration, could be a viable alternative to oral route (Chiu and Tsai, 2011; Subedial, 2010). In fact, the use of topical drugs relies on effective drug formulations, which maximize the drug concentrations in the target tissue. Some compounds must to stay on the skin surface, e.g., sunscreens or barrier creams (Hayden et al., 2005; Hayden et al., 1997; Zhai and Maibach, 1996), while others are intended to interact with the stratum corneum (SC), for example skin moisturizers. Other compounds should reach the viable epidermis, or dermis, like local analgesics, antifungal agents and drugs used in photodynamic therapy (Sawynok, 2003; Hadgraft, 2001). On the other hand we must consider that the protective function of human skin involves physicochemical limitations due to the type of substance that can traverse the barrier. A drug that needs to be delivered through the skin, it must have certain requirements, such as adequate lipophilicity and also a molecular weight <500 Da (Bhowmik and Kumar, 2013). In the last years various strategies are emerging to optimize the drug delivery across the skin. Formulations that improve the penetration of drugs through the stratum corneum, are essential to improve their efficacy as topical agents. Colloidal carriers are one of the approaches for controlled release of active substances and targeting to skin layers. Among these, the solid lipid nanoparticles (SLNs), are attracting much attention, in particular, for topical use (Figure 10.1) (Pople and Kamalinder, 2006). These spherical particles in the nanometer range, developed at the beginning of 1990s (Müller and Lucks, 1996), are composed of physiological and biodegradable lipids of low systemic toxicity and also low cytotoxicity, and are usually obtained by high pressure homogenization or microemulsion methods (Müller et al.,1997). Most of the used lipids are approved or are excipients used in commercially available topical cosmetic or pharmaceutical preparations (Souto et al., 2004). The SLNs represent an alternative drug delivery systems to colloidal carriers such as lipid emulsions, liposomes and polymeric nanoparticles because they combine the advantages of these colloidal drug delivery systems, avoiding some of their disadvantages and are very helpful, in particular, for the encapsulation of drugs with poor water solubility (Gasco, 1993; Müller, 2007; Trombino et al., 2009).

The SLNs are interesting, in particular for topical applications, both for the possibility to incorporate therein active ingredients chemically and physically unstable, protecting them against degradation (Dingler et al., 1999; Jenning and Gohla, 2001), both for the occlusive action that they exert on the skin (Figure 10.2). This last effect leads to an increase of hydration and improve the penetration of compounds in specific skin layers (Uner, 2006).

In addition it is the solid matrix-lipid which promotes a controlled release of the drug from SLNs. This is a very important aspect especially when it is necessary to administer the drug for a prolonged period of time, to reduce the systemic absorption, and when the drug is irritant at high concentrations (Souto et al., 2004). The present chapter intends to review the role of

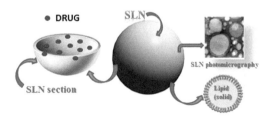

FIGURE 10.1   Solid lipid nanoparticles structure.

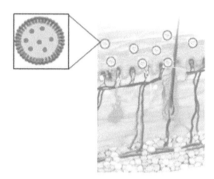

FIGURE 10.2   Interaction of SLNs with the skin.

the solid lipid nanoparticles as delivery systems for topical applications of antimicotic, antinflammatory, antibiotic and antioxidant drugs.

10.2   SLNs FOR TOPICAL DELIVERY OF ANTIFUNGAL DRUGS

The skin infections due to cutaneous mycoses, are triggered by the action of pathogenic fungi. It is fundamental to find an appropriate, adequate and rapid therapy. The oral route is reserved only to difficult cases, thus antimicotic creams-based on clotrimazole (CLT), miconazole (MN), etc. are often used. The antifungal administration of formulation-based on through SLNs could serve to obtain a better drug efficacy.

Therefore, Souto and Müller (2007) developed SLNs for the encapsulation of CLT, an imidazolic antifungal agent used in several dermatological creams, effective for the local treatment of cutaneous and mucosal infections. The aim of their study was to increase its dermal bioavailability and to control drug release, thereby potentially reducing its side effects. In particular Souto and Müller (2007), evaluated both the clotrimazole release, through Franz diffusion cells and both the stability of the lipid carriers after 3 months of storage. The entrapment efficiency and the drug release profile depended on the concentration and the lipid mixture employed: glyceryl tripalmitate (Dynasan®116). The release rate decreased for SLNs with a higher lipid concentration. The stability studies indicated that after 3 months of storage, at different temperatures, the mean diameter of SLNs was practically the same (<1 μm), which emphasized the physical stability of these lipid particles. Successively Souto and Müller developed SLNs containing a CLT concentration of 1% (m/m) identical to commercial Canesten and Fungizid-Ratiopharm creams (Souto and Müller, 2007). The release of clotrimazole from the two commercial creams, as well as from aqueous SLN dispersions was studied using Franz diffusion cells with a cellulose acetate membrane. After 6 h the amount of drug released was higher than 48% when delivered from both investigated commercial formulations and not higher than 25% when delivered from the aqueous SLN dispersion. After 24 h the percentage of released drug was more than 50% for Canesten cream and Fungizid-ratiopharm cream and not higher than 30% for the formulation-based on SLN that showing a prolonged release.

Sanna et al. (2006) realized SLNs for topical administration of econazole nitrate (ECN), by o/w high-shear homogenization method using different ratios of lipid Glyceryl Palmitostearate (Precirol ATO 5) and drug (5:1 and 10:1). Then they developed topical gels containing the SLNs dispersions loaded with ECN, rheological measurements were performed, and *ex-vivo* drug permeation tests were carried out using porcine SC. In vivo percutaneous absorption of ECN was simulated by tape-stripping technique, moreover penetration tests of the drug from a conventional gel were performed as comparison. Particles had a mean diameter of about 150 nm and the encapsulation efficiency values were about 100%. In vivo tests showed that SLNs were able to control the drug release through the SC. The release rate depended upon the lipid content on the nanoparticles. The SLNs promoted a rapid penetration of ECN through the SC after only 1 h and improved the diffusion of the drug in the deeper skin layers after 3 h of application compared with the reference gel.

Some years later Bhalekar et al. (2009) prepared miconazole nitrate (MN) loaded solid lipid nanoparticles (MN-SLNs) effective for topical delivery of MN, an imidazole antifungal agent, clinically administered both in oral and topical formulations. The SLNs were obtained using Compritol 888 ATO as lipid, propylene glycol (PG) to increase drug solubility in lipid, Polysorbate 80 (tween 80) and glyceryl monostearate as surfactants to stabilize SLNs dispersion. The particles exhibited average size between 244 and 766 nm. The drug entrapment efficiency resulted from 80% to

100%. The MN-SLNs dispersion showed good stability for a period of 1 month. In addition the lipid nanoparticles were incorporated in a gel for convenient topical application and the skin penetration was evaluated *ex-vivo* using cadaver skins and Franz diffusion cells. The results of these studies indicated that the MN-SLN formulations significantly increased the accumulative uptake of MN in the skin over the marketed gel, used as reference, and showed a significantly enhanced the delivery skin targeting effect of miconazole nitrate.

In 2010, Jain and his collaborators encapsulated MN in SLNs, for topical applications (Jain et al., 2010). In particular they prepared the particles by modified solvent injection method. In vitro drug release of MN-loaded SLNs-bearing hydrogel was compared with MN suspension and MN hydrogel. MN-loaded SLNs-bearing hydrogel showed a sustained drug release over a period of 24 h. Tape stripping experiments demonstrated 10-fold greater retention with MN-loaded SLNs-bearing hydrogel as compared to MN suspension and MN hydrogel. In vivo studies were performed by infecting rats with *candida* species. It was observed that MN-loaded SLNs-bearing hydrogel was more efficient in the treatment of candidiasis. The results indicated that MN-loaded SLNs-bearing hydrogel provides a sustaining MN topical effect as well as quicker relief from fungal infection.

More recently Cassano et al. (2015) reported on the preparation of SLNs-based on PEG-40 stearate and PEG-40 stearate acrylate containing antifungal drugs, useful in vaginal infections caused by *Candida albicans*, such as ketoconazole and clotrimazole. The goal was to obtain a carrier providing long-term therapeutic concentrations at the site of infection. For this reason, Cassano et al. (2015) used the polyethylene glycol (PEG), an inert polymeric support characterized by two terminal primary hydroxyl groups widely used in organic synthesis and as a conjugating agent for biologically active molecules, due to its unique characteristics of solubility and stability (Rhee et al., 1994). The prepared nanoparticles were submitted to the antifungal activity evaluation. The results of this test revealed that the particles characterized by a very high value of minimal inhibitory concentration (MIC), appeared to be potentially effective for a topical administration. Finally, all nanoparticles were submitted to the drug permeability studies that revealed significant effect, of obtained SLNs, on the transcellular transport.

10.3 SLNs FOR TOPICAL DELIVERY OF ANTIINFLAMMATORY DRUGS

Many time the skin is used also for the administration of antiinflammatory drugs. In an effort to capture the efficacy of these agents, decreasing their side effects, many clinicians are turning to their topical administration. In this context the SLNs could an ideal carrier also for topical delivery of antiinflammatory drugs. Maia et al. have incorporated within SLNs prednicarbate (PC) (Maia et al., 2000). It is a glucocorticoid drug, applied in chronic atopic eczema, superior to the halogenated glucocorticoids because of an improved benefit/risk ratio (Lange et al., 1997). The aim of the study was to obtain a further highly desired increase by drug targeting to viable epidermis. The behavior of PC incorporated within the SLNs has been compared to that of the same drug formulated in a conventional cream (Dermatop®). Maia and co-workers demonstrated that PC penetration into human skin increased by 30.0% as compared to PC cream possibly due to the small particle size and close interaction of SLNs with the stratum corneum. With reconstructed epidermis 8.0% of the applied drug (and metabolites) was recovered from the acceptor medium of Franz cells but only 2.8% with PC cream. PC penetration of excised skin following SLNs and cream was clearly less than the penetration of reconstructed epidermis. The penetration study of PC-SLNs showed that the biotransformation of PC did not changed by SLNs-incorporation. The results of this study proved that, the glucocorticoids entrapped into SLNs, could be well-tolerated by human skin.

In 2008, Attama et al. prepared SLNs with a combination of homolipid from goat (goat fat) and phospholipid (Phospholipon 90G®) for diclofenac sodium (DNa) delivery to the eye. The lipid nanoparticles, thanks to the presence of goat fat containing phospholipid, showed a good stability and crystal characteristics, which have favored the drug loading. The DNa delivery to the eye was evaluated using bio-engineered human cornea, produced from immortalized human corneal endothelial cells (HENC), stromal fibroblasts and epithelial cells (CEPI-17 -CL 4). The encapsulation efficiency was high and a sustained release of DNa and high permeation through the bio-engineered cornea were achieved. The results obtained by Attama and collaborators showed that the permeation of DNa through the cornea construct was improved by formulation-based on SLNs modified with phospholipid.

More recently Khalil et al. developed solid lipid nanoparticles for topical delivery of meloxicam (MLX), a nonsteroidal antiinflammatory drug, preferential inhibitor of cyclooxygenase 2 (Khalil et al., 2013) MLX-SLNs, made of glycerides (Geleol, Compritol 888 ATO or Precirol ATO 5) as solid core and poloxamer 188 as a surfactant, were obtained by high shear homogenization and ultrasonication technique. In particular Khalil and colleagues evaluated the influence of different formulation compositions, as lipid type and concentration, in addition to surfactant concentration on the physicochemical properties and drug release profile of MLX- SLNs. The results of the study revealed that MLX loaded SLNs showed extremely spherical shape with particle size in the range of 325 to 1080 nm and drug entrapment efficiency ranging from 61.94 to 85.33%. The zeta potential values indicated a good stability. In vitro release study showed a sustained release of MLX from the SLNs upto 48 h. Results of stability evaluation showed a long-term stability after storage at 4 C for 12 months. So the obtained SLNs could represent a promising carrier for topical delivery of meloxicam.

Dasgupta et al. (2013) prepared a xanthan gum loaded SLNs gel containing aceclofenac, nonsteroidal antiinflammatory/antirheumatic agent, and compared it to the SLNs dispersions. The lipid particles were prepared by high speed homogenization and ultra-sonication method using a fixed amount of aceclofenac (10%) and a pluronic F68 (1.5%) as nonionic surfactant. The particle size, zeta potential and span of developed formulations was in the range of 123 nm to 323 nm, −12.4 to −18.5 and 0.42 to 0.86, respectively as the lipid concentration was increased from 7.5% to 40%. The highest entrapment efficiency was found to be 75% with the formulation having lipid concentration of 30% and 0.85% of phospholipon 90G (Phospholipid as cosurfactant). Permeation rate and controlled release property of xanthan gum loaded SLNs gel formulations and SLNs dispersion was studied through excised pig skin for 24 h. The drug release of SLNs gel formulations was better controlled as compare to SLNs dispersions. In vivo antiinflammatory study showed that action of aceclofenac was enhanced for SLNs dispersion and gel formulations. The results indicated the improvement of SLNs-based formulations for topical delivery of aceclofenac.

Mei et al. (2005) developed solid lipid nanoparticles for topical application of Triptolide (TP). This drug has been shown to have antiinflammatory, antifertility, antineoplastic, and immunosuppressive activity, but its clinical usage is limited due to its poor water solubility and toxicity. In order to overcome these disadvantages Mei and co-workers prepared SLNs, based on tristearin glyceride, soybean lecithin, and Polyethylene glycol 400 Monostearate (PEG 400 MS), as delivery systems for TP. The transdermal delivery and antiinflammatory activity were also evaluated. The results indicated that SLNs could serve as an efficient promoter of TP penetrating into skin. Then the SLNs dispersion was incorporated into hydrogel, the nanoparticulate structure was maintained, and aggregation and gel phenomena of the particle was avoided. The cumulative transdermal absorption rate in 12 h was 73.5%, whereas the conventional TP hydrogel was 45.3%. The antiinflammatory effect was over two-fold higher than that of conventional TP hydrogel. Moreover, the SLNs hydrogel consisted of pharmaceutically acceptable ingredients, such as soybean lecithin and lipid. So the obtained results demonstrated that nanoparticle could improve safety and minimize the toxicity induced by TP.

10.4   SLNs FOR TOPICAL DELIVERY OF ANTIBIOTIC DRUGS

Topical antibiotics are used to treat infections of the skin and are generally applied as ointments and creams then solid lipid nanoparticles represent also a good strategy to favor their skin absorption. Cavalli et al. (2002) evaluated SLNs as carriers for topical ocular delivery of tobramycin (TOB), an aminoglycoside antibacterial with good activity against *Pseudomonas aeruginos* (Cheer et al., 2003). In particular, the aim of their study was to verify the topical instillation of TOB-loaded SLNs for improvement of the ocular bioavailability of the drug with respect to standard eye drops. The SLNs containing the ion-pair complex of TOB with hexadecyl phosphate (TOB-SLNs; drug: hexadecyl phosphate 1:2 molar ratio), were prepared by the warm o/w microemulsion method. The particles showed an average diameter below 100 nm and a polydispersity index below 0.2. The preocular retention of SLNs in rabbit eyes was tested using drug-free, fluorescent SLNs (F-SLNs): these were retained for longer times on the corneal surface and in the conjunctival sac when compared with an aqueous fluorescent solution. A suspension of TOB-loaded SLNs (TOB-SLNs) containing 0.3% w/v TOB was administered topically to rabbits, and the aqueous humor concentration of TOB was determined up to six hours. When compared with an equal dose of TOB administered by standard commercial eye drops, TOB-SLNs produced a significantly higher TOB bioavailability in the aqueous humor. On the basis of obtained results, Cavalli and co-workers proposed the SLNs as a promising vehicle for topical ocular administration of tobramycin replacing, with advantage, subconjunctival injections, which are necessary for treatment of 'resistant' pseudomonal keratitis.

Some years later Jain and Banerjee have developed different drug carriers, such as nanoparticles of albumin, gelatin, chitosan and between them also solid lipid SLNs, for ciprofloxacin hydrochloride delivery (Jain and Banerjee, 2008). Ciprofloxacin is a broad spectrum antibiotic having efficacy against Gram-positive and Gram-negative bacteria, used, in particular, for topical ocular infections like conjunctivitis and keratitis caused by *S. aureus* (Alonso, 2004; Dillen, 2004). Ciprofloxacin hydrochloride loaded-SLNs, prepared by a warm o/w microemulsion method, showed an average particle size in the range of 73 to 98 nm and a drug encapsulation of about 39%. SLNs were found to be capable to release the drug for prolonged durations up 80 h showing to overcome the burst effect of the free drug, which released in around 70 min. Thus, these particles were found to be promising formulations for prolonged release of ciprofloxacin, particularly, for local delivery in ocular and skin infections.

More recently, Butani et al. (2016) designed and developed SLNs to improve topical application of amphotericin B, a broad-spectrum antifungal and antiprotozoal macrolide polyene antibiotic. It possesses activity against experimental cutaneous leishmaniosis and used in clinical mucocutaneous and cutaneous leishmaniosis infections resistant to antimony treatment (Yardley and Croft, 2000). SLNs particles were prepared with the aim to overcome the problem of dermato pharmacotherapy such as limited local activity and enhance drug penetration into the skin. Amphotericin B loaded SLNs were prepared by novel solvent diffusion method and were characterized for particle size, zeta potential, drug entrapment, surface morphology, in vitro antifungal activity, *ex vivo* permeation, retention and skin-irritation. Different SLN formulations of amphotericin B were prepared but the best was that (SLN5) with drug: lipid ratio 1:10, and Pluronic F 127 0.25% as surfactant. They showed an average size of 111.1 ± 2.2 nm, zeta potential of −23.98 ± 1.36 mV and 93.8% of drug entrapment. The SLN5 formulation exhibited higher drug permeation as compared to plain drug dispersion and higher zone of inhibition in *Trichophyton rubrum* fungal species. The formulation was found to be stable at 2–8°C and about 25°C for the period of three months.

10.5   SLNs FOR TOPICAL DELIVERY OF ANTIOXIDANTS

Loading of drugs into the solid matrix of solid lipid nanoparticles can be one of effective means to protect them against chemical degradation which is shown for many antioxidant substances (Dingler et al., 1999; Jenning and Gohla, 2001).

In 2006 Jee et al. formulated SLNs for all-*trans* retinol (AR) to improve its stability, whose chemical instability has been a limiting factor in its clinical use (Jee et al., 2006). The physicochemical properties of AR-loaded SLNs, including mean particle diameter and zeta potential, were modulated by changing the total amount of surfactant mixture and the mixing ratio of eggPC and tween 80 as surfactant mixture. The AR-loaded SLNs formulation was irradiated with a 60-W bulb to investigate the photostability. The loading of AR in optimized SLNs formulations rather decelerated the degradation of AR, compared with AR dissolved in methanol. The photostability at 12 h of AR in SLNs was higher (43% approximately) than that in methanol solution (about 11%).

Pople and Singh (2006) investigated SLNs for topical application of vitamin A palmitate and studied their beneficial effects on the skin. The obtained SLNs were incorporated into polymeric gels of Carbopol, Pemulen, Lutrol, and Xanthan gum for suitable application on the skin. The solid lipid nanoparticulate dispersion showed mean particle size of 350 nm. Differential scanning calorimetry studies revealed no incompatibility between drug and excipient. In vitro release profile of vitamin A palmitate from nanoparticulate dispersion and its gel showed prolonged drug release upto 24 h, which could be owing to embedment of drug in the solid lipid core. In vitro penetration studies showed almost 2 times higher drug concentration in the skin with lipid nanoparticle-enriched gel as compared with conventional gel, thus indicating better localization of the drug in the skin. In vivo skin hydration studies in albino rats revealed increase in the thickness of the stratum corneum with improved skin hydration. The developed formulation was nonirritant to the skin with no erythema or edema.

Souto et al. realized SLNs containing α-lipoic acid. It is an anti-aging substance chemically labile in fact its degradation products possess an unpleasant odor (Souto et al., 2005). Therefore, the active substance was encapsulated in SLNs. A lipid with low melting point, Softisan®601, was selected for the preparation of active-loaded SLNs, after screening the solubility of α-lipoic acid in different lipids. An entrapment efficiency of 90% was obtained for all developed formulations using Miranol®Ultra C32 as emulsifying agent. Results of Differential Scanning Calorimetry (DSC) analysis confirmed that these systems were characterized by a solid-like behavior, although with a very low crystallinity index.

Some years later Mandawgade and Patravale focused their attention on the preparation of solid lipid nanoparticles from indigenous, natural solid lipids (refined, highly purified grade stearine fractions of fruit kernel fats) by using a simple microemulsion technique (Mandawgade and Patravale, 2008). Then their potential in the topical delivery of a lipophilic drug, tretinoin (TRN) was evaluated. TRN-loaded SLNs-based topical gels were formulated and the gels were evaluated comparatively with the commercial product through in vitro occlusivity and skin permeation tests. The particles with the novel lipids showed a mean particle size < 100 nm. Up to 46% of drug entrapment was attained. Lesser skin irritancy, greater skin tolerance, occlusivity and slow drug release were observed with the developed TRN-loaded SLN-based gels respect to the commercial product. So the studies effected by Mandawgade and Patravale have proved the safety, suitability and compatibility of indigenous natural lipids (stearine fractions) as a novel excipients for SLN formulations.

In 2009 Trombino et al. formulated, stearyl ferulate-based solid lipid nanoparticles (SF-SLNs), as vehicles for β-carotene and α-tocopherol (Trombino et al., 2009). Ferulic acid (FA) is a potent antioxidant having synergistic effects with other antioxidants and it is able to protect and stabilize them from degradation (Trombino et al., 2004; Lin et al., 2005). For such reasons, Trombino and co-workers linked ferulic acid to stearyl alcohol (SA) that differs from the essential component of traditional SA-SLNs, stearic acid, in the functional group. Then, both SF-SLNs and both the conventional SA-SLNs entrapping β-carotene and α-tocopherol, were submitted to stability, antioxidant and cytotoxicity studies and were characterized for entrapment efficiency, size and shape. Furthermore, SF-SLNs and SA-SLNs entrapping α-tocopherol and β-carotene were stored at room temperature (28–30°C) and exposed to sunlight for 3 months. After this period SF-SLNs entrapping both β-carotene both α-tocopherol showed only a small increase in size while SA-SLNs had a major increase. Also the content of the two antioxidants in SF-SLNs was similar after 3-months storage whereas the SA-SLNs showed a lower entrapment efficiency The obtained results indicated the SF-SLNs as a good vehicle for β-carotene and α-tocopherol as able to stabilize these antioxidants and preventing the degradation of both compounds. The cytotoxicity studies suggested that SF-SLNs are well tolerated at concentrations upto 5 mM, lower than SA-SLNs, in the formulation for short time exposure (no more than 6 h). Therefore, SF-SLNs could be an interesting carrier for chemically labile active ingredients, in particular, for what concerns topical application when they are used for all those formulations, either cosmetical or pharmaceutical, that are suitable with the low micromolar range of SF-SLNs or simply employed, at higher doses, to store photosensitive compounds before being used for chemical purposes.

10.6 CONCLUSIONS

In the last years, the attention on topical application of SLNs are increasing, because, these carriers combine the advantages of others colloidal drug delivery systems, avoiding some of their disadvantages and are very helpful, in particular, for the encapsulation of drugs with poor water solubility or to protect chemically and physically unstable compounds from degradation.

In addition, the occlusive effect of the same nanoparticles on the skin, induces an increase in hydration and, and enhances the penetration of compounds in specific skin layers.

In this chapter the potential of solid lipid nanoparticles for topical drug administration is highlighted. In particular the role of SLNs for the delivery of antifungal, antiinflammatory, antibiotic and antioxidant molecules were treated. Thus references of relevant literature published by various research groups are provided.

KEYWORDS

- antibiotic
- antifungal
- antiinflammatory
- antioxidant
- solid lipid nanoparticles
- topical delivery

REFERENCES

Alonso, M. J. Nanomedicines for overcoming biological barriers. *Biomed Pharmacother* **2004**, *58*, 168–172.

Attama, A. A.; Reichl, S.; Muller-Goymann, C. C. Diclofenac sodium delivery to the eye: In vitro evaluation of novel solid lipid nanoparticle formulation using human cornea construct. *Int. J. Pharm.* **2008**, *355*, 307–313.

Bhalekar, M. R.; Pokharkar, V.; Madgulkar A.; Patil, N.; Patil, N. Preparation and Evaluation of miconazole nitrate-loaded solid lipid nanoparticles for topical deliver. *AAPS Pharm Sci Tech.* **2009**, *10*, 289–296.

Bhowmik, D.; Sampath Kumar, K. P. Recent Trends in Dermal and Transdermal Drug Delivery Systems: Current and Future Prospects. *The Pharma Innovation Journal.* **2013**, *2*(6), 1–6.

Butani, D.; Yewale, C.; Misra, A. Topical Amphotericin B solid lipid nanoparticles: Design and development. *Coll. Surf, B: Biointerf.* **2016**, *139*, 17–24.

Cassano, R.; Ferrarelli, T.; Mauro, M. V.; Cavalcanti, P.; Picci, N.; Trombino, S. Preparation, characterization and in vitro activities evaluation of solid lipid nanoparticles-based on PEG-40 stearate for antifungal drugs vaginal delivery. *Drug Deliv.* **2015**, *21*, 1–10.

Cavalli, R.; Gasco, M. R.; Chetoni, P.; Burgalassi, S.; Saettone, M. F. Solid lipid nanopar-ticles (SLN) as ocular delivery system for tobramycin. *Int. J. Pharm.* **2002**, *238*, 241–245.

Cheer, S. M.; Waugh, J.; Noble, S. Inhaled tobramycin (TOBI): a review of its use in the management of Pseudomonas aeruginosa infections in patients with cystic fibrosis. *Drugs.* **2003**, *63*, 2501–20.

Chiu, H. Y. Tsai, T. F. Topical use of systemic drugs in dermatology: a comprehensive review. *J Am. Acad. Dermatol.* **2011**, *65*, 1048–1070.

Dasgupta, S.; Ghosh, S. K.; Ray, S.; Mazumder, B. Solid lipid Nanoparticles (SLNs) gels for topical delivery of aceclofenac in vitro and in vivo Evaluation. *Curr. Drug Del.* **2013**, *10*, 656–666.

Dillen, K.; Vandervoort, J.; Van den Mooter, G.; Verheyden, L.; Ludwig, A. Factorial design, physicochemical characterization and activity of ciprofloxacin-PLGA nanoparticles. *Int J Pharm.* **2004**, *275*, 171–187.

Dingler, A.; Blum, R. P.; Niehus, H.; Muller, R. H.; Gohla, S. Solid lipid nanoparticles (SLN/Lipopearls) a pharmaceutical and cosmetic carrier for the application of vitamin E in dermal products. *J. Microencapsul.* 1999, *16*, 751–767.

Fasinu, P.; Pillay, V.; Ndesendo, V. M. K.; Du Toit, L. C.; Choonara Y. E. Diverse approaches for the enhancement of oral drug bioavailability. *Biopharm. Drug Dispos.* **2011**, *32*, 185–209.

Gasco, M. R. US Patent 5250236 (1993).

Hayden, C. G. J.; Cross, S. E.; Anderson, C.; Saunders, N. A.; Roberts, M. S. Sunscreen penetration of human skin and related keratinocyte toxicity after topical application. *Skin Pharmacol. Physiol.* **2005**, *18*, 170–174.

Hayden, C. G. J.; Roberts, M. S.; Benson, H. A. Systemic absorption of sunscreen after topical application. *Lancet.* **1997**, *350*, 863–864.

Jain, D; Banerjee, R. Comparison of Ciprofloxacin Hydrochloride-Loaded Protein, Lipid, and Chitosan Nanoparticles for Drug Delivery. *J. Biomed. Mat. Res. Part B: Appl. Biomat.* **2008**, *86*, 105–112.

Jain, S.; Jain, S.; Khare, P.; Gulbake, A.; Bansal, D.; Jain, S. K. Design and development of solid lipid nanoparticles for topical delivery of an antifungal agent. *Drug Delivery*, **2010**, *17*, 443–451.

Jee, J. P.; Lim, S. J.; Park, J. S.; Kim, C. K. Stabilization of all-*trans* retinol by loading lipophilic antioxidants in solid lipid nanoparticles. *Eur. J. Pharm. Biopharm.,* **2006**, *63*, 134–139.

Jenning, V.; Gohla, S. Encapsulation of retinoids in solid lipid nanoparticles (SLN) *J. Microencapsul.* 2001, *18*, 149–158.

Khalil, R. M.; El-Bary, A.; Kassem, M. A.; Ghorab, M. M. *Eur. Sci. Journal. Special Edition (1st).* Annual International Interdisciplinary Conference, AIIC 2013, 24–26 April, Azores, Portugal – Proceedings, **2013**, 779–798.

Lange, K.; Gysler, A.; Bader, M.; Kleuser, B.; Korting, H. C.; Schäfer-Korting, M. Prednicarbate versus conventional topical glucocorticoids: pharmacodynamic characterizationin vitro. *Pharm. Res.* **1997**, *14*, 1744–1749.

Lin, F. H.; Lin, J. H.; Gupta, R. D.; Tournas, J. A.; Burch, J. A.; Selim, M. A.; Monteiro-Riviere, N. A.; Grichnik, J. M.; Zielinski, J.; Pinnel, R. S. Ferulic Acid Stabilizes a Solution of Vitamins C and E and Doubles its Photoprotection of Skin. *J. Invest. Dermatol.* **2005**, *125*, 826–832.

Maia, C. S.; Mehnert, W.; Schäfer-Korting, M. Solid lipid nanoparticles as drug carriers for topical glucocorticoids. *Int. J. Pharm.* **2000**, *196*, 165–167.

Mandawgade, S. D.; Patravale, V. B. Development of SLNs from natural lipids: Application to topical delivery of tretinoin. *Int. J. Pharm.* **2008**, *363*, 132–138.

Mei, Z.; Wu, Q.; Hu, S.; Li, X.; Yang, X. Triptolide loaded solid lipid nanoparticle hydrogel for topical application. *Drug. Dev. Ind. Pharm.* **2005**, *31*, 161–168.

Müller, R. H. Lipid nanoparticles: Recent advances. *Adv. Drug Deliv. Rev.* **2007**, *59*, 375–376.

Müller, R. H.; Lucks, J. S.; **1996**. Arzneistoffträger aus festen Lipidteilchen–Feste Lipid Nanosphären (SLNS). EP 0605497.

Müller, R. H.; Rühl, D.; Runge, S.; Schulze-Forster, K.; Mehnert, W. Cytotoxicity of solid lipid nanoparticles as a function of the lipid matrix and the surfactant. *Pharm. Res.* **1997**, *14*, 458–462.

Pang, K. S. Modeling of intestinal drug absorption: roles of transporters and metabolic enzymes. *Drug Metab. Dispos.* **2003**, *31*, 1507–19.

Pople, P. V.; Kamalinder, K. S. Development and Evaluation of Topical Formulation Containing Solid Lipid Nanoparticles of Vitamin A. *AAPS Pharm Sci Tech.* **2006**, *7*, 1–7.

Rhee, W.; Wallace, D. G.; Michaels, A. S. Biologically inert, biocompatible-polymer conjugates. **1994**. Patent Document No. 5324775.

Sanna, V.; Gavini, E.; Cossu, M.; Rassu, G.; Giunchedi, P. Solid lipid nanoparticles (SLNS) as carriers for the topical delivery of econazole nitrate: in-vitro characterization, ex-vivo and in-vivo studies. *J. Pharm. Pharmacol.* **2007**, *59*, 1057–1064.

Sawynok, J. Topical and peripherally acting analgesics. *Pharmacol. Rev.* 2003, 55, 1–20.9. Hadgraft, J. Skin, the final frontier. *Int. J. Pharm.* **2001**, *224*, 1–18.

Souto, E. B.; Müller, R. H. Rheological and in vitro release behavior of clotrimazole-containing aqueous SLN dispersions and commercial creams. *Pharmazie.* 2007, *62*, 505–509.

Souto, E. B.; Muller, R. H.; Gohla, S. A novel approach-based on lipid nanoparticles (SLN) for topical delivery of alpha-lipoic acid. *J. Microencapsul.* **2005**, *22*, 581–592.

Souto, E. B.; Wissing, S. A.; Barbosa, C. M.; Müller, R. H. Development of a controlled release formulation-based on SLNS and NLC for topical clotrimazole delivery. *Int. J. Pharm.* **2004**, *278*, 71–77.

Subedi, R. K.; Oh, S. Y.; Chun, M. K.; Choi, H. K. Recent advances in transdermal drug delivery. *Arch. Pharm. Res.* **2010**, 33, 339–351.

Trombino, S.; Cassano, R.; Muzzalupo, R.; Pingitore, A.; Cione, E.; Picci, N. Stearyl ferulate-based solid lipid nanoparticles for the encapsulation and stabilization of β-carotene and α-tocopherol. *Coll. Surf. B: Biointerfaces* **2009**, *72*, 181–187.

Trombino, S.; Serini, S.; Di Nicuolo, F.; Celleno, L.; Andò, S.; Picci, N.; Calviello, G.; Palozza, P. Antioxidant Effect of Ferulic Acid in Isolated Membranes and Intact Cells: Synergistic Interactions with α-Tocopherol, β-Carotene, and Ascorbic Acid. *J. Agric. Food Chem.* **2004**, *522*, 2411–2420.

Uner, M. Preparation, characterization and physicochemical properties of solid lipid nanoparticles (SLN) and nanostructured lipid carriers (NLC): their benefits as colloidal drug carrier systems. *Pharmazie.* 2006, *61*, 375–386.

Yardley, V.; Croft, S. L. A comparison of the activities of three Amphotericin B lipid formulations against experimental visceral and cutaneous leishmaniasis. *Int. J. Antimicrob. Agents* **2000**, *13*, 243–248.

Zhai, H.; Maibach, H. I. Effect of barrier creams: Human skin in vivo. *Contact Dermatitis.* **1996**, *35*, 92–96.

CHAPTER 11

HYDROPHOBIZED POLYMERS FOR ENCAPSULATION OF AMPHOTERICIN B IN NANOPARTICLES

YOSHIHARU KANEO

Laboratory of Biopharmaceutics, Faculty of Pharmacy and Pharmaceutical Sciences, Fukuyama University; Fukuyama, Hiroshima, 729-0292, Japan, Tel.: +81 849 362111; Fax: +81 849 362024; E-mail: kaneo@fupharm.fukuyama-u.ac.jp

CONTENTS

ABSTRACT

Amphotericin B (AmB) is a broad-spectrum fungicidal antibiotic used primarily in the treatment of life-threatening systemic fungal infections. Since AmB is sparingly soluble in water, the amphiphilic polymers were attempted to render it soluble and transportable by encapsulating it in macromolecular nanoparticles. In this short review, the author discussed the usefulness of the self-assembled nanoparticles using a variety of synthetic polymers and polysaccharides.

11.1 INTRODUCTION

One of the approaches for improving a drug's performance and reducing its toxicity involves the use of a macromolecular carrier system (Kwon, 2003). Polymeric nanoparticles have been somewhat successful for the delivery of sparingly water-soluble drugs into systemic circulation (Fahr and Liu, 2007; Thassu et al., 2007; Yallapu et al., 2007). In this short review, the author discussed the alternative formulations of amphotericin B need to be developed.

11.2 AMPHOTERICIN B: SPARINGLY WATER-SOLUBLE ANTIFUNGAL AGENT

Amphotericin is an antifungal agent; it was isolated in 1956 from a soil actinomycete, *Streptomyces nodosus*. It exists in two forms, A and B; the latter, amphotericin B (AmB) is more active and used primarily as a broad-spectrum fungicidal antibiotic in the treatment of life-threatening systemic fungal infections (Figure 11.1) (Brajtburg and Bolard, 1996; Gallis et al., 1990).

For many years, the classic Fungizone (AmB-desoxycholate) has been the mainstay of antifungal therapy; however, AmB treatment is frequently associated with numerous toxicities that limit its use, including infusion-related adverse effects, and particularly, nephrotoxicity (Deray, 2002; Laniado-Laborin and Cabrales-Vargas, 2009).

The strong lipophilic properties of AmB prompted investigations of its encapsulation in liposomes or of its binding to lipid complexes, in an

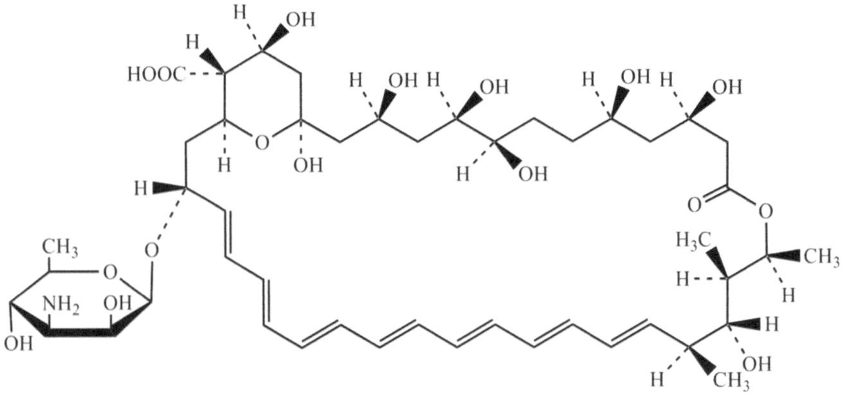

FIGURE 11.1 Chemical structure of amphotericin B (AmB).

effort to increase both its efficacy and safety. Subsequently, three lipid formulations, Amphotec (AmB colloidal dispersion), Abelcet (AmB lipid complex), and AmBisome (liposomal AmB), were developed and licensed. of these, AmBisome is the most widely used; however, it is associated with some toxicity and is not universally effective. Therefore, alternative formulations need to be developed (Guo, 2001; Herbrecht et al., 2003; Tomii, 2002).

11.3 STYRENE MALEIC ACID ANHYDRIDE COPOLYMER (SMA) FOR THE ENCAPSULATION OF AMB

11.3.1 SMA: A PROMISING POLYMERIC CARRIER

Styrene maleic acid anhydride copolymer (SMA) is an alternating copolymer that is easily formed by free radical copolymerization of maleic anhydride and styrene. S MA can be modified by reacting the functional anhydride groups with different nucleophiles. SMA is widely used in various industrial applications, such as in adhesives and coatings, because of its good durability and low cost. It has no teratogenic, acute, or chronic toxic effects (Winek and Burgun, 1977).

In 1979, Maeda et al., reported the first synthesis of the anticancer protein neocarzinostatin (NCS) conjugated to SMA, which they named SMANCS.

Later studies found that SMANCS accumulates in tumor tissues to a greater extent than NCS (Fang et al., 2011; Maeda et al., 2009). These studies reported that a micellar formulation of pirarubicin using SMA exhibited excellent tumor targeting capacity due to the enhanced permeability and retention (EPR) effect. The SMA-pirarubicin micelles contributed to prolonged survival in vivo in a murine liver metastasis model (Daruwalla et al., 2009, 2010). An antitumor compound zinc protoporphyrin IX (ZnPP)-SMA micelles also showed remarkable cytotoxicity against a variety of tumor cells. It demonstrated antitumor effects in Meth A fibrosarcoma and B16 melanoma tumor-bearing mice (Fang et al., 2011, 2012; Nakamura et al., 2011).

11.3.2 AMB-LOADED SMA NANOPARTICLES (YAMAMOTO ET AL., 2013A)

The anhydride form of SMA is insoluble in water; the maleic anhydride residue of the SMA was then hydrolyzed to the water-soluble maleic acid form by adding NaOH (Figure 11.2). The lyophilized powder of the sodium salt of SMA was extremely water soluble. The loading of sparingly water-soluble drugs was conducted using the hydrolyzed SMA solutions. AmB dissolved in dimethyl sulfoxide (DMSO) was added to SMA H_2O solution. The purification of each mixture solution was conducted using ultrafiltration, then lyophilized (Greish et al., 2004, 2005).

The loading of drugs into the nanoparticles occurred instantaneously when a small amount of DMSO containing the drug was added to SMA in water. High-performance size exclusion chromatography (HPSEC) was carried out using a liquid chromatography apparatus equipped with

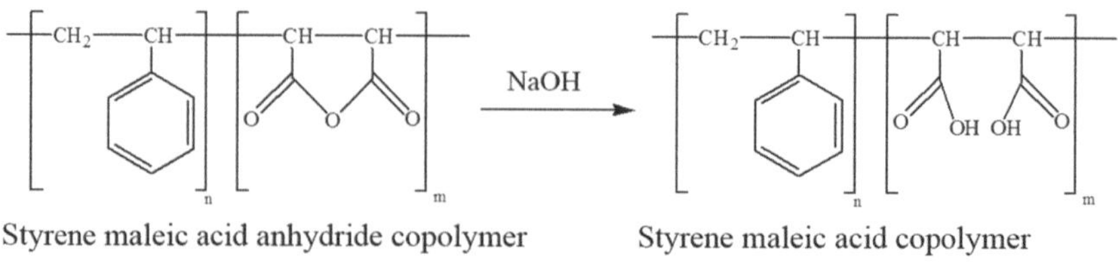

Styrene maleic acid anhydride copolymer   Styrene maleic acid copolymer

FIGURE 11.2 Preparation of the styrene maleic acid copolymer.

a differential refractometer. The drug-loaded SMA nanoparticles were eluted near the exclusion volume of the column. Elution peaks were also detected using a photodiode array detector. The peak retention time for AmB-loaded SMA was 6.2 min for each detection wavelength (405, 380, 360, and 335 nm), as observed in Figure 11.3. These results indicated that the introduction of hydrophobic drugs into the core of the nanoparticle makes the nanoparticle more rigid and more stable.

SMA is a useful material for medical applications demonstrated through the success of SMANCS. Only by simple hydrolysis of SMA without further chemical modification, amphiphilic carrier molecules could be obtained. Encapsulation in SMA nanoparticles could effectively increase the overall water solubility of sparingly water-soluble drugs. Furthermore, a relatively high retention of AmB in the blood circulation was also demonstrated in an in vivo animal experiment. Amphiphilic SMAs can give a AmB new pharmaceutical capabilities.

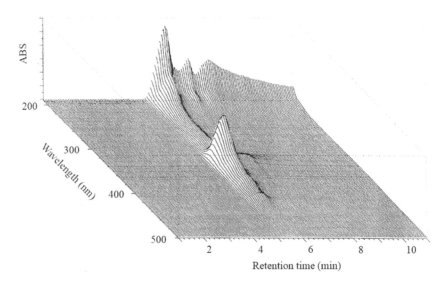

FIGURE 11.3 Three-dimensional chromatogram of AmB-loaded But-SMA. The peak retention time for the AmB-loaded But-SMA was 4.8 min for the 405, 380, 360, and 335 nm cross sections. High performance size-exclusion chromatography was carried out using an HPLC system equipped with a photodiode array detector. A 7.8 × 300 mm, TSKgel G4000PWXL column was used at 40°C. The mobile phase was water, and the flow rate was 1.0 mL/min.

11.4 HYDROPHOBIZED KOLLICOAT IR FOR THE ENCAPSULATION OF AMB

11.4.1 KOLLICOAT IR, A POLY(ETHYLENE GLYCOL)-POLY(VINYL ALCOHOL) GRAFT COPOLYMER (YAMAMOTO ET AL., 2013B)

Kollicoat IR (KOL) is a poly(ethylene glycol)-poly(vinyl alcohol) graft copolymer. KOL was developed as an excipient and a film coating polymer for instant release tablets by BASF Chemical Co. (Ludwigshafen, Germany) and is marketed under the trade mark Kollicoat IR. The polymer consists of 75 w/w% poly(vinyl alcohol) (PVA) units grafted on 25 w/w% poly(ethylene glycol) (PEG) units. The PVA moiety has good film-forming properties, and the PEG moiety acts as an internal plasticizer. The copolymer is hydrophilic and thus readily soluble in water. The copolymer's solubility is 40 w/w% in aqueous systems and 25 w/w% in a 1:1 ethanol-water mixture; the solubility in nonpolar solvents is low (Guns et al., 2010; Fouad et al., 2011).

Solubility is the key determinant of the oral bioavailability of drug components. Solid dispersion is among the most successful strategies to improve the dissolution rate of poorly soluble drugs. Synthetic polymers, including poly(vinyl pyrrolidone), PEG, and poly(methacrylate), and natural product-based polymers, such as cellulose derivatives, have been used to formulate solid dispersions. Recently, Janssens et al. (2007) reported that KOL is a promising excipient for the formulation of solid dispersions of itraconazole prepared by hot stage extrusion.

PEG, a constituent of KOL, is a water-soluble polymer composed of repeating ethylene oxide units flanked by alcohols. Modification of the nanoparticle surface with PEG (PEGylation) significantly improves circulation time (Jokerst et al., 2011). For example, a doxorubicin liposome formulation containing PEG (Doxil) shows increased circulation time in plasma, enhanced accumulation in murine tumors, and superior therapeutic activity over free doxorubicin (Gabizon et al., 1994). Two PEG-modified interferons (Pegintron and Pegasys) improved the treatment of chronic hepatitis C (Zeuzem et al., 2003).

11.4.2 AMB-LOADED KOL NANOPARTICLES (YAMAMOTO ET AL., 2013B)

Amphiphilic polymers were prepared with KOL and hydrophobic groups through ester bonds ($-[CH_2-CH (OCOR)]_n-$) (Figure 11.4). The degree of substitution was determined using ^{1}H-NMR to compare the integral of the peak at 0.677 ppm assigned to the 13-CH_3 of cholesteryl, the peak at 0.881 ppm assigned to the terminal CH_3 of stearoyl, or the peak at 0.881 ppm assigned to the terminal CH_3 of oleoyl, with the integral of the peaks at 4.26, 4.50, and 4.69 ppm being assigned to the H of the PVA chain. The degrees of substitution were 2.1 w/w% for KOL-cho, 5.0 w/w% for KOL-stearoyl, and 5.2 w/w% for KOL-oleoyl.

KOL is a new pharmaceutical excipient that was developed as a coating polymer for instant release tablets. Kaneo et al., first synthesized hydrophobized KOLs by a simple one-step chemical reaction. In general, hydrophobically modified polymers tend to lose their water solubility. However, KOL-cho, KOL-stearoyl and KOL-oleoyl maintained their water solubility and effectively solubilized a sparingly soluble drug, AmB, due to the KOLs' superior hydrophilicities.

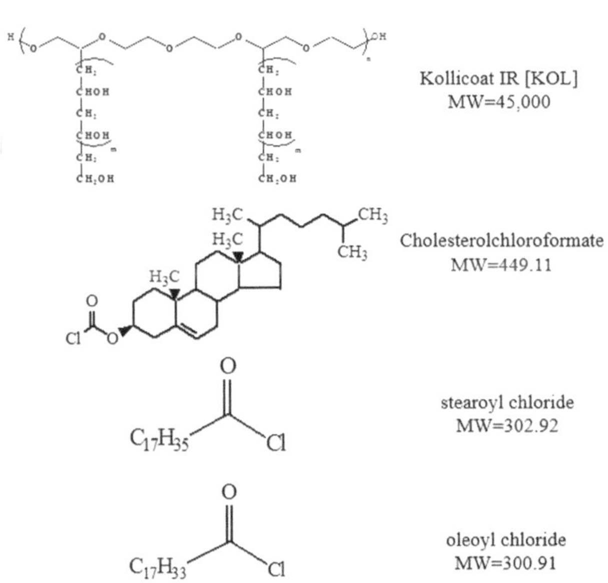

FIGURE 11.4   Chemical structures of Kollicoat IR and reagents.

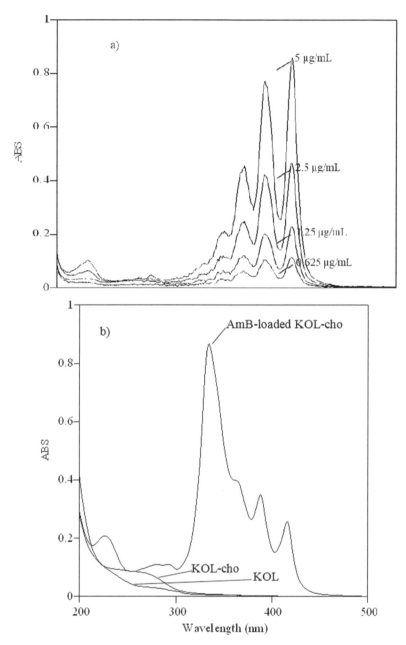

FIGURE 11.5 (a) UV absorption of AmB in 50% methanol solution at 5, 2.5, 1.25, and 0.625 µg/mL. (b) UV absorption of AmB loaded KOL-cho at 250 µg/mL, KOL-cho at 250 µg/mL, and KOL at 250 µg/mL.

UV spectroscopy was used to investigate the aggregation states of AmB when associated with the nanoparticles. The UV absorbance spectrum of AmB in 50 v/v% methanol, representing the monomeric form, showed high-intensity peaks at 405, 384, and 366 nm and a low-intensity peak at 348 nm (Figure 11.5a). The absorbance spectrum of AmB-loaded KOL-cho, which represented the self-aggregation state of AmB, showed a broad high-intensity peak at 331 nm and lower intensity peaks at 405, 384, and 366 nm (Figure 11.5b). This also suggested that AmB was encapsulated in the hydrophobic inner core of the nanoparticles (Adams and Kwon, 2003; Barwicz et al., 1992).

Nanoparticle formation and drug loading occurred simultaneously in the dialysis process when dimethyl sulfoxide (DMSO) was used as the solvent for AmB and the polymer. The amount of AmB loaded into the system reached 11.4 w/w%, depending both on the hydrophobic group itself and the feed-weight ratio of AmB to polymer. The water solubility of the AmB loaded in the KOL-cho was 4017.4 ± 24.4 µg/mL. KOL is a promising material for medical applications because of its biocompatibility and hydrophilic nature. By encapsulating AmB, the KOL-cho nanoparticle reduced AmB toxicity with respect to hemolysis. Although the AmB loaded in the KOL-cho showed the high retention in the plasma only in the early stage after injection, a relatively low accumulation in the liver was demonstrated in vivo in an animal experiment. Therefore, these results suggested that the KOL-cho nanoparticles could give AmB new pharmaceutical applications.

11.5 HYDROPHOBIZED POLY(VINYL ALCOHOL) FOR THE ENCAPSULATION OF AMB

11.5.1 POLY(VINYL ALCOHOL): ONE OF THE MOST IMPORTANT INDUSTRIAL POLYMERS (YAMAMOTO ET AL., 2013C)

Poly(vinylalcohol) (PVA) is used in a variety of applications, for textile fibers, in paper-coating agents, as an emulsion stabilizer, and in biomedical products. PVA's biocompatibility makes it excellent for medical applications such as use in soft contact lenses. Furthermore, it is used for long-term implants, including bioartificial pancreas, artificial cartilage, nonadhesive film, and esophageal or scleral buckling material (Hyon et al., 1994; Noguchi et al., 1991). Recently, PVA has been used as a macromolecular drug carrier (Orienti et al., 2005) and a surface modifier of liposomes (Takeuchi et al., 2000).

11.5.2 SYNTHESIS OF HYDROPHOBIZED PVA

Amphiphilic polymers were prepared by partial substitution of PVA with hydrophobic groups through ester bonds (-[CH$_2$-CH (OCOR)]$_n$-) (Figure 11.6). Table 11.1 shows the properties of the substituted polymers. The degree of substitution was determined using 1H-NMR, comparing the integral of the peak at 0.677 ppm assigned to the CH$_3$ of the cholesterol with that of the peaks at 4.26, 4.50, and 4.69 ppm assigned to the H of the PVA chain. The degree of substitution was 4.0 w/w% for PVA-cho.

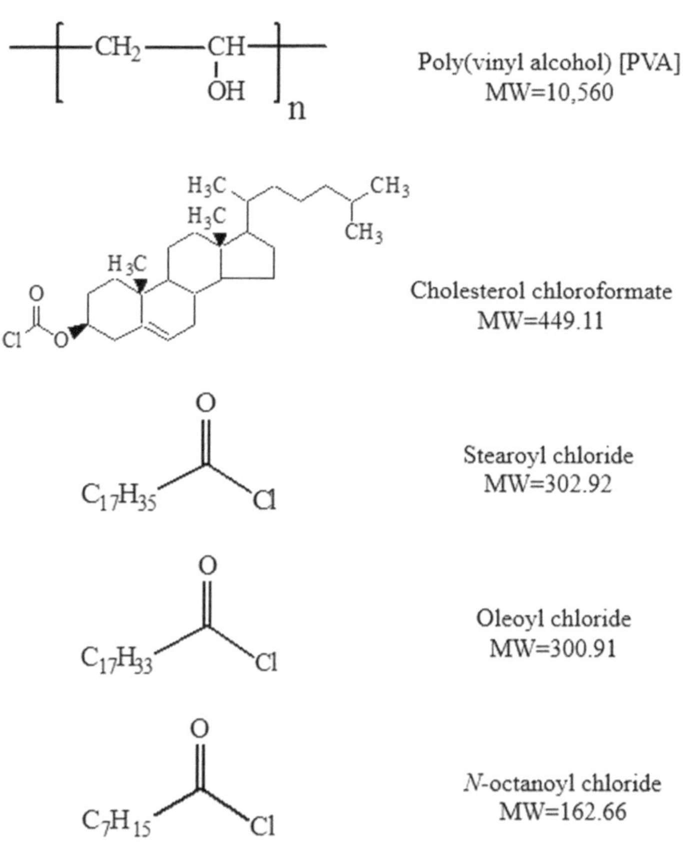

Poly(vinyl alcohol) [PVA]
MW=10,560

Cholesterol chloroformate
MW=449.11

Stearoyl chloride
MW=302.92

Oleoyl chloride
MW=300.91

N-octanoyl chloride
MW=162.66

FIGURE 11.6 Chemical structures of PVA and reagents.

TABLE 11.1 Properties of PVA Nanoparticles

	Unloaded particle				AmB-loaded particle	
	Diameter (nm)	Hydrophobic content (w/w%)	PEG content (w/w%)	CMC (μg/mL)	Diameter (nm)	AmB-loaded (w/w %)
PVA-octanoyl	19.5 ± 0.7	6.6	–	400	148.0 ± 14.8	0.8
PVA-stearoyl	22.7 ± 1.7	3.3	–	120	23.6 ± 1.6	24.0
PVA-oleoyl	26.0 ± 2.9	1.8	–	90	32.5 ± 1.4	22.1
PVA-cho	16.0 ± 0.8	4.0	–	80	25.3 ± 0.7	30.4
PVA-cho-sPEG	15.0 ± 0.4	4.0	46.9	60	31.5 ± 2.6	9.6
AmBisome	–	–	–	–	65.7 ± 0.8	4.0
Fungizone	–	–	–	–	17.3 ± 0.8	50.0

A cholesterol-grafted PVA, PVA-cho was synthesized by a simple one-step chemical reaction. Next, we substituted another hydrophilic polyethylene glycol (2 kDa) to the grafted PVA in order to retain water solubility using O-methyl-O'-succinyl polyethylene glycol (sPEG), since the high degree of substitution of the PVA with hydrophobic cholesterol groups reduced its water solubility (Opanasopit et al., 2006, 2007a,b).

The degree of PEGylation of the PVA-cho-sPEG was also determined using 1H-NMR to compare the integral of the peak at 4.12 ppm assigned to the CH_2 of the sPEG with the integral of the peaks at 4.26, 4.50, and 4.69 ppm assigned to the H of the PVA chain. The degree of PEGylation was 46.9 w/w% for PVA-cho-sPEG.

11.5.3 CRITICAL MICELLATION CONCENTRATION OF HYDROPHOBIZED PVA

The hydrophobized PVAs formed nanoparticles in an aqueous environment because of the aggregation of the grafted hydrophobized groups. The critical micellation concentration (CMC) of the self-assembled nanoparticles formed was determined by fluorescence probe techniques by using

N-phenyl-1-naphthylamine (PNA) (Akiyoshi et al., 1993). PNA, which is strongly hydrophobic, has very low solubility in water and is dissolved preferentially by the hydrophobic core of micelles, as shown in the chemical structure in Figure 11.7. The relative fluorescence intensity increased with increasing concentration of each hydrophobized PVA, as shown in Figure 11.8. The curves of relative fluorescence intensity versus the concentration of each hydrophobized PVA showed a sharp increase, from which the CMC values were estimated (Table 11.1). Furthermore, the shift of the emission maximum of the PNA to the lower wavelength, a so-called blue shift, was seen to be a function of the concentration of each hydrophobized PVA (Figure 11.8).

11.5.4 PREVENTION OF HEMOLYSIS OF AMB BY NANOPARTICLE INCLUSION

The ability of nanoparticle-encapsulated AmB to prevent the typical hemolysis provoked by AmB was examined. AmB itself caused 100% hemolysis above the concentration of 20 μg/mL due to membrane damage (Lavasanifar et al., 2001). Fungizone caused 50% hemolysis at 10 μg/mL as shown in Figure 11.9. In contrast, each AmB-loaded nanoparticle was less toxic than AmB itself. In particular, the AmB-loaded PVA-cho-sPEG was completely nonhemolytic upto 100 μg/mL. Furthermore, no significant difference in body weight change was seen between the mice treated with hydrophobized

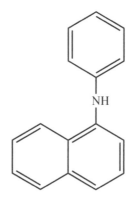

FIGURE 11.7   Chemical structure of *N*-phenyl-1-naphthylamine (PNA).

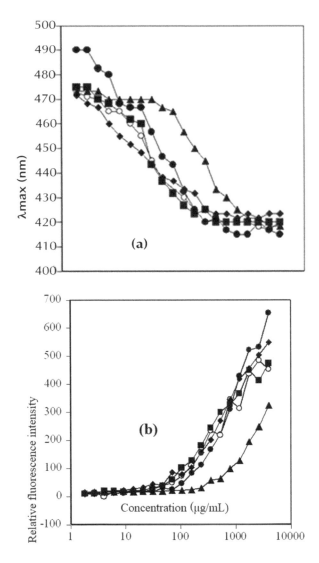

FIGURE 11.8 Emission maximum (λ max) and relative fluorescence intensity of PNA as a function of the logarithm of the concentration of PVA-cho-sPEG ($\circ$), PVA-cho ($\bullet$), PVA-oleoyl ($\blacksquare$), PVA-stearoyl ($\blacklozenge$), and PVA-octanoyl ($\blacktriangle$). Each point represents the mean of 4 to 5 measurements.

PVAs and those treated with 5% glucose. Therefore, AmB-loaded PVA-cho-sPEG is expected to be a very safe preparation for clinical treatment.

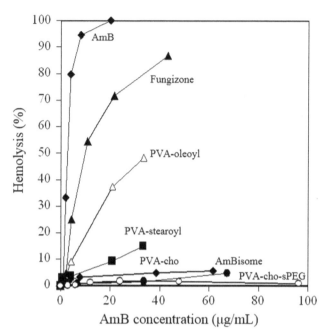

FIGURE 11.9 AmB hemolysis of mouse red blood cells in seven preparations: AmB (♦), Fungizone (▲), PVA-oleoyl (△), PVA-stearoyl (■), PVA-cho (◇), AmBisome (●), PVA-cho-sPEG (○).

11.5.5 BIODISPOSITION OF AMB LOADED TO HYDROPHOBIZED PVA NANOPARTICLES

The biodistribution of AmB loaded to the hydrophobized PVA nanoparticles was compared with that of AmBisome (liposomal AmB) after a single injection in mice. We demonstrated along persistence in the blood circulation and a low hepatic distribution of AmB after intravenous injection of AmB loaded to PVA-cho-sPEG nanoparticles (Figure 11.10). Van Etten also reported that the blood concentration of AmB after administration of liposomal AmB was consistently higher than that of the desoxycholate preparation (van Etten et al., 1995). The concentration in the liver after the injection of AmB-loaded PVA-cho-sPEG was lower than that of AmBisome and Fungizone.

When Fungizone is administered intravenously, a rapid disruption of the desoxycholate micelles may occur, and most of the liberated AmB is

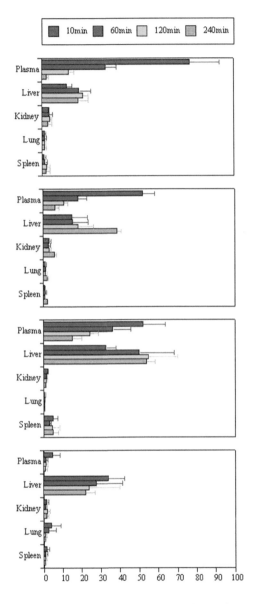

FIGURE 11.10   AmB levels in the blood and tissues after a single injection of AmB-loaded PVA-cho-sPEG nanoparticle, AmB-loaded PVA-cho nanoparticle, AmBisome, or Fungizone at an equivalent dose of 2 mg/kg in mice. Values given are mean ± SD for groups of 4–5 mice.

bound to plasma lipoproteins (Bekersky et al., 2002). As a result, AmB is distributed mostly in the liver, kidney, and spleen, revealing a relatively low plasma concentration after intravenous injection. A part of the AmB forms aggregates in the plasma, and these are sometimes distributed in the lung and kidney. They are considered a cause of the augmented nephrotoxicity (Deray, 2002; Laniado-Laborin and Cabrales-Vargas, 2009).

The RES in the liver and spleen takes up foreign substances. Surface PEGylation of the nanoparticles would allow them to avoid the RES. AmB-loaded PVA-cho-sPEG showed a greater persistence of AmB than that of PVA-cho (Figure 11.10). It was, therefore, suggested that PEGylation would allow the molecule to avoid being trapped by the RES. This feature was superior to that of liposomal AmB, AmBisome. The sites of fungal inflammation are characterized by increased vascular permeability, and as a consequence, smaller nanoparticles leak into the tissues with a plasma component. Therefore, greater therapeutic efficacy may be achieved with AmB-loaded PVA-cho-sPEG.

11.6 HYDROPHOBIZED CLUSTER DEXTRIN FOR THE ENCAPSULATION OF AMB

11.6.1 CLUSTER DEXTRIN AS A BIODEGRADABLE DRUG CARRIER (KANEO ET AL., 2014)

Cluster dextrin (CDex), a highly branched cyclic dextrin, is a novel glucose polymer that is produced from a waxy cornstarch by the cyclization reaction of a branching enzyme (Takata et al., 1996a; Takata et al., 1996b; Fujii et al., 2003). Despite its large molecular weight (462 kDa), CDex is highly soluble in water and has a relatively low propensity for retrogradation compared to that of commercial dextrins (Takata et al., 1997). As CDex is a safe substance consisting of glucose, a variety of applications, such as food additives for sports beverages or taste improvement, a spray drying aid for extracts of fruits and vegetables, and the powderization of fish oil, are proposed for the use of CDex.

One of the approaches for improving the performance of a drug and reducing its toxicity involves the use of a macromolecular carrier system (Kwon, 2003; Fahr and Liu, 2007; Thassu et al., 2007; Yallapu et al., 2007).

In particular, polysaccharide-based nanoparticles have been recognized as promising drug carriers because of their hydrophobic domain, which is surrounded by a hydrophilic outer shell, that serves as a preserver for various hydrophobic drugs (Chakravarthi et al., 2007; Liu et al., 2008; Namazi et al., 2012). Examples of hydrophobically modified polysaccharides used to prepare nanoparticles include pullulan, dextran, starch, dextrin, and cellulose (Akiyoshi et al., 1998; Aumelas et al., 2007; Daoud-Mahammed et al., 2007; Goncalves et al., 2007; Jeong et al., 1999; Jung et al., 2003, 2004; Namazi et al., 2012).

Most biodegradable polymers used in drug delivery are specifically intended for parenteral administration. Above all, the biodisposition of dextran has been widely investigated (deBelder, 1996; Molteni, 1979; Thoren, 1981; Schacht et al., 1985). Moreover, we have reported that certain neutral polysaccharides, such as dextran and pullulan, can accumulate in the liver for long periods after their injections because these materials are scarcely metabolized enzymatically in the circulating blood (Kaneo et al., 1996, 1997, 2000, 2001). Therefore, the higher molecular weight glucans are taken up into the hepatocytes by nonspecific fluid-phase endocytosis.

CDex is as digestible as commercial dextrins to form glucose by α-amylase and α-glucosidase, which are both present in the small intestine. However, we have also reported that intravenously administered FITC-labeled CDex is rapidly eliminated from the blood, followed by an appreciable excretion into the urine (Taguchi et al., 2009). High-performance size exclusion chromatographic analysis showed that the FITC-labeled CDex was quickly degraded into small molecules (–6 kDa) in the blood. Therefore, CDex alone would be expected to avoid long-term hepatic accumulation.

11.6.2 AMB LOADING INTO THE HYDROPHOBIZED CDEX NANOPARTICLE

The amphiphilic polymers were prepared by substituting CDex with hydrophobic groups through ester bonds. The degrees of substitution, the number of moles of hydrophobic groups per 100 moles of glucose units, were 1.00 mol/mol % for n-octanoyl CDex (CDex-octanoyl), 1.80 mol/

mol % for stearoyl CDex (CDex-stearoyl), 1.07 mol/mol % for oleoyl CDex (CDex-oleoyl), and 1.07 mol/mol % for cholesteryl CDex (CDex-cho). The linkages between CDex and hydrophobic groups were relatively stable because no disruption of the nanoparticles was detected in aqueous solution even after 1 week storage.

The AmB loadings (w/w %) within the hydrophobized CDex nanoparticles have values ranging from 7.7 to 19.1 w/w%, depending on the macromolecular vehicle. Free AmB was insoluble in water, whereas each AmB preparation—Fungizone, AmBisome, or the AmB-loaded hydrophobized CDex nanoparticles—yielded a colloidal solution. The sizes of the unloaded nanoparticles ranged from 16.4 to 21.7 nm, but those of the AmB-loaded nanoparticles varied from 19.0 to 278.0 nm, as determined by DLS. The values of polydisersity index of CDex nanoparticles were 0.08–0.10. A photograph of the AmB-loaded CDex-cho nanoparticles taken using SEM is presented in Figure 11.11, and the nanoparticles were observed to cluster together to form aggregates in the dry state.

11.6.3   ENZYMATIC DEGRADATIONS OF THE HYDROPHOBIZED CDEX NANOPARTICLES

The degradations of CDex and AmB-loaded CDex-cho by α-amylase were examined by HPSEC. The retention time of authentic CDex treated by α-amylase shifted from 8.0 min to 11.2 min, as shown in Figure 11.12. The AmB-loaded CDex-cho nanoparticles eluted much more rapidly (5.1 min) than CDex. Additionally, treatment by α-amylase changed the elution profile of the AmB-loaded CDex-cho to that of lower molecular weight glucans (11.2 min) (Figure 11.13). The reducing terminal of CDex was measured using the DNS method. Hydrolysis of both CDex and CDex-cho by α-amylase was fast at the initial stage and then proceeded slowly; however, the inclusion of AmB in the CDex-cho nanoparticle decreased the degradation of CDex by two-thirds, as shown in Figure 11.14.

11.6.4   BIODEGRADABLE CDEX AS A DRUG CARRIER FOR AMB

CDex is a promising material for medical applications because of its biocompatibility and hydrophilic nature. By encapsulating AmB, these

CDex-cho nanoparticles reduced AmB toxicity with respect to hemolysis and could effectively be used to increase the overall water solubility
of AmB. Furthermore, a relatively high retention of AmB in the plasma

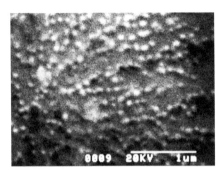

FIGURE 11.11   Scanning electron micrograph of the AmB-loaded CDex-cho nanoparticles.

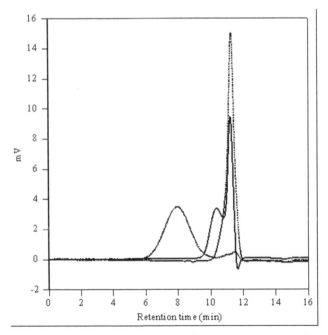

FIGURE 11.12   Degradation of CDex by α-amylase (0.1 U/mL) at 37°C. HPSEC was
performed using a HPLC system equipped with a differential refractometer. A 7.8×300-mm
TSKgel G4000PWXL column was used at 40°C. The mobile phase was water, and the flow
rate was 1.0 mL/min.

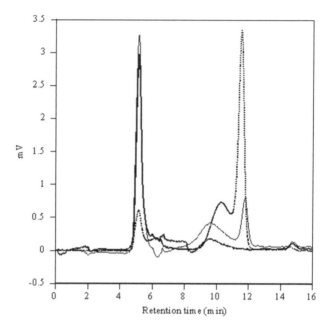

FIGURE 11.13   Degradation of AmB-loaded CDex-cho by α-amylase (0.1 U/mL) at 37°C. HPSEC was performed using a HPLC system equipped with a differential refractometer. A 7.8×300-mm TSKgel G4000PWXL column was used at 40°C. The mobile phase was water, and the flow rate was 1.0 mL/min.

was demonstrated in vivo in an animal experiment compared with that of Fungizone. It was suggested that these hydrophobized CDexs could give AmB new pharmaceutical applications.

ACKNOWLEDGMENTS

This work was supported in part by Grants in Aid (No. 13672406) for Scientific Research (C) from the Ministry of Education, Culture, Sports, Science and Technology, Japan. The author is also grateful to BASF Japan Ltd., Tokyo, to Japan Vam & Poval Co., Ltd., Osaka, and to Dr. H Takata of Ezaki Glico Co., Ltd., Osaka, for gifting the PVA samples, the KOL sample and the cluster dextrin samples, respectively. The author particularly wants to thank Professor T. Ishidzu, Professor Y. Yamaguchi and Professor H. Haraguchi for their technical guidance on 1H-NMR measurement, scanning electron microscopy observation, and assay of the

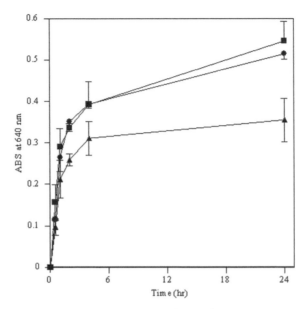

FIGURE 11.14 Appearance of reducing terminals of CDex (■), CDex-cho (●) and AmB-loaded CDex-cho (▲) after treatment with α-amylase (1 U/mL) at 37°C and pH 6.0. Quantitative analyzes of the reducing terminals were performed by the 3,5-dinitrosalicylic acid method.

antimicrobial activity, respectively. The author is grateful for Professor Y. Maitani, Professor H. Maeda, and Professor T. Tanaka for their valuable guidance. The author thanks to Dr. S. Yamamoto and Ms. K. Taguchi for their technical assistance.

KEYWORDS

- **amphotericin B**
- **biodisposition**
- **hemolysis**
- **nanoparticle**
- **polyethyleneglycol**
- **polysaccharides**
- **synthetic polymers**

REFERENCES

Adams, M. L.; Kwon, G. S. Relative aggregation state and hemolytic activity of amphotericin B encapsulated by poly(ethylene oxide)-block-poly(N-hexyl-L-aspartamide)-acyl conjugate micelles: effects of acyl chain length. *Journal of Controlled Release* **2003**, *87*, 23–32.

Akiyoshi, K.; Deguchi, S.; Moriguchi, N.; Yamaguchi, S.; Sunamoto, J. Self-Aggregates of Hydrophobized Polysaccharides in Water. Formation and Characteristics of Nanoparticles. *Macromolecules* **1993**, *26*, 3062–3068.

Akiyoshi, K.; Kobayashi, S.; Shichibe, S.; Mix, D.; Baudys, M.; Kim, S. W.; Sunamoto, J. Self-assembled hydrogel nanoparticle of cholesterol-bearing pullulan as a carrier of protein drugs: complexation and stabilization of insulin. *Journal of Controlled Release* **1998**, *54*, 313–320.

Aumelas, A.; Serrero, A.; Durand, A.; Dellacherie, E.; Leonard, M. Nanoparticles of hydrophobically modified dextrans as potential drug carrier systems. *Colloids Surf. B: Biointerfaces* **2007**, *59*, 74–80.

Barwicz, J.; Christian, S.; Gruda, I. Effects of the aggregation state of amphotericin B on its toxicity to mice. *Antimicrob Agents Chemother* **1992**, *36*, 2310–2315.

Bekersky, I.; Fielding, R. M.; Dressler, D. E.; Lee, J. W.; Buell, D. N.; Walsh, T. J. Plasma protein binding of amphotericin B and pharmacokinetics of bound versus unbound amphotericin B after administration of intravenous liposomal amphotericin B (AmBisome) and amphotericin B deoxycholate. *Antimicrob Agents Chemother* **2002**, *46*, 834–840.

Brajtburg, J.; Bolard, J. Carrier effects on biological activity of amphotericin B. *Clin Microbiol Rev* **1996**, *9*, 512–531.

Chakravarthi, S. S.; Robinson, D. H.; De, S. In *Nanoparticulate drug delivery systems*; Thassu, D.; Deleers, M.; Pathak, Y.; Eds.; Informa: Abingdon, UK, 2007, pp. 51–60.

Daoud-Mahammed, S.; Couvreur, P.; Gref, R. Novel self-assembling nanogels: stability and lyophilization studies. *Int J Pharm* **2007**, *332*, 185–191.

Daruwalla, J.; Greish, K.; Malcontenti-Wilson, C.; Muralidharan, V.; Iyer, A.; Maeda, H.; Christophi, C. Styrene maleic acid-pirarubicin disrupts tumor microcirculation and enhances the permeability of colorectal liver metastases. *Journal of Vascular Research* **2009**, *46*, 218–228.

Daruwalla, J.; Nikfarjam, M.; Greish, K.; Malcontenti-Wilson, C.; Muralidharan, V.; Christophi, C.; Maeda, H. In vitro and in vivo evaluation of tumor targeting styrene-maleic acid copolymer-pirarubicin micelles: Survival improvement and inhibition of liver metastases. *Cancer Sci.* **2010**, *101*, 1866–1874.

deBelder, A. N. In *Polysaccharides in Medical Applications*; Dumitriu, S.; Ed.; Marcel Dekker: New York, 1996, pp. 505–523.

Deray, G. Amphotericin B nephrotoxicity. *J Antimicrob Chemother* **2002**, *49* (Suppl 1), 37–41.

Fahr, A.; Liu, X. Drug delivery strategies for poorly water-soluble drugs. *Expert Opin. Drug Deliv.* **2007**, *4*, 403–416.

Fang, J.; Greish, K.; Qin, H.; Liao, L.; Nakamura, H.; Takeya, M.; Maeda, H. HSP32 (HO-1) inhibitor, copoly(styrene-maleic acid)-zinc protoporphyrin IX, a water-soluble micelle as anticancer agent: In vitro and in vivo anticancer effect. *European Journal of Pharmaceutics and Biopharmaceutics* **2012**, *81*, 540–547.

Fang, J.; Nakamura, H.; Maeda, H. The EPR effect: Unique features of tumor blood vessels for drug delivery, factors involved, and limitations and augmentation of the effect. *Adv. Drug. Deliv. Rev.* **2011**, *63*, 136–151.

Fouad, E. A.; El-Badry, M.; Neau, S. H.; Alanazi, F. K.; Alsarra, I. A. Technology evaluation: Kollicoat IR. *Expert Opin. Drug Deliv.* **2011**, *8*, 693–703.

Fujii, K.; Takata, H.; Yanase, M.; Terada, Y.; Ohdan, K.; Takaha, T.; Okada, S.; Kuriki, T. Bioengineering and application of novel glucose polymers. *Biocatal. Biotransformation* **2003**, *21*, 167–172.

Gabizon, A.; Catane, R.; Uziely, B.; Kaufman, B.; Safra, T.; Cohen, R.; Martin, F.; Huang, A.; Barenholz, Y. Prolonged circulation time and enhanced accumulation in malignant exudates of doxorubicin encapsulated in polyethylene-glycol coated liposomes. *Cancer Research* **1994**, *54*, 987–992.

Gallis, H. A.; Drew, R. H.; Pickard, W. W. Amphotericin B: 30 years of clinical experience. *Rev Infect Dis* **1990**, *12*, 308–329.

Goncalves, C.; Martins, J. A.; Gama, F. M. Self-assembled nanoparticles of dextrin substituted with hexadecanethiol. *Biomacromolecules* **2007**, *8*, 392–398.

Greish, K.; Nagamitsu, A.; Fang, J.; Maeda, H. Copoly(styrene-maleic acid)-pirarubicin micelles: high tumor-targeting efficiency with little toxicity. *Bioconjugate Chemistry* **2005**, *16*, 230–236.

Greish, K.; Sawa, T.; Fang, J.; Akaike, T.; Maeda, H. SMA-doxorubicin, a new polymeric micellar drug for effective targeting to solid tumors. *Journal of Controlled Release* **2004**, *97*, 219–230.

Guns, S.; Kayaert, P.; Martens, J. A.; Van Humbeeck, J.; Mathot, V.; Pijpers, T.; Zhuravlev, E.; Schick, C.; Van den Mooter, G. Characterization of the copolymer poly(ethyleneglycol-g-vinylalcohol) as a potential carrier in the formulation of solid dispersions. *European Journal of Pharmaceutics and Biopharmaceutics* **2010**, *74*, 239–247.

Guo, L. S. Amphotericin B colloidal dispersion: an improved antifungal therapy. *Adv. Drug Deliv. Rev.* **2001**, *47*, 149–163.

Herbrecht, R.; Natarajan-Ame, S.; Nivoix, Y.; Letscher-Bru, V. The lipid formulations of amphotericin B. *Expert Opin. Pharmacother.* **2003**, *4*, 1277–1287.

Hyon, S. H.; Cha, W. I.; Ikada, Y.; Kita, M.; Ogura, Y.; Honda, Y. Poly(vinyl alcohol) hydrogels as soft contact lens material. *J. Biomater. Sci. Polym. Ed.* **1994**, *5*, 397–406.

Janssens, S.; de Armas, H. N.; Remon, J. P.; Van den Mooter, G. The use of a new hydrophilic polymer, Kollicoat IR, in the formulation of solid dispersions of Itraconazole. *European Journal of Pharmaceutical Sciences* **2007**, *30*, 288–294.

Jeong, Y. I.; Nah, J. W.; Na, H. K.; Na, K.; Kim, I. S.; Cho, C. S.; Kim, S. H. Self-assembling nanospheres of hydrophobized pullulans in water. *Drug Dev Ind Pharm* **1999**, *25*, 917–927.

Jokerst, J. V.; Lobovkina, T.; Zare, R. N.; Gambhir, S. S. Nanoparticle PEGylation for imaging and therapy. *Nanomedicine (Lond)* **2011**, *6*, 715–728.

Jung, S. W.; Jeong, Y. I.; Kim, S. H. Characterization of hydrophobized pullulan with various hydrophobicities. *Int J Pharm* **2003**, *254*, 109–121.

Jung, S. W.; Jeong, Y. I.; Kim, Y. H.; Kim, S. H. Self-assembled polymeric nanoparticles of poly(ethylene glycol) grafted pullulan acetate as a novel drug carrier. *Arch Pharm Res* **2004**, *27*, 562–569.

Kaneo, Y.; Nakano, T.; Tanaka, T.; Tamaki, R.; Iwase, H.; Yamaguchi, Y. Characteristic distribution of polysaccharides in liver tissue. *J. Pharm. Sci. Technol., Jpn.* **2000**, *60*, 183–195.

Kaneo, Y.; Taguchi, K.; Tanaka, T.; Yamamoto, S. Nanoparticles of hydrophobized cluster dextrin as biodegradable drug carriers: solubilization and encapsulation of amphotericin B. *J. Drug Deliv. Sci. Technol.* **2014**, *24*, 344–351.

Kaneo, Y.; Tanaka, T.; Nakano, T.; Yamaguchi, Y. Evidence for receptor-mediated hepatic uptake of pullulan in rats. *Journal of Controlled Release* **2001**, *70*, 365–373.

Kaneo, Y.; Uemura, T.; Tanaka, T.; Kanoh, S. Polysaccharides as drug carriers: Biodisposition of fluorescein-labeled dextrans in mice. *Biological Pharmaceutical Bulletin* **1997**, *20*, 181–187.

Kaneo, Y.; Ueno, T.; Tanaka, T. In *23rd International Symposium on Controlled Release of Bioactive Materials*; Controlled Release Society: Kyoto, 1996, pp. 101–102.

Kwon, G. S. Polymeric micelles for delivery of poorly water-soluble compounds. *Crit Rev Ther Drug Carrier Syst* **2003**, *20*, 357–403.

Laniado-Laborin, R.; Cabrales-Vargas, M. N. Amphotericin B: side effects and toxicity. *Rev. Iberoam. Micol.* **2009**, *26*, 223–227.

Lavasanifar, A.; Samuel, J.; Kwon, G. S. Micelles self-assembled from poly(ethylene oxide)-block-poly(N-hexyl stearate L-aspartamide) by a solvent evaporation method: effect on the solubilization and haemolytic activity of amphotericin B. *Journal of Controlled Release* **2001**, *77*, 155–160.

Liu, Z.; Jiao, Y.; Wang, Y.; Zhou, C.; Zhang, Z. Polysaccharides-based nanoparticles as drug delivery systems. *Adv. Drug Deliv. Rev.* **2008**, *60*, 1650–1662.

Maeda, H.; Bharate, G. Y.; Daruwalla, J. Polymeric drugs for efficient tumor-targeted drug delivery-based on EPR-effect. *Eur. J. Pharm. Biopharm.* **2009**, *71*, 409–419.

Molteni, L. In: *Drug carriers in biology and medicine*; Gregoriadis, G.; Ed.; Academic Press: London, 1979, pp. 107–125.

Nakamura, H.; Fang, J.; Gahininath, B.; Tsukigawa, K.; Maeda, H. Intracellular uptake and behavior of two types zinc protoporphyrin (ZnPP) micelles, SMA-ZnPP and PEG-ZnPP as anticancer agents; unique intracellular disintegration of SMA micelles. *Journal of Controlled Release* **2011**, *155*, 367–375.

Namazi, H.; Fathi, F.; Heydari, A. In *The delivery of nanoparticles*; Hashim, A. A.; Ed.; InTech: Rijeka, Croatia, 2012, pp. 149–184.

Noguchi, T.; Yamamuro, T.; Oka, M.; Kumar, P.; Kotoura, Y.; Hyon, S.; Ikada, Y. Poly(vinyl alcohol) hydrogel as an artificial articular cartilage: evaluation of biocompatibility. *J Appl Biomater* **1991**, *2*, 101–107.

Opanasopit, P.; Ngawhirunpat, T.; Chaidedgumjorn, A.; Rojanarata, T.; Apirakaramwong, A.; Phongying, S.; Choochottiros, C.; Chirachanchai, S. Incorporation of camptothecin into N-phthaloyl chitosan-g-mPEG self-assembly micellar system. *Eur J Pharm Biopharm* **2006**, *64*, 269–276.

Opanasopit, P.; Ngawhirunpat, T.; Rojanarata, T.; Choochottiros, C.; Chirachanchai, S. N-phthaloylchitosan-g-mPEG design for all-trans retinoic acid-loaded polymeric micelles. *Eur J Pharm Sci* **2007a**, *30*, 424–431.

Opanasopit, P.; Ngawhirunpat, T.; Rojanarata, T.; Choochottiros, C.; Chirachanchai, S. Camptothecin-incorporating N-phthaloylchitosan-g-mPEG self-assembly micellar system: effect of degree of deacetylation. *Colloids. Surf. B Biointerfaces* **2007b**, *60*, 117–124.

Orienti, I.; Zuccari, G.; Carosio, R.; Montaldo, P. G. Preparation and evaluation of polyvinyl alcohol-cooleylvinyl ether derivatives as tumor-specific cytotoxic systems. *Biomacromolecules* **2005**, *6*, 2875–2880.

Schacht, E.; Ruys, L.; Vermeersch, J.; Remon, J. P.; Duncan, R. In *Macromolecules as Drugs and As Carriers for Biologically Active Materials*; Tirrell, D. A.; Donaruma, G.; Turek, A. B.; Eds.; The New York Academy of Science: New York, 1985; Vol. 446, pp. 199–212.

Taguchi, K.; Kaneo, Y.; Tanaka, T. Biodisposition of cluster dextrin as a biodegradable drug carrier. *J. Pharm. Sci. Technol., Jpn.* **2009**, *69*, 373–383.

Takata, H.; Takaha, T.; Nakamura, H.; Fujii, K.; Okada, S.; Takagi, M.; Imanaka, T. Production and some properties of dextrin with a narrow size distribution by the cyclization reaction of branching enzyme. *J. Ferment. Bioeng.* **1997**, *84*, 119–123.

Takata, H.; Takaha, T.; Okada, S.; Hizukuri, S.; Takagi, M.; Imanaka, T. Structure of the cyclic glucan produced from amylopectin by Bacillus stearothermophilus branching enzyme. *Carbohydr Res* **1996a**, *295*, 91–101.

Takata, H.; Takaha, T.; Okada, S.; Takagi, M.; Imanaka, T. Cyclization reaction catalyzed by branching enzyme. *J Bacteriol* **1996b**, *178*, 1600–1606.

Takeuchi, H.; Kojima, H.; Yamamoto, H.; Kawashima, Y. Polymer coating of liposomes with a modified polyvinyl alcohol and their systemic circulation and RES uptake in rats. *J. Control. Release* **2000**, *68*, 195–205.

Thassu, Deepak; Pathak, Yashwant; Deleers, Michel In *Nanoparticlate Drug Delivery Systems*; Thassu, D.; Ed. 2007, pp. 1–31.

Thoren, L. The dextrans – Clinical data. *Develop. Biol. Stand.* **1981**, 48, 157–167.

Tomii, Y. Lipid formulation as a drug carrier for drug delivery. *Curr Pharm Des* **2002**, *8*, 467–474.

van Etten, E. W.; Otte-Lambillion, M.; van Vianen, W.; ten Kate, M. T.; Bakker-Woudenberg, A. J. Biodistribution of liposomal amphotericin B (AmBisome) and amphotericin B-desoxycholate (Fungizone) in uninfected immunocompetent mice and leucopenic mice infected with Candida albicans. *J Antimicrob Chemother* **1995**, 35, 509–519.

Winek, C. L.; Burgun, J. J. Acute and subacute toxicology and safety evaluation of SMA 1440-H resin. *Clin. Toxicol.* **1977**, 10, 255–260.

Yallapu, M. M.; Reddy, M. K.; Labhasetwar, V. In *Biomedical Applications of Nanotechnology*; Labhasetwar, V.; Leslie-Pelecky, D. L.; Eds.; John Wiley & Sons, Inc.: Hoboken, NJ, USA, 2007, p 131–171.

Yamamoto, S.; Kaneo, Y.; Ishizu, T.; Yamaguchi, Y.; Haraguchi, H. Incorporation of amphotericin B into self-assembled hydrophobized Kollicoat IR nanoparticles. *J. Drug Deliv. Sci. Technol.* **2013b**, *23*, 591–596.

Yamamoto, S.; Kaneo, Y.; Maeda, H. Styrene maleic acid anhydride copolymer (SMA) for the encapsulation of sparingly water-soluble drugs in nanoparticles. *J. Drug Deliv. Sci. Technol.* **2013a**, *23*, 231–237.

Yamamoto, S.; Kaneo, Y.; Maitani, Y. Hydrophobized poly(vinyl alcohol) for encapsulation of amphotericin B in nanoparticles. *J. Drug Deliv. Sci. Technol.* **2013c**, *23*, 129–135.

Zeuzem, S.; Welsch, C.; Herrman'n, E. Pharmacokinetics of peginterferons. *Seminars in Liver Disease* **2003**, *23* (Suppl 1), 23–28.

INDEX

T - #0813 - 101024 - C398 - 229/152/18 - PB - 9781774631133 - Gloss Lamination